AF710216

CONSTRAINT AND INTEGER PROGRAMMING

OPERATIONS RESEARCH/COMPUTER SCIENCE INTERFACES SERIES

Series Editors

Professor Ramesh Sharda
Oklahoma State University

Prof. Dr. Stefan Voß
Universität Hamburg

Other published titles in the series:

Greenberg, Harvey J. / *A Computer-Assisted Analysis System for Mathematical Programming Models and Solutions: A User's Guide for ANALYZE*
Greenberg, Harvey J. / *Modeling by Object-Driven Linear Elemental Relations: A Users Guide for MODLER*
Brown, Donald/Scherer, William T. / *Intelligent Scheduling Systems*
Nash, Stephen G./Sofer, Ariela / *The Impact of Emerging Technologies on Computer Science & Operations Research*
Barth, Peter / *Logic-Based 0-1 Constraint Programming*
Jones, Christopher V. / *Visualization and Optimization*
Barr, Richard S./ Helgason, Richard V./ Kennington, Jeffery L. / *Interfaces in Computer Science & Operations Research: Advances in Metaheuristics, Optimization, & Stochastic Modeling Technologies*
Ellacott, Stephen W./ Mason, John C./ Anderson, Iain J. / *Mathematics of Neural Networks: Models, Algorithms & Applications*
Woodruff, David L. / *Advances in Computational & Stochastic Optimization, Logic Programming, and Heuristic Search*
Klein, Robert / *Scheduling of Resource-Constrained Projects*
Bierwirth, Christian / *Adaptive Search and the Management of Logistics Systems*
Laguna, Manuel / González-Velarde, José Luis / *Computing Tools for Modeling, Optimization and Simulation*
Stilman, Boris / *Linguistic Geometry: From Search to Construction*
Sakawa, Masatoshi / *Genetic Algorithms and Fuzzy Multiobjective Optimization*
Ribeiro, Celso C./ Hansen, Pierre / *Essays and Surveys in Metaheuristics*
Holsapple, Clyde/ Jacob, Varghese / Rao, H. R. / *BUSINESS MODELLING: Multidisciplinary Approaches — Economics, Operational and Information Systems Perspectives*
Sleezer, Catherine M./ Wentling, Tim L./ Cude, Roger L. / *HUMAN RESOURCE DEVELOPMENT AND INFORMATION TECHNOLOGY: Making Global Connections*
Voß, Stefan, Woodruff, David / *Optimization Software Class Libraries*
Upadhyaya et al/ *MOBILE COMPUTING: Implementing Pervasive Information and Communications Technologies*
Reeves, Colin & Rowe, Jonathan/ *GENETIC ALGORITHMS—Principles and Perspectives: A Guide to GA Theory*
Bhargava, Hemant K. & Ye, Nong / *COMPUTATIONAL MODELING AND PROBLEM SOLVING IN THE NETWORKED WORLD: Interfaces in Computer Science & Operations Research*
Woodruff, David L./ *NETWORK INTERDICTION AND STOCHASTIC INTEGER PROGRAMMING*
Anandalingam, G. & Raghavan, S. Raghu/ *TELECOMMUNICATIONS NETWORK DESIGN AND MANAGEMENT*
Laguna, Manuel & Martí, Rafael/ *SCATTER SEARCH: Methodology and Implementations in C*
Gosavi, Abhijit/ *SIMULATION-BASED OPTIMIZATION: Parametric Optimization Techniques and Reinforcement Learning*
Koutsoukis, Nikitas-Spiros & Mitra, Gautam/ *DECISION MODELLING AND INFORMATION SYSTEMS: The Information Value Chain*

CONSTRAINT AND INTEGER PROGRAMMING
Toward a Unified Methodology

Edited by
MICHELA MILANO
University of Bologna

Kluwer Academic Publishers
Boston/Dordrecht/London

Distributors for North, Central and South America:
Kluwer Academic Publishers
101 Philip Drive
Assinippi Park
Norwell, Massachusetts 02061 USA
Telephone (781) 871-6600
Fax (781) 871-9045
E-Mail: kluwer@wkap.com

Distributors for all other countries:
Kluwer Academic Publishers Group
Post Office Box 322
3300 AH Dordrecht, THE NETHERLANDS
Telephone 31 786 576 000
Fax 31 786 576 254
E-mail: services@wkap.nl

Electronic Services <http://www.wkap.nl>

Library of Congress Cataloging-in-Publication Data

A C.I.P. Catalogue record for this book is available from the Library of Congress.

CONSTRAINT AND INTEGER PROGRAMMING: Toward a Unified Methodology,
edited by Michela Milano

ISBN 1-4020-7583-9

Copyright © 2004 by Kluwer Academic Publishers.

All rights reserved. No part of this work may be reproduced, stored in a retrieval system, or transmitted in any form or by any means, electronic, mechanical, photocopying, microfilming, recording, or otherwise, without the written permission from the Publisher, with the exception of any material supplied specifically for the purpose of being entered and executed on a computer system, for exclusive use by the purchaser of the work.

Permission for books published in Europe: permissions@wkap.nl
Permissions for books published in the United States of America: permissions@wkap.com

Printed on acid-free paper.

Printed in the United States of America.

Contents

List of Figures xiii

List of Tables xvii

Preface xix

Contributing Authors xxiii

Foreword xxvii
John Hooker

1
Constraint and Integer Programming 1
Michela Milano and *Michael Trick*

1	Introduction		1
2	CP(FD) Basic Concepts		4
	2.1	Modeling	5
	2.2	Structure of a CP program	6
	2.3	Solving	8
	2.3.1	Constraint Propagation	8
	2.3.2	Search	10
	2.3.3	Branch and Bound in CP	12
	2.4	An example: the car sequencing problem	13
3	Integer Linear Programming Basic Concepts		15
	3.1	Modeling	16
	3.1.1	Logical Constraints	16
	3.1.2	Resource Constraints	17
	3.1.3	Routing Constraints	18
	3.1.4	Alternative Formulations	19
	3.2	Solving	20
	3.2.1	Relaxations	20
	3.2.2	Branch and Bound	22
	3.2.3	Improving the Effectiveness	23
	3.3	An example: the car sequencing problem revisited	26
4	Incomplete search strategies		27
5	Conclusion		28

References 29

2
Two Generic Schemes for Efficient and Robust Cooperative Algorithms 33
Emilie Danna and *Claude Le Pape*
1 Introduction 34
2 Operations Research Algorithms and Constraint Programming 37
3 Operations Research Algorithms and Mixed Integer Programming 39
4 Constraint Programming and Mixed Integer Programming 40
5 Operations Research Algorithms and Local Search 44
6 Mixed Integer Programming and Local Search 45
7 Constraint Programming and Local Search 48

References 53

3
Branch-and-Infer: A Framework for Combining CP and IP 59
Alexander Bockmayr and *Thomas Kasper*
1 Introduction 59
2 Modeling in CP and IP 60
3 An illustrating example: discrete tomography 63
3.1 Integer programming models 63
3.2 Constraint programming models 64
4 Branch and Infer 67
4.1 Primitive and non-primitive constraints 67
4.2 Inferring primitive from non-primitive constraints 68
4.3 Search 71
4.4 Pruning the search tree 73
5 Symbolic constraints in IP 74
5.1 Symbolic constraint abstractions 74
5.2 Extending expressiveness 75
5.3 Improving efficiency 76
5.4 Compositionality 77
6 Example: Symbolic constraints for supply chain planning 77
6.1 Stock-resource constraint 79
6.2 Finite capacity resource 80
6.3 State resource 81
6.4 Combined example 82
7 Summary 84

References 85

4
Global Constraints and Filtering Algorithms 89
Jean-Charles Régin
1 Introduction 89
2 Global Constraints 94

2.1 Preliminaries 94
2.2 Definition and Advantages 95
2.3 Examples 96
3 Filtering Algorithms 104
3.1 Algorithms Based on Constraints Addition 105
3.2 General Arc Consistency Filtering Algorithm 106
3.2.1 Preliminaries 106
3.2.2 A First Algorithm 107
3.2.3 A better general algorithm 107
3.2.4 Discussion and Example 109
3.3 Dedicated Filtering Algorithms 111
4 Two Successful Filtering Algorithms 112
4.1 Preliminaries 113
4.2 The Alldifferent Constraint 114
4.2.1 Consistency and Arc Consistency 114
4.2.2 Complexity 115
4.2.3 Some Results 116
4.3 The Global Cardinality Constraint 117
4.3.1 Consistency and Arc Consistency 117
4.3.2 Complexity 118
4.3.3 Some results 118
5 Global Constraints and Over-constrained Problems 120
5.1 Satisfiability Sum Constraint 121
5.2 Global Soft Constraints 122
5.2.1 General Definitions of Cost 123
5.2.2 Soft Alldifferent Constraint 124
6 Quality of Filtering Algorithms 125
7 Discussion 126
7.1 Incomplete Algorithms and Fixed-Point Property 126
7.2 Closure 127
7.3 Power of a Filtering Algorithm 128
8 Conclusion 129

References 131

5
Exploiting relaxations in CP 137
Filippo Focacci, Andrea Lodi and *Michela Milano*
1 Introduction and Motivation 138
2 Integer Linear Programming and Relaxations 139
2.1 Continuous Linear Relaxation 141
2.2 Structured Relaxations 142
2.2.1 Surrogate Relaxation 143
2.2.2 Lagrangean Relaxation 143
3 Integrating Relaxations in CP 144
3.1 Which relaxation to use 145
3.2 Which part of the problem 145
3.2.1 Relaxation of global constraints 147
4 Relax to propagate 150

4.1 Mapping 150
4.2 The Cost-Based Propagation 151
4.3 Triggering Events 152
5 Relax to guide the search 153
5.1 Synchronous update of the oracle 153
5.1.1 Select a branching object 153
5.1.2 Select promising values 154
5.2 Asynchronous update of the oracle 154
5.2.1 Generation of static heuristics 155
5.2.2 Defining sub-problems and incomplete methods 155
6 A case study: global optimization constraints for a *Path* constraint 156
6.1 Global constraint architecture 156
6.2 The `Path` constraint case 157
6.2.1 An ILP model of the `Path` constraint 158
6.2.2 Continuous (linear) relaxation 159
6.2.3 The linear Assignment Problem 159
6.2.4 The Minimum Spanning r-Arborescence 160
6.2.5 Adding cutting planes 160
6.2.6 Lagrangean relaxation of cuts 162

References 165

6
Hybrid Problem Solving in ECLiPSe 169
Farid Ajili and *Mark G. Wallace*
1 Introduction 170
1.1 Modelling Formalisms for Constraints and Mathematical Programming 170
1.2 Requirements on Hybrid Modelling Formalisms 170
1.3 ECLiPSe and Piecewise Linear Probing 171
2 Integration of Constraints and Operations Research 172
2.1 Constraint Programming 172
2.2 Operations Research 173
2.3 Hybridization context 173
3 Language Ingredients for Hybrid Solvers 174
3.1 Conceptual Models and Design Models 174
3.2 Context 175
3.3 Creating a Design Model - Introduction and Example 175
3.4 Information Passed between Solvers 177
3.5 Combining Solvers and Search Engine 178
3.5.1 Information Passed from Search Engine to Solvers 178
3.5.2 Information Passed from Solvers to Search Engine 178
3.6 Requirements on the Language used for Design Models 179
3.6.1 Variable Attributes 179
3.6.2 Control Requirements 179
3.7 Infeasibility Detection 180
4 ECLiPSe as a Platform for Building Hybrid Algorithms 181
4.1 Modelling in ECLiPSe 182
4.2 The Domain Solver: *ic* 183

4.3 The linear solver interface: *eplex* 184
4.4 The *repair* solver 185
4.5 Attributed Variables and Demons in ECLiPSe 185
5 Programming a Hybrid Search in ECLiPSe 186
5.1 An Illustrative Example 187
5.2 Problem Modelling 189
5.3 Hybrid Probe-based Algorithm 189
5.3.1 The Algorithm Design Model 191
5.3.2 Inference phase 191
5.3.3 Probing phase 192
5.3.4 Resource feasibility 192
5.4 Probing Strategies 193
5.5 Mixed Integer Programming based Probing 194
5.6 Linear Relaxation based Probing 196
5.7 Evaluation and Performance Analysis 197
5.7.1 Setting up the benchmark instances 198
5.7.2 Computational results 199
6 Conclusion 200

References 203

7
CP Based Branch-and-Price 207
Kelly Easton, *George Nemhauser* and *Michael Trick*
1 Introduction 207
2 Three Illustrative Examples 210
2.1 The Generalized Assignment Problem 210
2.2 The Traveling Tournament Problem 212
2.3 The Social Golfers Problem 216
2.4 Other Applications 218
3 Implementation Issues 219
3.1 Set Partitioning Versus Set Covering 219
3.2 Initial Solution 220
3.3 Column Management 221
3.4 LP Relaxation 222
3.5 Branching 223
3.6 CP as a Subproblem Solver 224
3.7 Column Generation Heuristics 224
3.8 Combining Column and Row Generation 225
3.9 Parallel Implementation Issues 226
4 Future Directions for CP Based Branch-and-Price 226

References 229

8
Randomized Backtrack Search 233
Carla P. Gomes
1 Introduction 234

2	Randomization of Backtrack Search Methods	237
3	Formal Models of Heavy-Tailed Behavior	241
	3.1 Random Walk	241
	3.2 Tree Search Model	242
4	Heavy and Fat-Tailed Distributions	245
	4.1 Heavy-Tailed Distributions	248
	4.1.1 Definition	249
	4.1.2 Visualization of Heavy-Tailed Behavior	249
	4.1.3 Estimation of Index of Stability (α)	250
	4.2 Fat *vs.* Heavy-Tailed Distributions	252
5	Heavy and Fat-Tailed Distributions in Backtrack Search	255
	5.1 CSP Formulations	255
	5.2 Mixed Integer Programming Formulations	257
	5.3 Boolean Satisfiability Formulations	260
	5.4 Graph Coloring Formulations	262
	5.5 Discussion	263
6	Restart Strategies	269
	6.1 Elimination of Heavy-Tails	269
	6.2 Cutoff Value for Restart Strategy	271
	6.3 Formal Results on Restarts	273
	6.4 Restart Results on a Range of Problem Instances	274
	6.5 Learning Dynamic Restart Strategies	276
	6.6 Variants of Restart Strategies	277
7	Portfolio Strategies	278
	7.1 Portfolio Design	279
	7.2 Portfolio Results	279
	7.3 Variants of Portfolios	282
8	Conclusions	283

References 285

9

Local Search and Constraint Programming 293

LS and CP illustrated on a transportation Problem
Filippo Focacci, Francois Laburthe, Andrea Lodi

1	Introduction	294
2	A didactic transportation problem	297
3	A CP approach for dTP	298
	3.1 A CP model for dTP	298
	3.1.1 Basic model	300
	3.2 Propagation	301
	3.2.1 Disjunctive Relations	301
	3.2.2 Linking trucks and bins	302
	3.2.3 Propagating costs	302
	3.3 A redundant routing model	303
	3.3.1 Propagation	304
4	Constructive Algorithms	306
	4.1 Insertion algorithms	306
	4.2 Greedy insertion	307

	4.3	Restricted Candidate Lists and GRASP	309
	4.4	Discrepancy-based search	310
5	LS as Post-Optimization		312
	5.1	LS + constraint checks	312
	5.2	Constraint checks within the neighborhood iteration	314
	5.3	Freezing Fragments	316
	5.4	CP models for the Neighborhoods	318
6	Metaheuristics		320
	6.1	Control strategies from metaheuristics	320
	6.2	Managing several neighborhoods	322
7	LS during construction		323
	7.1	Single Route Optimization	323
	7.2	Ejection Chains	324
8	Conclusions		325

References 327

10
Open Perspectives 331
Mark Wallace, *Yves Caseau* and *Jean-Francois Puget*

1	Motivations, Challenges and Applications		331
	1.1	Overview	331
	1.2	Challenges	332
	1.3	Focus on Hard to Solve Problems	333
	1.3.1	Global Constraints	333
	1.3.2	Scalability Limitations of Tree Search	334
	1.3.3	Limitations of Global "Soft" Constraints	335
	1.4	Problem Analysis vs. Resolution	335
	1.4.1	Multi-criteria problems	335
	1.4.2	Uncertain Problems	336
	1.4.3	Probabilistic Data	337
	1.5	Supporting the Problem Solving Process	337
	1.6	Software Engineering Issues	339
	1.6.1	Software Engineering in Support of Combinatorial Problem Solving	339
	1.6.2	Combinatorial Problem Solving in Support of Software Engineering	340
2	Transforming Models to Algorithms		341
	2.1	Conceptual and Design Models	341
	2.2	Decompositions	342
	2.3	Transformations	344
	2.3.1	Separating Modeling for Performance Improvements	344
	2.3.2	Motivations and Examples	345
	2.3.3	Transforming Conceptual Models into Design Models	346
	2.4	Search	347
	2.4.1	Local and Constructive Search	347
	2.4.2	Combining Different Search Methods	348
	2.4.3	Concurrent Search	348
	2.5	Inference	349
	2.5.1	Global and Local Control of Inference	349

2.5.2 Controlling the Local Inference Procedure 350
2.5.3 Controlling Communication of Inferences 350
2.6 Symmetries 350
3 New Techniques 351
3.1 Stochastic Optimisation 351
3.2 Overconstrained problems and robustness 353
3.3 User Support 354
3.3.1 A Simple Environment for Solving LSICO Problems 354
3.3.2 Automating Algorithm Development and Testing 356
3.4 Packaging 357
4 New Application Areas 359
4.1 Computer-Aided decision analysis based on simulation 359
4.2 Cooperative Problem Solving 360
4.3 Interleaved Planning and Execution 360

References 363

Index 367

List of Figures

1.1 Interaction among constraints 10
3.1 `Cumulative` constraint in discrete tomography 66
3.2 Two- and three-dimensional `diffn` constraint in discrete tomography 66
3.3 Primitive constraints in CP 68
3.4 Primitive constraints in IP 69
3.5 Architecture for constraint-based solving 71
3.6 Inference in IP and CP 71
3.7 Supply Chain Example 78
4.1 An Assignment Timetable. 92
4.2 An example of a global cardinality constraint. 93
4.3 Rules of the ad-hoc filtering algorithm for the n-queens problem. 112
5.1 Linearization of all constraints in a single store 147
5.2 Optimization component in each constraints 148
5.3 Merging of two constraints 149
5.4 Global constraint architecture 157
5.5 Addition of cuts in optimization constraints. 161
6.1 The input schedule and the PL cost function ϕ of an activity. 187
6.2 Results of running H(relax) and H(lambda) on the four test classes. 199
8.1 Two different executions of a backtrack search methods looking for an assignment to the variables A, B, C that satisfies the formula $(A \vee \neg B \vee \neg C) \wedge (B \vee \neg C) \wedge (A \vee C)$ 238
8.2 Top panel: symmetric random walk (10,000 steps). Bottom panel: the heavy-tailed nature of a random walk contrasted with the exponential decay of the normal distribution. 243

8.3 Top panel: imbalanced tree model; p is the probability that the branching heuristic will make a wrong decision, and b is the branching factor (in this case, $b = 2$). This tree model has a finite mean and an infinite variance when $1/b^2 < p < 1/b$; both the mean and the variance are infinite when $p \geq 1/b$. Bottom panel: example distributions the imbalanced and bounded imbalanced models, contrasted with the balanced model. Parameters: $b = 2$, $n = 20, p = 0.5$ and 0.75 (n is the number of variables). 244

8.4 Quasigroup Completion Problem (QCP). 246

8.5 Erratic behavior of mean cost of completing a quasigroup (order 11, 30% pre-assignment) vs. stabilized behavior of mean for a standard distribution (gamma) 247

8.6 The phenomenon of long tails with randomized complete search procedures (QCP instance, order 11, 30% pre-assignment) 248

8.7 Left panel: log–log plot of heavy-tailed behavior (QCP instance, order 11, 30% pre-assignment; uncensored data). Right panel: no heavy-tailed behavior. 250

8.8 Log-log plot of heavy-tailed behavior. 251

8.9 Standardized normal, Cauchy, and Lévy densities 253

8.10 Comparison of the *fatness* of tails: normal (1,1), lognormal(1,1), exponential(1), and pareto(1,1). 254

8.11 Comparison of the tail *fatness*: normal (1,1), lognormal(1,1), exponential(1), and pareto(1,1). 254

8.12 Log-log plot of the complement to one of the cumulative distribution of the runtime for the round-robin sports scheduling (12 teams; CSP formulation); (a) full distribution; (b) tail ($X > 10,000$). 257

8.13 (a) Cost profiles for a logistics planning problems for depth-first and best-bound search strategies. (b) Hevay-tailed behavior of depth-first search. 259

8.14 Log-log plot of heavy-tailed behavior for SAT formulations of (a) logistics planning, and (b) register allocation 260

8.15 Log–log plot of the tail of the runtime distributions of an exact backtrack search coloring algorithm on three different graph topologies: random ($log_2(p) = 0$); small-world ($log_2(p) = -4$); and more structured ($log_2(p) = -8$). p is the probability of randomly rewiring an edge of a regular graph. Size of the regular graph: 100 nodes and degree 8 (Walsh, 1999). 262

8.16 Log–log plot of heavy-tailed behavior for the timetabling problem 263
8.17 Left panel: heavy tail for quasigroup completion with extensive propagation (CSP formulation, order 30, with 55% pre-asssignment; uncensored data). Right panel: comparison of Satz to Relsat on logistics planning (SAT formulations). 266
8.18 Survival function $(1 - F(x))$ of the number of backtracks needed to solve different instances of model E with 20 variables and a domain size of 10. The parameter p captures the constrainedness of the instances; it is the probability of choosing one constraint (out of the total number of possible constraints). Two different regimes in the decay of the survival function can be identified: a heavy-tailed regime (curves with linear behavior) and a non–heavy-tailed regime. 267
8.19 Absence of heavy-tailed behavior for unsolvable QCP instances (CSP formulation) 268
8.20 Left panel: comparison of runtime profiles for proving optimality of `misc07` from MIPLIB. Right panel: absence of heavy-tailed behavior in proving optimality of `misc07` 268
8.21 Restarts: (a) Tail (of $(1 - F(x))$ as a function of the total number of backtracks for a QCP instance (CSP formulation), log–log scale; (b) same as (a), but for a logistics instance (SAT formulation). In (b) the left-most curve is for a cutoff value of 16, the middle curve is for a cutoff of 250, and the right-most curve is without restarts. 270
8.22 The effect of random restarts on solution cost for the logistics.d planning problem (SAT formulation) 272
8.23 Portfolio results for logistics planning: left panel, 2 processors; right panel, 20 processors 280
8.24 A range of portfolios for the MIP formulation of logistics planning (expected runtime). 280
8.25 Expected runtime (a) and expected total cost (b) of optimal portfolio for different numbers of processors. 281
9.1 Example of dTP with $N = 10, M = 3, C = 10$. 299
9.2 Example of 3-opt move. 314
9.3 Example of node-transfer move. 315
9.4 Example of ejection chain. 319

List of Tables

3.1	Complexity of the polyomino reconstruction problem	63
3.2	Primitive and non-primitive constraints	68
3.3	Inference in IP and CP	70
6.1	Definition of the four classes of experiments.	198
8.1	Estimates of the index of stability (α). k is the sample size, and u is the number of runs that failed to terminate before reaching the cutoff in the number of backtracks. The values within parentheses are the estimated asymptotic standard deviations.	252
8.2	Comparison of tail probabilities, $\Pr\{X > c\}$, for standard normal, Cauchy, and Lévy distributions	253
8.3	Estimates of the index of stability (α). k is the sample size. The values within parentheses are the estimated asymptotic standard deviations (SAT encoding).	261
8.4	Estimate of the index of stability (α) for the school-timetabling problem. k is the sample size. The value within parentheses is the estimated asymptotic standard deviation	263
8.5	Index of stability: impact of look-back strategies on heavy-tailed behavior (SAT encodings).	266
8.6	Solving (a) a 16-team round-robin scheduling problem (CSP formulation) and (b) the logistics.d instance (SAT formulation) for a range of cutoff values.	271
8.7	Restarts using a hybrid CSP/LP randomized backtrack search on a balanced QCP instance of order 50, at the phase transition.	272
8.8	Randomized rapid restarts (RRR) versus deterministic versions of backtrack search procedures (Satz solver used on SAT encodings; Ilog solver on CSP encodings).	275

Preface

During his invited talk at the first Workshop on *Integration of Artificial Intelligence and Operations Research in Constraint Programming for Combinatorial Optimization*, CP-AI-OR99, Jean-Francois Puget claimed that "Constraint Programming (CP) can be seen as *Software Engineering applied to Operations Research (OR)*". This statement is very strong, but indeed "almost" true. To my mind, CP is both something more and something less than this. It is something more since CP has its roots not only in Mathematical Programming, but also in Logic Programming and in Constraint Solving. From Logic Programming, CP inherits the declarative semantics, flexibility, the relational form which enables the definition of constraints without specifying input and output parameters, and the non-determinism which leads to the exploration of a search tree with a backtracking scheme. From Constraint Solving and Artificial Intelligence, CP has inherited powerful filtering algorithms, sophisticated search techniques and nogood recording. However, CP is something less than this: OR is so wide, and well studied since the fifties that so far CP could only cover a small, yet effective, portion of OR. The integration of areas of OR such as game theory, decision theory and stochastic programming into a CP framework has yet to be addressed.

However, experience with current implementations has already shown that CP is a powerful framework where Mathematical Programming concepts, and also local search techniques can be smoothly integrated and easily used.

The OR community has strengths both in breadth and in depth. The techniques of Integer and Linear programming and Local search are applicable to a huge swathe of (combinatorial) optimisation problems. Yet OR researchers have explored certain problem classes in great depth. These problems (the Travelling Salesman Problem, Set Partitioning and Covering, the Knapsack Problem, to name a few) are *pure problems*. No side constraints are considered. This simplification has an enormous advantage: pure problems are structured. The geometrical structure of the problem can help to reveal important problem properties. Geometric and algebraic structures enable practitioners to define ad hoc super-efficient (often polynomial) solving algorithms. Although pure problems rarely appear in real life applications, their investigation leads to the development of algorithms that can be effectively exploited to solve variants of these problems. These algorithms can be considered as *software components* providing results exploited in a large variety of applications.

The exploitation of bounds, reduced costs, optimal solutions of subparts of the original problems, heuristic suggestions coming from relaxations or problem decompositions dramatically enhance CP solver performance. Linear programming solvers are nowadays integrated in all commercial CP solvers.

When it was first introduced, Constraint Logic Programming was a declarative, general framework for solving Constraint Satisfaction Problems. It was more efficient than Logic Programming, but for combinatorial optimization problems, it was not powerful enough to compete with OR methods.

Now, things have changed. CP is closer to software engineering applied to OR. Firstly global constraints have been introduced into CP solvers, embedding complex pruning techniques. They are a powerful, fundamental tool for modelling and solving constraint (optimization) problems. Secondly, the optimization side, that was naively treated at the beginning, is now becoming an important aspect of CP solvers exploiting OR techniques.

Many steps need to be taken. CP researchers need to study, learn and think to discover which is the most general way to incorporate OR components in CP solvers. It is a big challenge.

This book was conceived during the CP-AI-OR School on Optimization I organized in Le Croisic (France) in 2002. The School was a successful event with more than fifty participants, and the speakers covered almost all aspects of the integration in an exhaustive and stimulating way. After the school, I was solicited to collect papers concerning the talks into a book. Therefore, the book is mainly devoted to students, researchers and practitioners who are entering this evolving research field. The book covers a wide range of perspectives on the field. Beside chapters based on talks from the school, the book contains some additional papers covering aspects which were not treated during the school for reasons of time.

The book is organized as follows: Chapter 1 is devoted to an introduction that provides a high level overview of the main concepts of Constraint Programming (CP) and Integer Programming (IP) used in the book. Chapter 2 informally introduces integration methods describing and classifying the main works in the field. Chapter 3 presents a unifying framework that presents under a uniform perspective the main concepts of CP and IP, underlining similarities and differences and stating the basis for possible integrations. In Chapter 4 global constraints are described as a vehicle for integrating IP concepts in CP in a transparent way for the user. Chapter 5 presents various ways to integrate relaxations in Constraint Programming focussing on global constraints. Then, Chapter 6 describes hybrid solvers and Chapter 7 concerns Column Generation and its integration in Constraint Programming. Chapter 8 concerns randomization and problem structure as a basis for understanding the intrinsic difficulty of the combinatorial problems. Many incomplete methods have been proposed mixing incomplete search (like local search and metaheuristics) in CP. Thus, Chapter 9 is devoted to a survey on the subject. Finally, the last chapter is devoted to open perspectives and future directions.

The authors are eminent and well known researchers and have significantly contributed to the field. They come from Universities, Research Centers and industries.

Finally, I would like to thank all the people who have helped me in the realization of this book. Andrea Lodi and Jean-Francois Puget suggested me to write this book. Krzysztof Apt spared his precious time to help me at the beginning of the process with invaluable advice. Mark Wallace who kindly proofread some parts of this book. John Hooker introduced me to Kluwer and pointed me out the CS/OR series. Gary Folven provided me with constant assistance. François Laburthe and Narenda Jussien helped me in the organization of the School of Optimization in Le Croisic.

Warm thanks go to the authors of the papers presented here for their work and their constant commitment in respecting the numerous requests I have asked of them, among which were requests to review other chapters in the book. Besides the authors, I would like to thank external reviewers: Torsten Fahle, Marco Gavanelli, Willem Jan van Hoeve, Ulrich Junker, Meinolf Sellman and Louis Martin Rousseau. I would also like to thank my colleagues Luca Benini and Andrea Roli for their help in the editing of several parts of the manuscript.

MICHELA MILANO

Contributing Authors

Farid Ajili
IC-Parc, Imperial College London
London SW7 2AZ, United Kingdom.
f.ajili@imperial.ac.uk

Alexander Bockmayr
Université Henri Poincaré, LORIA, B.P. 239
54506 Vandœuvre-lès-Nancy, France
bockmayr@loria.fr

Yves Caseau
BOUYGUES TELECOM
20, quai du Point du Jour
92640 Boulogne Billancourt, France
yves@caseau.com

Emilie Danna
ILOG and LIA,
Université d'Avignon,
CNRS-FRE 2487
edanna@ilog.fr

Kelly King Easton
Department of Industrial and Manufacturing Engineering
Kansas State University
237 Durland Hall - Manhattan, KS 66506
keaston@isye.gatech.edu

Filippo Focacci
ILOG S.A.

9, rue de Verdun
BP 85, 94253 Gentilly, France
ffocacci@ilog.fr

Carla P. Gomes
Dept. of Computer Science
Cornell University
Ithaca 14853
gomes@cs.cornell.edu

Thomas Kasper
GBU Supply Chain Management, IBD3 Optimization
SAP AG, Neurottstr. 16, 69190 Walldorf, Germany
thomas.kasper@sap.com

François Laburthe
BOUYGUES S.A.
1, avenue E. Freyssinet
78061 St-Quentin-en-Yvelines Cedex, France
flaburthe@bouygues.com

Claude Le Pape
ILOG S.A.
9 rue de Verdun
F-94253 Gentilly Cedex
clepape@ilog.fr

Andrea Lodi
DEIS, University of Bologna
Viale Risorgimento 2, 40136 Bologna, Italy
alodi@deis.unibo.it

Michela Milano
DEIS University of Bologna
V.le Risorgimento 2, 40136 Bologna, Italy
mmilano@deis.unibo.it

George L. Nemhauser
School of Industrial and Systems Engineering

Georgia Institute of Technology
Atlanta, GA 30332-0205, USA
gnemhaus@isye.gatech.edu

Jean-Francois Puget
ILOG S.A.
9, rue de Verdun
BP 85, 94253 Gentilly, France
puget@ilog.fr

Jean-Charles Régin
ILOG
1681, route des Dolines,
Sophia Antipolis, 06560 Valbonne, France
jcregin@ilog.fr

Michael Trick
Graduate School of Industrial Administration
Carnegie Mellon University
Pittsburgh, PA 15213 - USA
trick@cmu.edu

Mark G. Wallace
IC-Parc, Imperial College London
London SW7 2AZ, United Kingdom.
mgw@imperial.ac.uk

Foreword

Constraint Programming and Mathematical Programming are beginning to merge, despite their very different origins. Constraint programming grew out of the logic programming community as part of an effort to embed constraints in a programming language. Mathematical programming is a much older field, rooted in the mathematics of optimization. Yet recent years have seen a series of efforts to integrate the two.

Why should these two technologies be combined? The most obvious reason is that they have complementary strengths. Constraint programming solvers are adept at logical reasoning, particularly constraint propagation, while mathematical programming solvers use sophisticated relaxation techniques. Constraint programming may therefore excel at combinatorial problems with weak continuous relaxations, such as sequencing and scheduling problems, or problems with disjunctive and other logical conditions. Mathematical programming may be the method of choice for resource allocation and other planning problems.

Constraint programming can focus on highly structured portions of a constraint set, while mathematical programming can take a global view, particularly in its continuous relaxations. When side constraints complicate a problem, constraint programming can accommodate them gracefully and actually use them to advantage. When the problem structure is pure, without side constraints, mathematical programming's analysis of polyhedral structure can yield powerful methods. Constraint programming may be effective when constraints contain only a few variables, since such constraints tend to "propagate well." Mathematical programming is more suited for constraints (or objective functions) with many variables, such as constraints that measure cost, since they tend to "relax well."

Constraint programming brings to the table a large modeling vocabulary that can result in succinct and natural problem formulations. Mathematical programming provides the ability to translate a wide variety of models into an economical language of inequalities, which may yield a useful continuous relaxation.

While some problems seem suited for constraint programming and some for mathematical programming, many others have elements amenable to both. For instance, some constraints in the problem may propagate well, while others relax well. It seems only reasonable to call on the resources of both fields, as needed, in a "hybrid" or integrated solver.

Integration can be accomplished in a fairly natural way, if one recognizes that constraint programming and mathematical programming use basically the same overall solution strategy. Focusing for the moment on exact solution, they have four points of commonality—similar search methods, their use of inference, their use of relaxation, and the analogous ways that inference and relaxation interact. More specifically: (a) Both constraint programming and mathematical programming typically rely on branching methods to search the solution space exhaustively. (b) *Inference* in constraint programming takes the form of domain reduction, constraint propagation and constraint learning (otherwise known as *nogood* generation). Mathematical programming systems, on the other hand, infer cutting planes, separating cuts, and Benders cuts, the last two being forms of nogood generation. (c) *Relaxation* appears in constraint programming in the form of a "constraint store," which keeps track of the domain (set of feasible values) of each variable. Continuous relaxation is of course the work horse of mathematical programming. (d) In both technologies, inference is often used to strengthen the relaxation. Inference in constraint programming strengthens the constraint store by reducing its variable domains, and inference in mathematical programming strengthens the continuous relaxation by adding cutting planes.

Thus constraint programming and mathematical programming are special cases of the same *search-infer-and-relax* solution strategy. They combine naturally in this common framework. At the inference stage of the algorithm, for example, one can apply inference methods from either or both fields, depending on the characteristics of the problem. One can do the same at the relaxation stage. There can be mutual reinforcement, as when reduced domains inferred for the constraint store help tighten the continuous relaxation, and cutting planes inferred for the continuous relaxation propagate to reduce domains further.

The respective strengths of constraint programming and mathematical programming complement each other to accelerate the search. The inference technology of constraint programming strives to maintain various types of "consistency," which in the simplest case amounts to reducing the variable domains and thereby the number of values on which one must branch. The relaxation technology of mathematical programming accelerates the search by providing bounds on the optimal value, which can be used to prune the search tree as part of a branch-and-bound strategy (known as "branch and relax" in the constraint programming community).

We are arriving at a deeper and more satisfying answer to the question posed above: why should constraint programming and mathematical programming be combined? They should be combined because they are already the same. They are instances of the same general method.

The modeling style of this general method is likely to favor that of constraint programming rather than mathematical programming, but only because it already represents a certain kind of integration. The declarative modeling style of mathematical programming helped to inspire constraint programming's research goal of embedding constraints in a programming langauge. Yet constraint programmers, rooted as they are in a computer science tradition, continue to regard constraints as procedures—specifically, as procedures that operate on the solution space by removing solutions that cannot satisfy the associated constraints. Constraint programming formulations are therefore a blend of declarative and procedural modeling.

The simple idea of viewing constraints as procedures opens the door to exploiting substructure in a problem. Constraint programmers typically write a group of constraints with a specific overall structure as a single "global constraint," such as the alldifferent constraint, which requires a set of variables to take distinct values, or the cumulative constraint, which is used for resource-constrained scheduling. In so doing the modeler communicates substructure in the formulation to the solver, which contains domain reduction algorithms tailored to each type of global constraint. Thus constraint programming exploits problem structure primarily through its inference methods, which turn out to be the key to its success. Mathematical programming, by contrast, exploits problem structure primarily through its relaxations, which are likewise essential to its success.

The idea of global constraints can of course be used in a hybrid setting, where they can invoke specialized relaxations as well as domain reduction procedures. For instance, constraints that require a hamiltonian path (used in the traveling salesman problem) can be written as a single global constraint (the "cycle" constraint), which can activate specialized domain reduction procedures developed by constraint programmers and generate cutting planes developed by integer programmers.

Viewing constraint and integer programming as special cases of a single search-infer-and-relax method poses the interesting intellectual challenge of defining just what this single method is, and how much it can be usefully generalized. A good deal of generalization turns out to be useful, particularly when taken far enough to encompass decomposition methods. For instance, the search can be viewed as the enumeration of Benders subproblems, and the relaxation as the Benders master problem. If the master problem is solved by mixed integer programming and the subproblems by constraint programming, the result is an integration scheme based on decomposition. (Benders cuts can be obtained from the constraint programming solution by extending classical concepts of duality.) This scheme turns out to be intuitive and useful in manufacturing applications, in which the master problem typically assigns jobs to machines, and the subproblem schedules the jobs that are assigned to each machine—thus exploiting the mathematical programming advantage in resource allocation, and the constraint programming advantage in scheduling. In fact, decomposition approaches have so far delivered the most impressive computational speedups among hybrid methods.

In general, a search-infer-and-relax method moves from one problem restriction to another, while applying various types of inference and relaxation to each problem restriction. In integer and constraint programming, the problem restrictions are obtained at the nodes of the search tree, and the relaxations are the constraint store and the continuous relaxation. In a Benders approach, the restrictions are the successive subproblems, and the relaxation is the master problem. In each case, the search direction is suggested by the solution of a relaxation. For instance, in integer programming one branches on a variable with a fractional value in the solution of the continuous relaxation, and in Benders decomposition the next subproblem is obtained by solving the master problem.

Local search methods also fit into this scheme, since they proceed by searching a sequence of neighborhoods, each of which can be viewed as the feasible set of a problem restriction. Local search has long been studied in Operations Research, but the constraint programming community has recently provided a fresh perspective. It has experimented with "large neighborhood" search, where constraint programming methods are used to search neighborhoods, as well as quasi-declarative languages for specifying local search algorithms. There have also been efforts to incorporate randomized search and other heuristic ideas into an exhaustive enumeration procedure.

At this writing there is no consensus on the best architecture for an integrated method. Some commercial software products support a certain amount of integration, such as the ECLiPSe system, ILOG's OPL Studio, and Dash Optimization's Mosel language, but no system comes close to exploiting the full advantage of integration. Nonetheless, the situation will quickly change. Research proceeds apace, computational successes accumulate, and more powerful software is under development. Integrated methods have already proved themselves in industrial applications. Hybrid solvers and modeling systems may become the norm in a very few years. The very term "hybrid" may become obsolete, as we increasingly see today's technologies as parts of a seamless whole.

It is therefore an ideal time to familiarize oneself with the basic ideas of what are now known as hybrid methods. The present volume serves this end. Michela Milano has performed a great service by collecting a set of tutorial essays written by some of the top researchers in the field. They present some of the basic ideas of constraint programming and mathematical programming, explore approaches to integration, bring us up to date on heuristic methods, and attempt to discern future directions in a fast-moving field.

JOHN HOOKER

To my parents

Chapter 1

CONSTRAINT AND INTEGER PROGRAMMING

Basic concepts

Michela Milano
DEIS University of Bologna
V.le Risorgimento 2, 40136 Bologna, Italy
mmilano@deis.unibo.it

Michael Trick
Carnegie Mellon University
trick@cmu.edu

Abstract The purpose of this introductory chapter is to provide the basic concepts behind Constraint Programming (CP) and Integer Programming (IP). These two fields cover a variety of aspects and have been widely studied. Therefore, here we do not intend to give a deep insight of the fields, but to provide the definitions and concepts for understanding the rest of this book. We explain CP and IP modelling aspects and solving strategies. We ground our discussion on an example: the car sequencing problem. The chapter provides references to relevant biography which can be referred to for a deeper understanding.

Keywords: Constraint Programming, Integer Linear Programming, Modeling, Constraint Propagation, Relaxation, Cutting Planes, Search, Branch and Bound

1. Introduction

Combinatorial Optimization problems appear in many real life applications, such as production scheduling, DNA sequencing, hardware design, protocol simulation and testing, option trading. This class of problems is, in general, extremely difficult to solve and has been traditionally tackled with Integer Linear Programming (referred to as IP or sometimes ILP) techniques, coming from Operations Research. Integer Programming is based on a model where variables represent problem decisions and can assume integer values. IP variables are linked by linear constraints. If only some of the variables are required to be integral while some other are continuous, we have a Mixed Integer Linear Program (referred to as MIP or MILP). If all variables are continuous, we have a Linear Program Mixed Integer Linear Program (LP). In this chapter we consider IPs, but the concepts explained can be easily extended also to MIPs.

The modeling phase in IP requires high experience since it strongly influences the performances of the solver. The solving technique generally used is branch-and-bound which is heavily based on relaxations, i.e., easier subproblems obtained by removing or relaxing some constraints. Relaxations provide bounds and other important results which can be used to prune provably suboptimal parts of the search tree. Another important aspect of IP solving concerns the use of cutting planes, i.e., linear inequalities which can be added to the relaxation to better approximate the geometrical structure of the original problem. For a long time, IP has been the only technique for solving combinatorial optimization problems.

Since the end of the eighties, a different and declarative framework coming from logics, i.e., Constraint Logic Programming (CLP), has been developed with the hope of reducing the development time while preserving the efficiency of procedural languages. CLP has been proposed as a general scheme which has been specialized on different domains such as reals, rationals, boolean (and pseudo-boolean) and finite domains. Each specialization results in the corresponding constraint language with its declarative and operational semantics.

We will focus in this book on Constraint (Logic) Programming on Finite Domains where variables represent problem decisions, ranging on finite domains of values and are linked by mathematical and symbolic constraints. Nowadays, not all solvers are based on logics. For this reason, the name we will use throughout this chapter and in most part of this book will be Constraint Programming (CP) on finite domains (CP(FD)), including also Constraint Logic Programming on finite domains. In the following, when we use the acronym CP we mean CP(FD). In CP the problem is easily modeled as a set of subproblems each represented by a constraint. Operationally, the solving part is composed by two steps: constraint propagation and search. Constraint propagation is an inference mechanism aimed at reducing provably inconsistent configurations. Each constraint can be seen as a software component containing either a general purpose or a specialized *constraint-dependent* filtering algorithm. These constraints are triggered, and the filtering algorithm run, each time an event happens on a variable involved in the constraint.

Unfortunately, in general, constraint propagation is not complete, i.e., values which are not removed are not guaranteed to be consistent. Thus, search is needed to explore the remaining search tree. Constraint propagation and search are interleaved, since after each branch in the tree search process, propagation is again triggered.

Beside the feasibility side, if an objective function should be optimized, CP implements a naive form of branch-and-bound by adding each time a solution is found, a constraint imposing that only improving solutions are accepted.

The first Constraint Programming (CP) solvers on finite domains were not competitive with Integer Programming techniques. They were based on general purpose propagation algorithms, namely arc and bound consistency. Moreover, the optimality aspect of the problem was naively treated by simply adding each time a solution is found a new constraint requiring improving solutions. Relaxations were never incorporated in CP solvers. Thus, the *optimality-based* pruning was very weak.

Now things are changing and, in a sense, have already changed. Global constraints have been integrated in all constraint solvers and embed powerful *constraint-dependent* filtering algorithms, see section "Dedicated Filtering Algorithms" in Chapter 4. As far as optimality is concerned, Mathematical Programming techniques are now integrated in all commercial constraint solvers. Results coming from relaxations can be exploited in a CP setting to prune also sub-optimal parts of the search tree and not only infeasible parts.

Therefore, now CP solvers are, for certain applications, competitive with Integer Programming solvers. Of course, the application tackled strongly influences the choice of the appropriate solver. Some applications are more suitable for Constraint Programming, while for some other CP obtains poor results. Pure problems, i.e., problems with an underlying structure and no side constraints, are in general more effectively solved with IP solvers exploiting problem dependent relaxations and cutting planes. However, when side constraints appear in the problem, when the structure is even slightly modified, when problem requirement change rapidly, Constraint Programming, for its high flexibility, can be the technique of choice (see, e.g., (Smith et al., 1996), (Darbi-Dowman and Little, 1998)). In most cases, researchers have shown that the integration of both techniques gives better results then the two techniques separately. In fact, hybrid solvers in many cases are able to overcome the weakness of both techniques.

The purpose of this chapter is to introduce the reader to CP and IP basic concepts used throughout the book. Clearly, this chapter is far from being comprehensive, but it provides some pointers to the literature that can be used for a deeper study.

This paper is organized as follows: in Section 2, we describe Constraint Programming on Finite Domain, addressing both the modelling (Section 2.1) and solving aspect (Section 2.3). An example on car sequencing and its modelization with CP is presented in Section 2.4. In Section 3 Integer Linear Programming basic concepts are presented, again addressing modelling (Section 3.1) and solving (Section 3.2). The IP modelization of the car sequencing problem is presented in Section 3.3. To conclude this overview, incomplete search is presented in Section 4.

2. CP(FD) Basic Concepts

Constraint Programming on Finite Domain (referred to as CP(FD)) has been recognized as a suitable modelling and solving tool to face combinatorial (optimization) problems. In this section, we provide some basic notions on modeling aspects, constraint propagation, search and optimization.

The CP modeling and solving activity is highly influenced by the Artificial Intelligence area on Constraint Satisfaction Problems, CSPs (see, e.g., the book by (Tsang, 1993)).

A CSP is a triple $\langle V, D, C \rangle$ where V is a set of variables $X_1, \ldots, X_n$, D is a set of finite domains $D_1, \ldots, D_n$ representing the possible values that variables can assume, and C is a set of constraints $C_1, \ldots, C_k$. Each constraint involves a set of variables $V' \subseteq V$ and defines a subset of the cartesian product of the corresponding domains containing feasible tuples of values. Therefore, constraints limit the values that variables can simultaneously assume.

A solution of a CSP is an assignment of values to variables which is consistent with constraints.

Usually, research on CSPs is concerned with binary CSPs where only binary constraints are considered, i.e., constraints linking pairs of variables, e.g., $X_i > X_j$. One of the general propagation methods applicable to binary CSPs is arc-consistency. A binary constraint $c(X_i, X_j)$ is arc-consistent iff for each value in the domain of X_i there exists at least one value in the domain of X_j consistent with the constraint and viceversa.

However, practical applications often require non binary constraints. Therefore, many efforts (Bèssiere and Régin, 1997), (Bèssiere and Régin, 1999), (Bèssiere et al., 1999b), (Bèssiere et al., 1999a) have been performed in order to extend results obtained for binary CSPs to general CSPs, where non-binary constraints appear. These algorithms are general and can be applied to any constraint. However, efficiency is an issue when generality is achieved. Thus, in CP, symbolic global constraints, embedding special purpose *constraint-dependent* filtering algorithms have been developed. Global constraints will be presented in Chapter 4. In the next section, we will concentrate on problem modeling and solving in CP along with an example on the car-sequencing problem. We will focus on advantages and limitations of Constraint Programming.

2.1 Modeling

There exists a one to one mapping between CSP concepts and CP on finite domains CP(FD) syntactic structures. Thus, CP(FD) benefits from and extends results achieved for CSPs.

For modeling a problem in a CP(FD) language, we define variables ranging on a finite domain of values and a set of constraints involving those variables. Values can be objects of arbitrary type, but often these languages manage finite domains of integers.

One of the characterizing aspects of CP(FD) is the facility of modelling problems which can be declaratively stated, extended and modified: expressivity and flexibility have always been considered important features of CP.

In CP(FD) languages, variables range over finite domains of integers. Variables' domains represent the values that variables can assume during the computation. For example, in a scheduling application, if $Start$ is a variable representing the starting time of an activity, its domain can be the schedule horizon, e.g., $Start :: [1..100]$. This (unary) constraint states that variable $Start$ can assume one of the integer values from 1 to 100. As another example, if a given variable Pos represents the position in a sequence of a given object, and positions available for the object are only 3, 5 and 9, we can express this constraint with a domain variable $Pos :: [3, 5, 9]$. Thorough the book, other notations can be used for associating a domain to a variable. For instance the domain can be represented as a set like $Pos \in \{3, 5, 9\}$ or $D(Pos) = \{3, 5, 9\}$, but they all mean the same. In addition, the notation of representing variables with a capital letter as initial is derived from the Logic Programming roots of CP. Indeed, in the book also non capital letters are used for indicating variables.

Variables are linked by constraints that can be either mathematical constraints or symbolic constraints. Mathematical constraints have the form: $t_1 \ R \ t_2$ where t_1 and t_2 are finite terms, i.e., variables, finite domain objects and usual expressions, and R is one of the constraints defined on the domain of discourse (e.g., for integers we have the usual relations: $>, \geq, <, \leq, =, \neq$). For example, if two activities i and j characterized by starting times $Start_i$ and $Start_j$ and durations d_i and d_j are linked by a precedence constraint stating that activity i should be executed before activity j, the following mathematical constraint can be imposed, $Start_i + d_i <= Start_j$.

Symbolic constraints, called also global constraints, are predicates involving finite domain variables. They are expressive and powerful constraints (which can also be defined by the user) embedding constraint-dependent filtering algorithms. A typical global constraint is the[1]

$$alldifferent([X_1, \ldots, X_n])$$

available in most CP(FD) solvers like CHIP (Dincbas et al., 1988), ECL^{i}PS^e (IC-Park, 2003), and ILOG Solver (Puget, 1994). Declaratively, the constraint *alldifferent*($[X_1, \dots, X_n]$), holds iff all variables are assigned to a different value. Thus, it is declaratively equivalent to a set of $n * (n-1)/2$ binary inequality constraints. However, its compact representation allows more concise models and embeds a specialized efficient filtering algorithm, described in the next section and extensively detailed in Chapter 4.

An extension of the *alldifferent* is the *global cardinality constraint*:

$$gcc([X_1, \dots, X_n], [v_1, \dots, v_m], [l_1, \dots, l_m], [u_1, \dots, u_m])$$

which holds iff the number of variables in $[X_1, \dots, X_n]$ which assume value v_i is within l_i and u_i.

A further extension of the global cardinality constraint, which will be used in the car sequencing example in Section 2.4 is the *global sequencing constraint*:

$$gsc([X_1, \dots, X_n], [v_1, \dots, v_m], min, max, q, [l_1, \dots, l_m], [u_1, \dots, u_m]).$$

This constraint holds iff each v_i appears a number of times between l_i and u_i and for each sequence S_i of q consecutive variables, each v_i appears a number of times between min and max.

Other important constraints in CP have the form:

$$atmost(i, [X_1, \dots, X_n], v)$$

$$atleast(i, [X_1, \dots, X_n], v)$$

$$exactly(i, [X_1, \dots, X_n], v)$$

meaning respectively that at most, at least and exactly i variables in the list assume the value v.

Another fundamental constraint in CP is the *element constraint*. It is very useful since with the element we can *link* variables. The declarative semantics of the element is the following:

$$element(I, [X_1, \dots, X_n], V)$$

holds iff the element of index I of the list $[X_1, \dots, X_n]$ has the value V.

Chapter 4 will describe global constraints in a wider perspective.

2.2 Structure of a CP program

The basic structure of a CP(FD) program describing a CSP is the following:

- definition of variables and domains;

- posting of the constraints among variables;
- definition of the search strategy.

When variables and domains are introduced in the model, they are collected in a store. Then constraints among variables are stated. As soon as they are stated, they start propagating to remove values inconsistent with the problem definition. Constraints can be seen as software components suspended on variables appearing in the constraint itself and embedding a filtering algorithm. The filtering algorithm can be either a general or a specialized one. Propagation is an iterative process: each constraint is triggered as soon as *something* happens on a domain of a variable involved. For example, if a variable X is instantiated or its domain reduced, all the constraints containing X are triggered, propagated and suspended again. When propagation ends, either a solution has been computed, or a failure has been found or search should be triggered. Search is nothing else than a series of more or less intelligent guesses. Each time a guess is performed, instantiating for example a variable, or partitioning its domain, propagation is again triggered.

For optimization problems the structure of the program should be extended to take into account the objective function.

The basic structure of a CP(FD) program describing a optimization problem is the following:

- definition of variables and domains;
- posting of the constraints among variables;
- definition of the objective function and its link with problem decision variables;
- definition of the search strategy.

The objective function in a CP program is represented by a domain variable. As an example, consider a scheduling problem. If we have to minimize the makespan, i.e., the duration of the whole schedule, we have to minimize the ending time of the last activity. Therefore, we have to minimize $Z = max_i\{Start_i + D_i\}$.

As another example, if we have a cost associated to each variable-value assignment and we have to minimize the overall cost, in the model we introduce a cost variable C_i associated to each problem variable. The link between a cost variable and the corresponding problem variable is achieved through an element constraint. The objective function is $Z = \sum_i C_i$.

On the objective function, a minimization or a maximization predicate should be imposed. This predicate automatically posts a constraint, called *bounding constraint* each time a solution is found. The bounding constraint imposes that next solutions should be better than the best one found so far. Therefore, solving an optimization problem in CP amounts of solving a set of feasibility problems obtained by the original problem plus the bounding constraint. In this way CP solvers implement a simple, indeed weak, form of branch-and-bound that is discussed in Section 2.3.3.

2.3 Solving

The solving process of constraint programming is characterized by two interleaved steps: constraint propagation and search.

2.3.1 Constraint Propagation. In general Constraint Satisfaction Problems can be solved by enumeration, but this approach is obviously not practical since the search space grows exponentially with the dimension of the problems. Thus, these problems greatly benefit from the reduction of the search space performed by constraint propagation, and some applications, where the feasibility side it very tight, are solved more effectively by using CP with respect to Integer Programming approaches (see, e.g., (Smith et al., 1996), (Darbi-Dowman and Little, 1998)).

During the computation, constraints are propagated in order to reduce variable domains by removing inconsistent values. If a domain becomes empty, a failure is raised and a backtrack forced. On of the basic propagation algorithms used for binary constraints is the one achieving arc-consistency (see the seminal paper by (Mackworth, 1977)).

As an example, consider variables $X :: [1..10]$ and $Y :: [4..18]$ linked via a constraint $X > Y$. This constraint is not arc-consistent: in fact, values $1, 2, 3, 4$ for variable X do not have a *support* in Y. Similarly, for all values between 10 and 18 for Y there is no consistent value in the domain of X. Therefore, they should be removed in order to make the constraint arc-consistent and obtain $X :: [5..10]$ and $Y :: [4..9]$.

Arc-consistency has been extended for non binary constraints, and called Generalized Arc-Consistency ((Bèssiere and Régin, 1997), (Bèssiere and Régin, 1999), (Bèssiere et al., 1999b), (Bèssiere et al., 1999a)). The interesting aspect of this algorithm is that it is general and problem-independent. The drawback is that its efficiency is indeed quite poor for large problems.

For this reason, symbolic constraints in general embed propagation algorithms which exploit the semantics of the constraint itself (see, e.g., (Beldiceanu and Contejean, 1994), (Régin, 1994), (Baptiste et al., 1995), (Régin, 1999b)). Consider, for example, the symbolic constraint:

$$\textit{alldifferent}([X_1, \ldots, X_n]).$$

it declaratively holds if all variables are assigned to a different value and it is equivalent to a set of binary inequality constraints connecting each pair of values in the list $[X_1, \ldots, X_n]$ (corresponding to $n(n-1)/2$ binary constraints $\neq$). However, we can perform more global and informed reasoning on the set of variables. In (Régin, 1994), a filtering algorithm based on network flow has been defined. We here make an intuitive example on how the algorithm prunes the search space.

Suppose for example that we have an *alldifferent* constraint among three variables $[X_1, X_2, X_3]$ whose domains are $D_1 = D_2 = [1, 2]$ and $D_3 = [1..10]$. While a set of binary inequalities are arc-consistent and cannot infer any value removal, the *alldifferent* constraint can reason globally on the cardinality of the sets of variables and values. We have two variables, i.e., X_1 and X_2, whose common domains $D_1 = D_2 = [1, 2]$ contain exactly two values. Thus, values 1 and 2 are *reserved* for variables X_1 and X_2 (no matter which value is assigned to which variable), and are no longer feasible for variable X_3 whose domain is reduced to $[3..10]$.

Interaction among constraints. An interesting feature of Constraint Programming is the interaction among constraints that cooperate through shared variables. The propagation algorithm is part of the constraint itself, and is triggered when an event is raised due to a domain modification of one variable involved in the constraint. The event can be one of the following:

- reduction of a domain bound;
- removal of a value;
- instantiation of a variable (the domain is reduced to a single value).

As soon as one constraint produces a modification on the domain of one variable, say X, all constraints involving X are triggered, and perform propagation on the basis of the current state of the variables' domains. At the end of the propagation, they are suspended until another event occur. A constraint is considered solved when it is always satisfied, i.e., $X_1 < X_2$ where $D_1 = [1..5]$ and $D_2 = [7..10]$ is always satisfied for all pairs of values in the domains of X_1 and X_2. Thus, if a constraint is solved it does not need to be triggered again.

In order to explain the interaction among constraints, let us consider the example depicted in Figure 1.1. The first propagation of $X = Y + 1$ yields to the following domain reduction $X :: [2..5]$, $Y :: [1..4]$, and $Z :: [1..5]$. Since $X = Y + 1$ is not solved it is suspended, since it could be triggered again after modifications of the domain of some variables. Then, propagation of $Y = Z + 1$ reduces domains as $X :: [2..5], Y :: [2..4], Z :: [1..3]$. The domain of Y has been changed, thus the first constraint is awakened and removes value 2 from the domain of X. Finally, the propagation of $Z = X + 1$ removes all values from the domain of X and a failure is detected. The order in which constraints are triggered and propagated does not affect the result, but can possibly affect the performances of the propagation process.

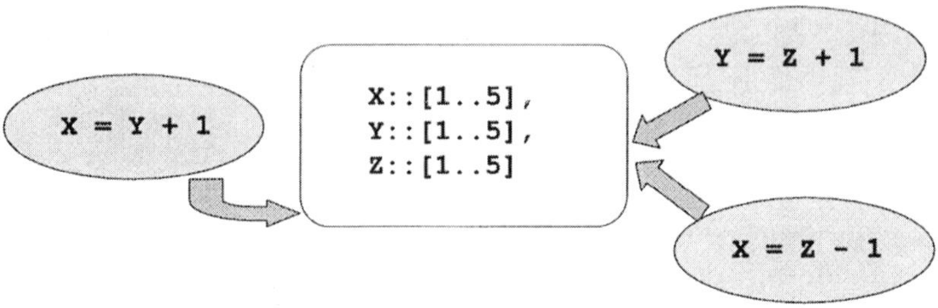

Figure 1.1. interaction among constraints

2.3.2 Search. From the previous section, we have explained that constraint propagation is a process that reduced a problem P in a problem P' equivalent to P, i.e., P and P' have the same solutions, where domains are possibly smaller. If P' is infeasible, so is P, but if P' has no domain empty, we don't know anything about P. Thus we should perform search.

Therefore, at the end of the constraint propagation process, we have three possible scenarios: (i) a domain becomes empty and a failure occurs; (ii) a solution is found, i.e., all variables are assigned to one value; (iii) some domains contain more than one value. In this third case, since constraint propagation is not complete, we need a search strategy in order to explore the remaining search tree.

The most commonly used search strategy is depth-first search with chronological backtracking but other choices are possible. At each node N of the search tree, we have to partition the corresponding problem P_N in a set of subproblems $P^1_{N+1}, \ldots, P^b_{N+1}$ so that a feasible solution to P_N is feasible to at least one of $P^1_{N+1}, \ldots, P^b_{N+1}$. One widely used way to partition a problem is called *labelling*, which means that one variable is selected and one value belonging to its domain is assigned to the variable itself. In other words, if the selected variable is X_i and its domain D_i contains values $v^i_1, \ldots, v^i_b$. Labelling partitions the problem in b subproblems where $P^1_{N+1} = P_N \cup \{X_i = v^i_1\}, \ldots, P^b_{N+1} = P_N \cup \{X_i = v^i_b\}$. The constraints $X_i = v^i_k$ are called *branching constraints*. The way we select the variable and we order its values greatly affects the performances of the tree search.

There are some general, problem independent strategies which can be effective. As an example, for variable selection, a widely used strategy is called *first-fail*. In this strategy, the selected variable is the one with the smallest domain since it is the most difficult to instantiate. General purpose search strategies are often not effective for real life applications. Thus,S it is important to study the problem and define specialized heuristics. For example, consider a scheduling problem. We have to allocate tasks to limited resources in time windows minimizing the makespan, i.e., the duration of the whole schedule. The model can be defined using variables $Start_i$ for each activity i representing their starting times. In this case, an effective strategy is to choose the variable $Start_i$ which has the earliest start time, i.e., the smallest domain lower bound. On backtracking, instead of choosing the next value in the domain of $Start_i$ as the labelling would do, the decision on this activity is postponed and another activity is selected. The reason why we postpone the decision and do not perform labelling is very simple: it is unlikely that we have a very *different* schedule if we shift one activity of a single time tick.

Another possibility often used in optimization problems, when the assignment cost should be minimised, is to choose the variable-value assignment associated with the lowest cost.

Different kind of branching constraints can be used in order to split the problem at a given node in simpler subproblems representing a partition of the original one. Domain partitioning can be used instead of labelling: for example, if a variable X has the domain $[1..10]$, a possible branching constraint can be of the type $X > 5$ and $X \leq 5$ upon backtracking.

Binary constraints can also be used for branching: in a scheduling application if we have two activities, associated with start variables $Start_i$ and $Start_j$ and durations d_i and d_j, sharing a unary resource, they cannot be scheduled at the same time. So, we can branch imposing $Start_i + d_i \leq Start_j$ and upon backtracking $Start_j + d_j \leq Start_i$.

When a problem relaxation is available, strategies based on the relaxation can be used. In a relaxation the solution violates some original problem constraints. A search strategy can be conceived so as to remove one or more violations at each branch. This technique will be discussed in Chapter 5. This technique is also used in Integer Programming branch-and-bound on the basis of the solution of the linear relaxation. In fact, a fractional variable is selected and two branching constraints imposed (the search tree is thus binary) both removing the fractional value (see Section 3.2.2).

In addition, variants of the depth-first search have been encapsulated in almost all CP solvers. One of the most frequently used and effective is Limited Discrepancy Search (see (Harvey and Ginsberg, 1995)) (LDS). LDS is a search strategy *trusting* the heuristics. At each node of the search tree, the heuristic is supposed to provide the *good* choice (corresponding to the leftmost branch) among possible alternative branches. Any other choice is considered a *discrepancy*. In LDS, one tries to explore first the branches with low discrepancy. In fact, a perfect heuristic would provide us the (optimal) solution immediately. Since this is not often the case, we have to increase the number of discrepancies so as to make it possible to find the optimal solution after correcting the mistakes made by the heuristic. However, the goal is to use only few discrepancies since in general good solutions are provided soon. Many variants of of LDS have been proposed like Depth Bounded Discrepancy Search, and Discrepancy-Bounded Depth First Search (see (Walsh, 1997) and (Beck and Perron, 2000)).

Eventually, discrepancy-based techniques can be stopped after a limited number of discrepancies and become incomplete strategies. These and other incomplete strategies will be extensively discussed in Chapter 9 and sketched in this Chapter in Section 4.

An important point to be clarified is that during the search, constraints are always taken into account and propagated in order to prune, as much as possible, the search space. In fact, a variable instantiation raises an event awaking all the constraints involving that variable and a propagation process starts again.

The way the search space is explored greatly influences the performances of the overall constraint-based computation. All CP languages have high level predicates allowing to easily write search heuristics (branching methods). This leads to the development, within the CP community of sophisticated branching methods for many type of problems allowing to solve them effectively (Perron, 1999).

2.3.3 Branch and Bound in CP. Finally, in some applications we are not looking just for a feasible solution, but for an optimal one with respect to some objective function f defined on problem variables. With no loss of generality, we restrict our discussion to minimization problems. CP systems usually implement a sort of branch-and-bound algorithm to find an optimal solution. The idea is to solve a set of decision (feasibility) problems (i.e., a feasible solution is found if it exists), leading to successively better solutions. In particular, each time a feasible solution z^* is found (whose associated cost is $f(z^*)$), a constraint $f(x) < f(z^*)$, where x is any feasible solution, is added to each subproblem in the remaining search tree. The purpose of the added constraint, called *upper bounding constraint*, is to remove portions of the search space which cannot lead to better solutions than the best one found so far. The problem with this approach is twofold: (i) CP does not rely on sophisticated algorithms for computing lower and upper bounds for the objective function, but derives their values starting from the variable domains; (ii) in general, the link between the objective function and the problem decision variables is quite loose and does not produce effective domain filtering.

As already mentioned, the objective function is represented by a domain variable C ranging over $[C_{min}..C_{max}]$ which is a function of the problem decision variables, i.e., $C = f(X_1, \ldots, X_n)$. Thanks to constraint propagation, the domain reduction of C can be reflected on decision variable domains and vice versa. C_{min} represents the problem lower bound, and C_{max} the problem upper bound. In CP the bounds of the domain of C are computed starting from domain variables. Suppose, for example, that C represents the makespan of a schedule. C_{max} at the beginning of the constraint solving process represents the schedule horizon, and it is updated each time a solution is found. Value C_{min} is set to the minimal ending time of tasks. If the earliest ending time of the tasks changes, C_{min} is updated accordingly. The fact that C bounds are updated during search is important since they provide information on the range where the optimal solution can be found. If the connection between C and problem decision variables were tight, the information provided by the upper bound (the best solution found) would be enough to find the optimal solution and prove optimality rather efficiently. Unfortunately, it is in general not the case in CP. Thus, a good information on the problem lower bound could be useful in order to speed up the search. In addition, it may be less computationally expensive to compute a good lower bound C_{min} than it is to perform effective constraint propagation from a given upper bound C_{max}.

Therefore, recently some efforts have been performed in order to embed in CP some techniques that take into account some form of *optimality reasoning*. Many parts of the book are indeed devoted to this aspect. For example, bounds deriving from the optimal solution of a (linear) relaxation have been used in CP based algorithms, (e.g., (Refalo, 1999), (Refalo, 2000), (Rodosek et al., 1997), (Beringer and Backer, 1995), (Hajian et al., 1996), (Pesant et al., 1998)). Different forms of relaxations have been also incorporated in symbolic constraints ((Régin, 1999a), (Focacci et al., 1999b), (Focacci et al., 1999a), (Caseau and Laburte, 1997)), thus achieving a form of pruning based on bounds and reduced costs.

2.4 An example: the car sequencing problem

This example is aimed at showing an application which has been widely faced with CP: the car sequencing problem. Cars in production are placed on an assembly line that moves through productions units that install options such as radios, air conditioning, etc. There are c classes of cars to produce and for each class a number of cars n_i. Therefore the total number of cars to produce is $nc = \sum_{i=1}^{c} n_i$.

Each class needs a fixed set of options and each car in the same class requires the same set of options. The assembly line is composed of nc slots and each car is allocated to a single slot. There are m possible options. Options cannot be installed on all cars of a production line, there are setup times that create capacity constraints. Capacity constraints are in the form *r out of p* meaning that only r cars in a sequence of p can be installed with that option. The car sequencing problem amounts to finding an assignment of cars to slots that satisfies the capacity constraints.

As mentioned in section 2.1, to define the CP model we should defined variables, domain values and constraints. For modelling the car sequencing problem, we introduce nc variables, one for each slot, that are assigned to the type of car to produce:

$$S_i :: [1, \ldots, c], \ \forall i \in \{1, \ldots, nc\}$$

Let m be the number of options, we also introduce $nc \times m$ variables

$$O_i^o :: [0, 1], \ \forall i \in \{1, \ldots, nc\}, o \in \{1, \ldots, m\}$$

if $O_i^o = 1$ the option o is installed on the slot i, 0 otherwise.

To ensure that the quantities required are indeed produced, we state the global cardinality constraint, introduced in Section 2.1, where the lower and upper bound for each assignment coincide.

$$gcc([S_1, \ldots, S_{nc}], [1, 2, \ldots, c], [n_1, \ldots, n_c], [n_1, \ldots, n_c])$$

To model the capacity constraints, we have alternative formulations. The first possibility is to use the following set of linear constraints: for each option o only r_o cars out of p_o can be produced, thus

$$O_i^o + \cdots + O_{i+r_o}^o \leq p_o \ for \ i \in \{1, \ldots, nc - r_o\}$$

Each of the previous linear constraints constraint can be replaced with a symbolic constraint of the form:

$$atmost(p_o, [O_i^o, \ldots, O_{i+r_o}^o], 1)$$

In both cases, we have a local propagation. Instead, we can use the global sequencing constraint, already introduced in section 2.1 and extensively explained in Chapter 4: for each option o, the number of cars nc_o installing that option can be easily computed as the number of classes requiring that option multiplied for the number of cars to produce in that class.

$$gsc([O_1^o, \ldots, O_{nc}^o], 1, nc_o, nc_o, p_o, r_o, r_o).$$

Now suppose that, in terms of cost, it is not equivalent producing a given option with different slots. So we have a cost matrix where cost c_{ij} is the cost of producing option i with slot j. The model of the problem should not change since the feasibility part is equal. We simply have to add an objective function to minimize, which is the sum of costs for producing all the options.

3. Integer Linear Programming Basic Concepts

Integer Programming is an alternative to Constraint Programming for solving many hard combinatorial problems. In fact, integer programming is a much older method, with roots that date back to the late 1950s with the pioneering work of Gomory and Land and Doig. In the last 40 years, there have been thousands of papers published on integer programming methods and modeling, and dozens of books. This section provides an overview of some of the most critical issues. Many more details can be found in survey papers (Johnson et al., 2000) and (Bixby et al., 2000) and in books such as (Wolsey, 1998).

Integer Programming can be thought of as a restriction of Constraint Programming. Instead of the variety of possible variables and domains of Constraint Programming, (Mixed) Integer Programming has two types of variables: integer variables with a domain consisting of the non-negative integers and continuous variables with the domain of nonnegative real values. More critically, instead of the rich and varied constraint types of constraint programming, integer programming allows only one type of constraint: linear inequalities. Finally, the objective function must be linear in the variables.

These restrictions make it seem that integer programming is much narrower than constraint programming. It is true that many problems have somewhat "unnatural" formulations in integer programming. However, the restrictions on integer programs still allow a great many problems to be modeled effectively, and algorithms for integer programs can find optimal solutions quickly for many applications.

The strengths of integer programming include

- Constraints are handled simultaneously through the *linear relaxation*, allowing arbitrary sets of linear constraints to be treated as a global constraint.
- Continuous variables can be handled naturally and efficiently along with discrete variables.
- Bounds are generated to give deviation from optimality even when optimal solutions are not proven.

For many applications, it is worth working within the limits of integer programming to get these advantages. Even better would be to combine constraint programming and integer programming to get the best of both worlds: that is the subject of the rest of this volume.

To fix notation, let x be the vector of variables, $x = [x_1, x_2, \ldots, x_n]$. A set of these variables I are required to take on integer values, while the remaining variables can take on any real value. Each variable can have a range, represented by vectors l and u such that $l_i \leq x_i \leq u_i$.

A *linear constraint* on the variables is a vector of coefficients $a = [a_1, \ldots, a_n]$ and a scalar right-hand-side b. The constraint is then the requirement that

$$\sum_j a_j x_j = b$$

or

$$ax = b$$

The "=" in the constraint can also be $\leq$ or $\geq$ (but cannot be $<$ or $>$). A linear objective is formed by a vector of coefficients $c = [c_1, c_2, \ldots, c_n]$, with the objective of minimizing (or maximizing) cx. An integer program consists of a single linear objective and a set of constraints. If we create a matrix $A = [a_{ij}]$, where a_{ij} is the coefficient for variable j in the ith constraint, then an integer program can be written:

$$\min cx \tag{1.1}$$
$$s.t \quad Ax = b \tag{1.2}$$
$$l \leq x \leq u \tag{1.3}$$
$$x_j \text{ integer for all } j \in I \tag{1.4}$$

In the next section, we will show how many problems of practical interest can be modeled in the above form. We will then outline some of the major algorithms for finding optimal solutions to integer programs and discuss methods for making those algorithms work better. We conclude by modeling the car sequencing problem given in the previous section as an integer program.

3.1 Modeling

While it seems that integer programming is quite restrictive, that is not the case. Many structures of practical interest can be modeled as linear constraints (though the model might not be immediately obvious). Here we outline some non-obvious uses of integer variables in modeling. For more examples, see the book by (Williams, 1999).

3.1.1 Logical Constraints. For many problems, it is important to identify when certain conditions occur. For instance, in a warehouse location problem, there may be a fixed cost incurred when a particular warehouse is opened. Many logical constraints can be modeled through the use of *binary* integer variables: integer variables that can only take on the values 0 or 1. The binary variable x_j might be used to represent opening ($x_j = 1$) or not opening ($x_j = 0$) warehouse j.

It is possible to combine binary variables into more complicated functions. Take binary variables x_1 and x_2. How can we force variable x_3 to be the function (x_1 OR x_2)? In constraint programming, every system has a function to represent this directly as something like $x_3 = x_1 \vee x_2$. In integer programming, we need to represent this as linear constraints as follows:

$$x_3 \geq x_1$$
$$x_3 \geq x_2$$
$$x_3 \leq x_1 + x_2$$
$$0 \leq x_3 \leq 1$$

In this formulation, the value 1 for a variable represents TRUE; 0 represents FALSE. If either of x_1 or x_2 is TRUE (1), then one of the first two constraints forces x_3 to be TRUE. If both are FALSE, then the third constraint forces x_3 to be FALSE.

Similarly, to represent $x_3 = x_1$ AND x_2, we can use the constraints

$$x_3 \leq x_1$$
$$x_3 \leq x_2$$
$$x_3 \geq x_1 + x_2 - 1$$
$$0 \leq x_3 \leq 1$$

Often, some of these constraints are not needed if the objective function "forces" variables in certain directions. For instance, for the AND function, if we are minimizing and there is a positive cost associated with x_3, then the first two constraints may not be needed, since x_3 will naturally remain 0 unless forced to take the value 1 by the third constraint.

Using these formulations as building blocks, it is possible to model arbitrary logical constraints using linear functions. Often it is necessary to add auxiliary variables to correctly handle intermediate values, which can result in a large and difficult-to-understand formulation, but conceptually, logical requirements are well-represented by integer programs.

3.1.2 Resource Constraints. Integer programming can also be used to represent sequencing constraints, such as those that arise in scheduling applications. Typically, a scheduling application will result in two types of variables: s_j, a continuous variable representing the start time for job j, and x_{ij} a binary variable representing whether job i occurs before job j or not.

As an example of the sort of constraints that occur, suppose we have a constraint along the lines of "If job 2 occurs after job 1, then it starts at least 10 time units after the end of job 1". Let d_1 be the duration of job 1 (this may be either data or a decision variable). Then the constraint

$$s_2 \geq s_1 + d_1 + 10 - M(1 - x_{12})$$

(where M represents a "suitable" large number) is a linear constraint that represents the requirement. "If job 2 occurs after job 1" means $x_{12} = 1$. When x_{12} is 1, then we get

$$s_2 \geq s_1 + d_1 + 10$$

as required. When $x_{12} = 0$, then we get

$$s_2 \geq s_1 + d_1 + 10 - M$$

As long as M is a sufficiently large number, this implies no restriction on s_2, again, as required.

In these applications, we also need to ensure that the x values are consistent, representing the concept of "coming before". The details vary by application, but are generally of the form:

$$x_{ij} + x_{ji} = 1$$

(either i is before j or the reverse, but not both), and

$$x_{ij} + x_{jk} - x_{ik} \leq 2$$

(if i is before j and j is before k, then i is before k).

3.1.3 Routing Constraints. Another common type of constraint occurs in routing applications where a set of sites must be visited. For instance, in the Traveling Salesman Program, we need to visit n sites. There is a distance c_{ij} from i to j, and we wish to travel the minimum distance while visiting all the sites. We can formulate it with binary variables x_{ij}, with a value of 1 representing travel from i to j and 0 representing no such travel. Our formulation becomes:

$$\min \sum_{ij} c_{ij} x_{ij} \tag{1.5}$$

$$s.t \sum_{j} x_{ij} = 1 \tag{1.6}$$

$$\sum_{j} x_{ji} = 1 \tag{1.7}$$

The first set of constraints enforces the requirement that we leave every site; the second forces us to enter every site. Unfortunately, this set of linear constraints is not sufficient. For instance, for 6 cities, consider the solution with $x_{12}, x_{23}, x_{31}, x_{45}, x_{56}, x_{64}$ equal to 1 and the rest equal to zero. It is easy to see that this satisfies the above formulation but the solution consists of two disjoint cycles: 1-2-3 and 4-5-6, which is not desired.

Enforcing a single cycle is not easy to do. One way to do this is to have a constraint for every proper subset of sites. Let S be the set of all proper subsets of sites, and let s be a generic member of S. Then we need

$$\sum_{i \in s, j \notin s} x_{ij} \geq 1$$

forcing us to leave every proper subset of sites. The above solution would violate the constraint for $s = \{1, 2, 3\}$.

This formulation is, of course, very large for all but the smallest of instances. We will discuss how to handle such huge formulations when we discuss algorithms.

3.1.4 Alternative Formulations. Formulations for integer programs are not unique. Normally there are many choices for variables, objective, and constraint that represent the same problem. For instance, consider a warehouse location example, where we have a set $1 \ldots n$ of customers who need to be assigned to one of warehouses, numbered $1 \ldots m$. Customers can only be assigned to an open warehouse, with there being a cost of c_j for opening warehouse j. Once open, a warehouse can serve as many customers as it chooses (with differing costs d_{ij} for each (customer,warehouse) pair). We can formulate this with binary variables x_j, representing the decision to open warehouse j, and y_{ij} representing the decision to assign customer i to warehouse j. This gives a partial formulation:

$$\min \quad \sum_j c_j x_j + \sum_i \sum_j d_{ij} y_{ij} \tag{1.8}$$

$$s.t \quad \sum_j y_{ij} = 1 \text{ for all } i \tag{1.9}$$

This forces every customer to be assigned to exactly one warehouse. How can we force a customer to be assigned only to an open warehouse? Here are two approaches:

If a warehouse is open, it can serve all the customers (all n of them). So, we get $\sum_i y_{ij} \leq n x_j$ for all j. Alternatively, if a customer i is assigned to warehouse j, then it must be open, leading to the constraints $y_{ij} \leq x_j$ for all

i and j. We can use either of these sets of constraints (or both) in our integer programming formulation to correctly model the problem. Algorithmically, however, the choice of formulation can have a strong effect on the effectiveness of the solution method.

3.2 Solving

The previous section gave a few examples of the strength of integer programming in formulating problems. Such formulations would not be useful if there were no algorithmic method for finding optimal solutions. There are a variety of methods for solving integer programs, and it is through these methods that we see the real reasons for formulating problems as integer programs.

3.2.1 Relaxations. If "constraint propagation" and "domain reduction" define constraint programming, the key concept in integer programming is "relaxation". A relaxation of a problem removes constraints in order to get an easier-to-solve problem. This easier problem is then used repeatedly to get a solution to the original problem.

Formally, a relaxation R(IP) to an integer program IP is a problem with the following characteristics (for a minimization problem):

- the optimal solution value to R(IP) is no more than that of IP
- if the solution to R(IP) is feasible to IP, then it is optimal for IP
- if R(IP) is infeasible, then so is IP.

For any integer program, there are a large number of possible relaxations. For instance, for our general IP:

$$\min cx \tag{1.10}$$
$$s.t \quad Ax = b \tag{1.11}$$
$$l \le x \le u \tag{1.12}$$
$$x_j \text{ integer for all } j \in I \tag{1.13}$$

we could create a relaxation by deleting the $Ax = b$ constraints, to form the problem (denoted BR(IP) for bounds relaxation):

$$\min cx \tag{1.14}$$
$$s.t. \quad l \le x \le u \tag{1.15}$$
$$x_j \text{ integer for all } j \in I \tag{1.16}$$

It is easy to see that this problem satisfies the relaxation requirements listed above: the objective is less than or equal to that of the original IP; if the optimal solution satisfies $Ax = b$ then it is optimal for IP; and if it is infeasible, so is the IP.

Over the years, researchers have studied many relaxations. This is particularly true for specially structured integer programs, for which there are a plethora of relaxations, including lagrangian relaxations, lagrangian decompositions, combinatorial relaxations, and many more. For many problems, and for general integer programs, the most common relaxation is the linear relaxation, which is formed by removing the integrality constraints (denoted LR(IP)):

$$\min cx \tag{1.17}$$
$$s.t \quad Ax = b \tag{1.18}$$
$$l \leq x \leq u \tag{1.19}$$

There are a number of reasons why the linear relaxation is used so often. Most important, the linear relaxation is simply a linear program, and there are fast computer codes to solve large linear programs (see, for instance (ILOG, 2002) (Associates, 2002)). Solvability of the relaxation is critical, and, as we shall see, it is important to be able to solve many relaxations quickly. Linear programming codes provide that capability.

Second, the linear relaxation embeds much critical information for many problems. The relaxation BR(IP) ignores much of the information about the problem embedded in the $Ax = b$ constraints. In contrast, the linear relaxation includes all of that information. We say that one relaxation R1 is stronger than another relaxation R2 if:

- The optimal objective value for R1 is at least that of R2, and
- R1 is infeasible whenever R2 is infeasible.

LR(IP) is always a stronger relaxation than BR(IP), for integral l and u. In our warehouse example from the previous section, the relaxation with the constraints $y_{ij} \leq x_j$ is stronger than the relaxation with $\sum_i y_{ij} \leq nx_j$.

The strength of this relaxation is related to the "global" view of the linear relaxation. By combining all of the constraints into a single linear program, the linear relaxation is able to identify infeasibilities that are not generated by any single constraint. This makes the linear relaxation a powerful method for uniting integer and constraint programming approaches.

A third advantage of the linear relaxation is its ability to generate a "dual value" for each constraint. Dual values are a natural side value generated by every linear programming algorithm. A technical introduction to linear programming duality is beyond this chapter (see any of the previously cited books for details), but intuitively, a dual value gives the marginal change in objective per unit of change of the right-hand-side of a constraint. So, if a constraint has a dual value of 10, changing the right-hand-side of the constraint by δ will change the objective value by 10δ. If another constraint has a dual value of -1, and we simultaneously change the first constraint by γ and the second by 2γ, the overall effect will be a change of $10\gamma + 2(-1)\gamma = 8\gamma$.

These dual values are of great use algorithmically. For instance, in a variable generation (or branch-and-price) approach (see Chapter 7), not every variable is included in the linear relaxation. Instead, the linear relaxation is solved with a subset of variables. Then, for each remaining variable, the effect of adding that variable is calculated through the dual values. Only those variables that can reduce the objective are added to the problem: the rest can be ignored. This method is enhanced by not checking each variable one-by-one but instead creating a *pricing subproblem* to find one or more variables whose addition will improve the overall objective. This turns out to be a very effective mechanism for solving problems with a huge number of variables.

Duality theory can also be used to identify variables that can be fixed at particular values, offering a simple domain reduction method based on reduced costs (see Chapter 5).

3.2.2 Branch and Bound. Armed with a relaxation, we can now present the general branch-and-bound algorithm for solving integer programs. This, like the tree search methods for constraint programming, is a recursive algorithm that generates multiple subproblems, each with additional variables fixed from is predecessor.

Beginning with integer program IP and relaxation R(IP), branch and bound works as follows:

1 Solve R(IP) to get optimal relaxation solution x^*.

2 If R(IP) is infeasible, then so is IP; else if x^* is feasible to IP, then x^* is optimal to IP; else create new problems IP1, IP2, ..., LPk by *branching* and solve each recursively. Stop if proved subproblem cannot be optimal to IP (*bounding*).

Branching. Branching is the process of creating new subproblems based on on an x^* that is infeasible to IP, but optimal to R(IP). These subproblems IP1, IP2, ... IPk must have the following properties:

- Every feasible solution to IP is feasible to at least one of IP1, IP2, ... IPk
- x^* is infeasible in each of R(IP1), R(IP2), ... R(IPk).

The first condition assures we do not make any feasible solution infeasible; the second condition prohibits the regeneration of x^* in the subproblems.

For the linear relaxation, x^* is infeasible if and only if there is a $j \in I$ such that x_j^* is fractional. In this case, we can create two new problems: one with the additional constraint $x_j \leq \lfloor x_j^* \rfloor$, and one with the constraint $x_j \geq \lfloor x_j^* \rfloor + 1$. This branching clearly meets our conditions.

Bounding. Bounding is the final use of the relaxation. Suppose during the branch-and-bound process, we have generated a feasible solution to some subproblem with objective value z^*. While z^* may be optimal to a subproblem, we do not yet know whether it is optimal to IP. But suppose we have a subproblem with relaxation objective value $z' \geq z^*$. Further branching can only increase the objective value, so it is clear that this subproblem cannot be the source of a better solution than z^*. We can cease examining this subproblem. It is through bounding that integer programming gets much of its effectiveness.

3.2.3 Improving the Effectiveness. Provided we continue branching until every subproblem is either infeasible, feasible to IP, or bounded, the branch-and-bound algorithm is guaranteed to find an optimal solution to IP. Of course, many difficult problems can be formulated as integer programs, so the number of subproblems might be very large. In these cases, there are methods to improve the computational features of branch and bound. We can divide these methods into formulation, cutting plane, branching, and feasible solution approaches. This short tutorial can only provide an overview of some of the major issues: for a more technical exposition, see, in particular, the survey papers (Johnson et al., 2000) and (Bixby et al., 2000).

Formulations. As we have seen, different formulations can have different linear relaxations. Which formulation should be used? Often, this is a computational issue: it is difficult to tell except by trying both.

If the choice is between two formulations where one has a provably stronger relaxation (like our warehouse example), it is almost always better to go with the one with the stronger relaxation. There are tradeoffs, however. Many formulations with stronger relaxations are also larger in terms of the number variables and/or constraints. This makes the linear relaxation slower to solve. The resulting decrease in number of subproblems in the branch and bound algorithm often overwhelms the increase in time per problem, however.

One sign of a "poor" formulation, or one unlikely to do well, is the appearance of multiple "Big-M" values in the formulation. Both our scheduling example and our warehouse example had constraints where a large value was used to make a constraint inactive in certain situations. Such constraints often result in very poor linear relaxations with linear relaxation objectives significantly different than integer programming values.

Another sign of difficulty comes from highly-symmetric formulations. Any problem with a symmetric relaxation runs the risk of excessive branching exploring alternative equivalent solutions. In such a case, it is important to break symmetry. This can be done either by fixing variables or by reformulating so that the variables do not exhibit the symmetries.

The general point to make is that formulating an integer program is a process: with many alternative formulations it may take some work to find one with good computational characteristics.

Cutting Planes. Cutting planes are constraints added to a relaxation to "cut off" (make infeasible) the optimal solution x^*. These methods automatically generate improved relaxations in the vicinity of the linear relaxation optimum. There are three major types of cutting planes:

- constraints embedded in the base formulation (like our connectivity constraints in our routing example)
- constraints associated with specific structures in the formulation
- constraints applicable to every integer program.

In all cases, to develop a cutting plane approach, we need to solve a "separation problem": given a class of constraints and a solution x^*, find a constraint in the class that is violated by x^*. If we can solve the separation problem, we can add the violated constraint to the relaxation and resolve, automatically getting a stronger relaxation.

To illustrate this idea, consider an integer program that contains one or more *knapsack constraints*. A knapsack constraint is a single linear constraint of the form

$$\sum_j a_j x_j \leq b$$

where the x_j associated with the nonzero a_j are all required to be binary and all the a_j are nonnegative. We say that C is a *cover* for this knapsack if

$$\sum_{j \in C} a_j > b$$

For any cover C, it is clear that we can deduce the following *cover constraint*:

$$\sum_{j \in C} x_j \leq |C| - 1$$

We can treat the set of all cover constraints as our constraint class, and ask the question: Given a fractional solution x^*, find a cover constraint that x^* violates.

This separation problem can be solved via an auxiliary knapsack problem, (Crowder et al., 1983). This problem will generate a violated cover constraint (if one exists) that can be added to strengthen the relaxation. In fact, (Crowder et al., 1983) showed that this approach often finds optimal solutions without branching, and greatly strengthened relaxations in other cases.

In addition to specially structured constraints like knapsack covers (and there are literally thousands of constraint classes known for various problems), there are some constraints that can be applied to any integer program. The most well-known such constraint is the Gomory Cut. Gomory gave a constraint that separates a fractional solution to any linear relaxation, (Gomory, 1958). In theory, such a constraint class obviates the need for any branching: repeated application of Gomory Cuts will eventually lead to the optimal integer solution. In practice, this approach is ineffective for most real problems, but the judicious addition of Gomory cuts has greatly decreased the solution times for many instances (Balas et al., 1996).

Cutting planes offer the possibility of automatically improving relaxation values, and have been a very fertile area of research.

Branching. Another area that can be exploited to make branch-and-bound more effective is the branching rule. Standard branch-and-bound has a very simple rule: find a fractional variable and limit it to no more than its round-down in one subproblem and at least its round-up in the other. More complicated branching rules are possible. For instance, suppose there is a constraint

$$x_1 + x_2 + x_3 + x_4 \leq 1$$

and in our current solution at least one of x_1^*, x_2^*, x_3^* and x_4^* is fractional. Rather than branching on the onc variable, it is possible to create five subproblems: one for each variable taking on the value 1, and one for having all variables take on value 0. These subproblems completely cover the feasible solutions relative to this constraint, and can decrease the amount of branching needed relative to standard branching.

Branching can even be used to model aspects that are difficult to embed in the integer programming formulation. An example is a constraint class known as Specially Ordered Sets of Type 2 (SOS-Type 2). In this class, there is an ordered set of continuous variables $x_1, x_2, \ldots, x_n$. The constraint requires that at most two of these variables are non-zero, and that if two are non-zero, they are consecutive. This constraint is used for modeling piece-wise linear functions.

Formulating SOS-Type 2 constraints is difficult in integer programming and the resulting formulation has a very weak linear relaxation. Instead, it is possible to embed these constraints in the branching. Consider a variable x_k. It is clear that there cannot be a non-zero variable among both $x_1, x_2 \ldots, x_{k-1}$ and $x_{k+1}, x_{k+2}, \ldots, x_n$. So a branching rule which adds a constraint

$$x_1 + x_2 + \ldots + x_{k-1} = 0$$

to one problem and

$$x_{k+1} + x_{k+2} + \ldots + x_n = 0$$

to the other meets our needs for a branching rule.

There are many other aspects of branching that might be addressed, including variable selection, choice of branch to first explore, and so on. Many of these issues are well explored in (Bixby et al., 2000).

Feasible Solutions. One final area that can be used to make branch-and-bound more effective is the generation of feasible solutions. In the simple outline given, feasible solutions were generated only when the relaxation solution happened to be feasible. But a good feasible solution can decrease computation time by increasing opportunities for bounding. Feasible solutions can be generated in a number of ways:

- Heuristic solutions can be generated outside of the branch-and-bound algorithm, and included as bounding values.
- The optimal relaxation solution x^* can be rounded to a feasible solution in some problems.
- Dual values and reduced costs can be used to guide a heuristic approach to generating solutions in every subproblem.

Many of these approaches are problem-specific, but there are some general methods available, including LP-Diving and Pivot-and-Complement. See (Johnson et al., 2000) for a description of these.

3.3 An example: the car sequencing problem revisited

We conclude by revisiting the car sequencing problem for which a constraint programming model was given. This problem can also be modeled within the limits of integer programming.

Recall that we have c classes of cars to produce and we need to produce n_i cars of class i. Each car is allocated to a single slot.

One choice of variables is to use binary variables x_{ij}, where a 1 is interpreted as producing a car of class i in slot j. We would then have constraints:

Produce enough of each car class: $\sum_j x_{ij} = n_i \; \forall i$.

One car per slot: $\sum_i x_{ij} = 1 \; \forall j$.

The other major constraint requires that no more than r_o out of p_o cars have option o installed. Let a_{io} be data that is 1 if class i has option o, and 0 otherwise. Then this constraint can be written as

$$\sum_{j_1=j}^{j_1 \leq j+p_o} \sum_i a_{io} x_{ij_1} \leq r_o \; \forall j, o$$

In words, for every time slot j and option o, take that timeslot and the p_o following slots, and count the number of cars whose class has option o. The result must be less than r_o.

In this formulation, it is easy to apply costs of having a car class in a particular time slot: this can provide an objective function $\sum_i \sum_j c_{ij} x_{ij}$.

4. Incomplete search strategies

We have seen that both CP and IP rely on tree search to explore the search space. However, this is not the only possibility. When the problem dimension becomes large, it is sometimes not possible to explore it completely. Therefore, either we stop the tree search at a given point, or we exploit a different search strategy, namely local search. Local search starts from a simple, but effective idea. Build a starting solution (either randomly or following a heuristics) then improve the solution with some given moves. Starting from a given solution S, all states reachable from it by the application of some given moves belong to the neighborhood of S. In this case, instead of a search tree, we explore a landscape called*fitness landscape*. A *Fitness Landscape* is defined by a triple:

$$\mathcal{L} = (S, \mathcal{N}, F)$$

where:

- S is the set of solutions (or states);
- $\mathcal{N}$ is the neighborhood function $\mathcal{N} : S \rightarrow 2^S$ that defines the neighborhood structure, by assigning to every $s \in S$ a set of states $\mathcal{N}(s) \subseteq S$.
- F is the objective function, in this specific case called *fitness function.*

The neighborhood function $\mathcal{N}$ implicitly defines an *operator* φ which takes a state s_1 and transforms it into another state $s_2 \in \mathcal{N}(s_1)$. Conversely, given an operator φ, it is possible to define a neighborhood of a variable $s_1 \in S$:

$$\mathcal{N}_\varphi(s_1) = \{s_2 \in S \setminus \{s_1\} \mid s_2 \text{ can be obtained by one application of } \varphi \text{ on } s_1\}$$

Usually, the operator is *symmetric*: if s_1 is a neighbor of s_2 then s_2 is a neighbor of s_1.

The local search process stops as soon as no improving solutions are found in the neighborhood, i.e., when a local optima has been found.

To escape from a local optima, many techniques have been devised going under the term metaheuristics. Metaheuristics are approximate algorithms which encompass and combine constructive methods, local search strategies, local optima escaping strategies and population-based search. Surveys and current research on metaheuristics can be found in (Aarts and Lenstra (eds.), 1997), (Corne et al., 1999).

The use of incomplete search strategies in Constraint Programming will be discussed in Chapter 9.

5. Conclusion

In this Chapter we have introduced the basic concepts underlying CP and IP. Our discussion is however far from being comprehensive. Many aspects of CP and IP have not been considered, and for details we remind to related bibliography. However, the main modelling aspects and solving mechanisms have been explained and can serve as a basis for understanding the rest of this book where some the integration of CP and IP is thoroughly explained.

Notes

1. Often in the rest of the book, this constraint is also referred to as *alldiff*, and the list or array of variables simply as X.

References

Aarts, E. and Lenstra, J.K. (eds.), (1997). *Local Search in Combinatorial Optimization.* Princeton University Press.

Associates, D. (2002). *XPRESS-MP User's Manual.*

Balas, E., Ceria, S., Cornuejols, G., and Natraj, N. (1996). Gomory cuts revisited. *OR Letters*, 19:1–9.

Baptiste, P., Pape, C. L., and Nuijten, W. (1995). Efficient operations research algorithms in constraint-based scheduling. In *Proceedings of the International Joint Conference on Artificial Intelligence, IJCAI'95.*

Beck, J.C. and Perron, L. (2000). Discrepancy-Bounded Depth First Search, *International Workshop on Integration of AI and OR Technologies for Combinatorial Optimization Problems*, CP-AI-OR'00.

Beldiceanu, N. and Contejean, E. (1994). Introducing global constraints in CHIP. *Mathematical Computer Modelling*, 20(12):97–123.

Beringer, H. and Backer, B. D. (1995). Combinatorial problem solving in constraint logic programming with cooperating solvers. In Beierle, C. and L.Plumer, editors, *Logic Programming: formal Methods and Practical Applications.* North Holland.

Bèssiere, C., Freuder, E., and Régin, J. (1999a). Using constraint metaknowledge to reduce arc-consistency computation. *Artificial Intelligence*, 107:125–148.

Bèssiere, C., Meseguer, P., Freuder, E., and Larrosa, J. (1999b). On forward checking for non binary constraint satisfaction. In *Proceedings of the International Conference on Principles and Practice of Constraint Programming, CP'99.*

Bèssiere, C. and Régin, J. (1997). Arc consistency for general constraint networks: preliminary results. In *Proceedings of Proceedings of the International Joint Conference on Artificial Intelligence, IJCAI'97.*

Bèssiere, C. and Régin, J. (1999). Enforcing arc consistency for global constraints by solving subproblems on the fly. In *Proceedings of the International Conference on Principles and Practice of Constraint Programming, CP99.*

Bixby, R., Fenelon, M., Gu, Z., Rothberg, E., and Wunderling, R. (2000). MIP: Theory and practice — closing the gap. In Powell, M. and Scholtes, S., editors, *System Modelling and Optimization: Methods, Theory and Applications*. Kluwer.

Caseau, Y. and Laburte, F. (1997). Solving small TSPs with constraints. In *Proceedings of the Fourteenth International Conference on Logic Programming - ICLP'97*, pages 316–330.

Corne, D., Dorigo, M., and (eds.), F. G. (1999). *New Ideas in Optimization - Advanced Topics in Computer Science Series*. McGraw-Hill.

Crowder, H., Johnson, E., and Padberg, M. (1983). Solving large-scale zero-one linear programming problems. *Operations Research*, pages 803–834.

Darbi-Dowman, K. and Little, J. (1998). Properties of some combinatorial optimization problems and their effect on the performance of integer programming and constraint logic programming. *INFORMS Journal of Computing*, 10(3):276–286.

Dincbas, M., Hentenryck, P. V., Simonis, M., Aggoun, A., Graf, T., and Berthier, F. (1988). The constraint logic programming language CHIP. In *Proceedings of the International Conference on Fifth Generation Computer System*, pages 693–702.

Focacci, F., Lodi, A., and Milano, M. (1999a). Cost-based domain filtering. In *Proceedings of the International Conference on Principles and Practice of Constraint Programming, CP99*, pages 189–203.

Focacci, F., Lodi, A., and Milano, M. (1999b). Solving tsp with time windows with constraints. In *ICLP'99 International Conference on Logic Programming*.

Gomory, R. (1958). Outline of an algorithm for integer solutions to linear programs. *Bulletin of the Americal Mathematical Society*, 64:275–278.

Hajian, M., El-Sakkout, H., Wallace, M., Lever, J., and Richards, E. (1996). Towards a closer integration of finite domain propagation and simplex-based algorithms. *In Proceedings of the Fourth International Symposium on Artificial Intelligence and Mathematics* 1996.

Harvey, W. D. and Ginsberg, M. L. (1995). Limited Discrepancy Search. In Mellish, C. S., editor, *Proceedings of the International Joint Conference on Artificial Intelligence IJCAI95*.

IC-Park (2003). *ECL^iPS^e User Manual.*

ILOG (2002). *ILOG Cplex Users Manual 7.0.*

Johnson, E., Nemhauser, G., and Savelsbergh, M. (2000). Progress in linear programming based branch-and-bound algorithms: An exposition. *INFORMS Journal on Computing*, 12:1–48.

Mackworth, A. (1977). Consistency in networks of relations. *Artificial Intelligence*, 8:99–118.

Perron, L. (1999). Search procedures and parallelism in constraint programming. In *Proceedings of the International Conference on Principles and Practice of Constraint Programming, CP99.*

Pesant, G., Gendreau, M., Potvin, J., and Rousseau, J. (1998). An exact constraint logic programming algorithm for the travelling salesman problem with time windows. *Transportation Science*, 32(1):12–29.

Puget, J. (1994). A C++ implementation of CLP. Technical Report 94-01, ILOG Headquarters.

Refalo, P. (1999). Tight cooperation and its application in piecewise linear optimization. In *Proceedings of the International Conference on Principles and Practice of Constraint Programming, CP99.*

Refalo, P. (2000). Linear formulation of constraint programming models and hybrid solvers. In *Proceedings of the International Conference on Principles and Practice of Constraint Programming, CP2000.*

Régin, J. (1994). A filtering algorithm for constraints of difference in CSPs. In *Proceedings of the American Conference on Artificial Intelligence, AAAI94.*

Régin, J. (1999a). Arc consistency for global cardinality constraints with costs. In *Proceedings of the International Conference on Principles and Practice of Constraint Programming, CP99.*

Régin, J. (1999b). The symmetric Alldiff constraint. In *Proceedings of the International Joint Conference on Artificial Intelligence, IJCAI'99.*

Rodosek, R., Wallace, M., and M.T.Hajian (1997). A new approach to integrating mixed integer programming and constraint logic programming. *Annals of Operational Research.* Recent Advances in Combinatorial Optimization.

Smith, B., Brailford, S., Hubbard, P., and Williams, H. (1996). The progressive party problem: Integer linear programming and constraint programming compared. *Constraints*, 1:119–138.

Tsang, E. (1993). *Foundation of Constraint Satisfaction.* Academic Press.

Walsh, T. (1997). *Depth-Bounded Discrepancy Search.* In *Proceedings of the International Joint Conference on Artificial Intelligence, IJCAI97.*

Williams, H. (1999). *Model Building in Mathematical Programming.* John Wiley and Sons, 4 edition.

Wolsey, L. (1998). *Integer Programming.* Wiley.

Chapter 2

TWO GENERIC SCHEMES FOR EFFICIENT AND ROBUST COOPERATIVE ALGORITHMS

Emilie Danna
ILOG S.A. and
LIA, Université d'Avignon, CNRS-FRE 2487
edanna@ilog.fr

Claude Le Pape
ILOG S.A.
9 rue de Verdun
F-94253 Gentilly Cedex
clepape@ilog.fr

Abstract Many examples of hybrid algorithms have been described in the recent optimization literature and shown to provide better results than pure algorithms based on only one technology. Yet, it is difficult to understand what makes a cooperative strategy work, hence it is difficult to design a good cooperative strategy. In this paper, we try to convey our understanding of the performance of hybrid algorithms with respect to two criteria: efficiency and robustness. We first present the strengths and weaknesses of each possible component of hybrid algorithms: polynomial operations research algorithms, constraint programming, mixed integer programming, and local search. We then give an overview of hybrid algorithms, looking at each of the six possible combinations of two of the techniques enumerated above. We try to abstract from hybrid examples the two following generic cooperative schemes. The decomposition scheme consists in applying one technique to a sub-problem in order to gain information that is then used by another technique to solve the overall problem. The multiple search scheme combines several techniques that solve in turn or in parallel the full problem and exchange information which allows for better diversification and intensification.

1. Introduction

In the recent past, many examples of hybrid applications have been described and the adopted hybrid algorithms shown to provide better results than pure algorithms based on only one technology. Yet the number of cases in which cooperative strategies have been applied remains very small compared to the number of cases in which they could potentially prove useful.

There might be several explanations for this fact. First, most work on cooperative optimization strategies has been focused either on the resolution of a particular problem at hand or on the design of generic software integration principles. A finer understanding of what makes a cooperative strategy work on a particular application and of the variants of the problem that could be solved well with the same cooperative strategy would be needed to allow an efficient generalization to similar yet different applications. Second, cooperative strategies are hard to develop:

- Appropriate sub-models or alternative models have to be designed and implemented, which often takes more time than the design and implementation of a unique model. One could actually argue that most experimental results from the literature showing the superiority of a cooperative strategy are biased by the fact that the cooperative strategy took much longer to develop than the non-cooperative strategies against which it was tested.

- The design of the cooperative strategy requires some understanding of what each component does, of the strengths and weaknesses of different technologies for the resolution of different types of problems. Very few people if any have a deep understanding of the four main classes of optimization techniques that can be used: (1) efficient (generally, polynomial or pseudo-polynomial) operations research algorithms, (2) constraint programming, (3) linear and mixed integer programming, and (4) local search.

- Cooperative strategies are hard to tune, probably just because the number of parameters that can be tuned is much higher than in the case of a "pure" strategy. Good tools and introductions to these tools are necessary to allow more developers of optimization applications to design, implement, and tune cooperative problem-solving strategies.

This chapter relies on examples from the literature to convey some understanding, or at least some intuition, of what types of cooperative strategies may be worth developing for a given optimization application. Given an optimization application, two complementary issues must be considered. First, the application must be "efficient", *i.e.*, provide a "good" (ideally, optimal) solution within reasonable time and resource limits (CPU, memory). Second, the application must be "robust" with respect to three types of variations in the problem instances: variations in problem size, variations in numerical characteristics, and addition of side constraints. It shall in most cases be easy to include additional constraints without re-designing the overall problem-solving strategy. The goal of hybridization is to improve either efficiency or robustness (or both) without sacrificing the other.

The four main classes of optimization techniques listed above present different characteristics with respect to the efficiency versus robustness compromise:

- Polynomial operations research algorithms exploit the problem structure of well-defined problems to solve them to optimality in time bounded by a polynomial function of the problem size. When the degree of the polynomial function is low, the algorithm is efficient and in general robust to changes in problem size and numerical characteristics. But in most cases the algorithmic efficiency is based on a set of compatible dominance properties $\{p_1, p_2, \ldots, p_n\}$, which state that at least one of the optimal problem solutions satisfies p_1, p_2, $\ldots$, and p_n. The use of such dominance rules strongly reduces the set of alternative solutions to be considered. However, dominance properties are in general not stable against the addition of side constraints to the original problem, *i.e.*, when the problem definition is extended, the dominance rules are no longer correct and the algorithm is likely to miss the optimal solutions or even all admissible solutions to the extended problem.

- Constraint programming (*cf.* Chapter 1) is a problem solving paradigm which establishes a distinction between, on the one hand, a precise definition of the constraints that define the problem to be solved and, on the other hand, the algorithms and heuristics enabling the selection and cancellation of decisions to solve the problem. In constraint programming, a purely deductive process referred to as "constraint propagation" is used to propagate the consequences of problem constraints and decisions. This process is applied each time a new decision is made, and is clearly separated from the decision-making algorithm (usually some form of tree search) *per se*. Most importantly, the overall constraint propagation process results from the combination of several local and incremental processes, each of which is associated with a particular constraint or a particular constraint class. This means that in constraint programming, any added side constraint can propagate and play a role in the resolution of the problem. Constraint programming is particularly efficient when constraint propagation is fast and effective, *i.e.*, if the critical constraints propagate well and, in an optimization context, if a tight bound on the optimization criterion results (by propagation) in constraints that effectively guide the search towards a good solution. Compared to other techniques, constraint programming tends to be more robust against the addition of side constraints, but it is not that robust against changes in problem size and numerical characteristics.

- Mixed integer programming (*cf.* Chapter 1) also establishes a distinction between the definition of the problem and its resolution. But it gets much of the guidance towards good solutions from the continuous relaxation of the problem. Mixed integer programming is especially efficient whenever the continuous relaxation is a good approximation of the convex envelope of the solutions (at least around the optimal solution) or when the relaxation can be iteratively tightened (by adding cuts) to improve this approximation. Much work has been done to make commercial implementations rather robust with respect to changes in problem size and numerical characteristics, but there are some problems for which the size of the mixed integer programming formulation grows much faster than the size of the problem. Linear side constraints are easy to accomodate, but the addition of non-linear or disjunctive side constraints generally enforces the addition of new integer variables in the linear model. In both cases, the continuous relaxation may become a bad approximation of the mixed integer problem.

- Various forms of local search (*cf.* Chapter 1), operating either on individual solutions (*e.g.*, simulated annealing, tabu search) or on populations of solutions (*e.g.*, genetic algorithms, path relinking), provide excellent results on some problems. Roughly speaking, local search tends to be efficient when good problem solutions share characteristics (*e.g.*, the ordering of two tasks) that can be compactly represented and that are likely to be kept when local search operators proceed from a solution or a set of solutions to the next. The robustness of a local search algorithm depends on the representation, operators, and overall strategy that are used. For example, some local search operators may become too expensive to apply systematically when problem size grows, or even useless when a side constraint is added (*e.g.*, a constraint relating the ordering of two tasks to the ordering of two other tasks, in a complex manner). If the effect of adding a side constraint is that the density of feasible solutions is reduced or that simple problem characteristics are less often shared by good solutions, the local search algorithm needs to be revised.

Looking at examples from the literature, two generic cooperative schemes emerge. The **decomposition** scheme consists in applying one of the above techniques (the "slave") to a sub-problem of the overall problem, in order to gain information. This information is then used by another of the above techniques (the "master"), to solve the overall problem. The sub-problem given to the slave may be a reduced problem with less variables, an approximation, a relaxation, or a tightened version of the overall problem. In some cases, the results given by the slave may be used by the master to focus the overall search, without any risk of missing the optimal solution. In other cases, the results given by the slave may be used as a guide towards good solutions, but do not strictly allow to reduce the search space.

In the **multiple search** scheme, two or more optimization techniques are used in turn or in parallel to provide solutions to the full problem. In the simplest case, each optimization component operates independently from the others. In most cases, however, the solutions found by one component and, more generally, information generated by one component as part of its own optimization process, can be used by other components to better focus subsequent search. Gains in efficiency and robustness stem from introducing diversification in the methods used to explore the search space, as well as some intensification on some promising parts of the search space.

The remainder of this chapter provides examples from the literature which illustrate the use of these two schemes. It is organized in six sections. Each section is dedicated to examples of cooperation involving two of the four classes of techniques discussed above. It is not intended to be exhaustive, but attempts to expose the most significant ways we are aware of to apply the above schemes.

2. Operations Research Algorithms and Constraint Programming

Obviously, the use of a cooperative scheme is not likely to be worthwhile when an efficient operations research algorithm can solve the complete problem under consideration. Hence, operations research algorithms either work alone or appear as slaves in the **decomposition** scheme, working on sub-problems or on simplified versions of the problem to provide guidance to another algorithm.

In the case of constraint programming, two different roles can be taken by the operations research algorithm: either it is used to reinforce constraint propagation, *e.g.*, deduce better bounds for variables of the problem, including the variable representing the cost of a solution, or it is used to heuristically guide the constraint-based tree search to promising regions. The main overall advantage of the combination is that the specific operations research algorithm brings efficiency to the constraint programming framework, while constraint programming is far more robust than the specific algorithm with respect to the addition of side constraints.

Chapter 4 provides many examples of so-called "global" constraints for which polynomial algorithms can be used to reinforce propagation. The improvement in efficiency, compared to more "naive" constraint propagation methods can be significant. The most striking examples come from the scheduling domain. For example, the "edge-finding" algorithm, first presented in (Pinson, 1988; Carlier and Pinson, 1989; Carlier and Pinson, 1990), was adapted by various researchers to the constraint programming framework (*e.g.*, (Nuijten, 1994; Caseau and Laburthe, 1994; Baptiste and Le Pape, 1995)), and generalized by these researchers to more general cases (*e.g.*, (Nuijten and Aarts, 1994; Caseau and Laburthe, 1996; Baptiste et al., 1999)). Results obtained on an industrial project scheduling problem (Baptiste and Le Pape, 1995) show how the use of edge-finding enabled the resolution in at most five minutes of problems which were not solved in one hour, using standard constraint propagation techniques.

Another typical example is the use of flow or assignment algorithms within global constraints. For example, (Régin, 1994) describes an algorithm, based on matching theory, to achieve the global consistency of the "all-different" constraint. This constraint is defined on a set of variables and constrains these variables to assume pairwise distinct values. Régin's algorithm maintains arc-consistency on the n-ary "all-different" constraint, which is more powerful than achieving arc-consistency for the $n*(n-1)/2$ corresponding binary "different" constraints. Both (Caseau and Laburthe, 2000) and (Focacci et al., 1999) have proposed extensions of such a global constraint to the case in which the cost to minimize is defined by summing individual costs assigned to each selected assignment.

In the scheduling domain, the case in which the cost function to minimize is a sum (like the weighted number of late jobs or the sum of setup times between activities) is also of particular interest. Propagating an upper bound on the cost, each time a new improving solution is found, is not a problem when the optimization criterion is a "maximum" such as the makespan (maximal completion time) or the maximal tardiness of a given set of activities. Indeed, an upper bound on the optimization criterion is directly propagated on the completion time of the activities under consideration, *i.e.*, the latest end times of these activities are tightened efficiently. On the contrary, when the cost function is a sum, an upper bound on the cost does not directly provide good upper bounds on the different terms of the sum because it is not clear which terms actually have to be reduced. In such situations, better results can be obtained by combining constraint programming with deductive algorithms targeted towards specific cost functions. See, for example, (Baptiste et al., 1998) for the weighted number of late jobs and (Focacci et al., 2000) for the sum of setup times.

A good example in which an operations research algorithm is used to guide the search is presented in (Le Pape et al., 2002). The goal is to route commodities over a network and dimension the links of the network so as to minimize the dimensioning costs. A shortest path algorithm is used as a heuristic. It determines the route with the smallest marginal cost for a given commodity, given the commodities that have already been routed. Search proceeds by imposing this route and backtracking on it if necessary. An analysis of the explored search trees showed that during the search almost all the improving solutions questioned one of the first routing decisions that had been taken in order to build the previous solution, *i.e.*, that in the middle and close to the leaves of the search tree, this heuristic almost always led to the best solution, given the choices made close to the root of the search tree. Experimental results with limited CPU time showed that using graph-oriented algorithms, both for heuristic choices and constraint propagation, enabled to save 14% of the costs on average over seven small instances provided by France Telecom. Without these graph-oriented algorithms, constraint programming would not be a viable option for the resolution of these problems.

3. Operations Research Algorithms and Mixed Integer Programming

As in the previous section, operations research algorithms are used as slaves in a **decomposition** scheme, working on sub-problems or on simplified versions of the problem to provide guidance to a MIP solver which acts as a master.

A typical example is column generation or branch-and-price (Barnhart et al., 1998). Chapter 7 explains in detail how column generation works, but let us describe the idea on a specific problem: the vehicle routing problem with time windows. The problem consists in assigning to each vehicle of a given pool an ordered list of customers to visit, within allowed time windows. Each "column" corresponds to one feasible route, starting from and ending at the vehicle depot and visiting several customers in their allowed time windows. The master problem is a set covering problem, that selects a number of such columns to cover all customers and to minimize the overall traveling cost. The number of possible columns is exponential so they cannot all be present at the beginning like in mixed integer programming, but they are generated on the fly. At the beginning, a set of naive columns compose a gross approximation of the convex hull of the integer solutions. At each iteration, this approximation is refined by solving a sub-problem to introduce new columns with negative reduced cost (*cf.* Chapter 1 for an introduction to dual values and reduced costs): according to the continuous relaxation of the set covering problem, these columns represent a better way to visit customers than the already existing columns. The sub-problem is a shortest path with time windows and it can be solved with any kind of optimization method. It is usually solved with a dynamic programming algorithm (Desrochers, 1988) that has pseudo-polynomial complexity but is not too expensive in practice. This algorithm is based on labels: each partial path from the depot to node j is associated to a cost and a label, that is a multidimensional vector corresponding to the quantity of each resource used by this path. Partial paths are progressively extended and a dominance relation allows to remove dominated partial paths following Bellman's optimality principle:

all non-dominated paths are the extension of a non-dominated path. In other words, if at node j a partial path arrives sooner and is shorter than another path, then this latter path can be removed because all of its extensions will be dominated by the extensions of the former partial path. Additional constraints can easily be taken into account if they are accumulation-type constraints, that is if the state of the constrained resource at node j only depends on the state of the resource at the predecessor of node j in the partial path considered: one only needs to add to the labels one dimension for each additional constrained resource. For example, a capacity constraint limiting the amount of goods that can be carried by each vehicule can be taken into account in that way. This label-based algorithm can also be adapted to a more complex additional constraint: generate only elementary routes, as reported in (Feillet et al., 2001) by adding to each label n 0-1 dimensions (if n is the number of customers), and independently in (Chabrier et al., 2002) by modifying the dominance relation.

Another example is Lagrangean relaxation. It consists in relaxing some constraints by penalizing the violation of these constraints in the cost function. The Lagrangean relaxation often yields lower bounds of good quality (when minimizing). This technique is especially interesting if the constraints matrix is structured in k independent sub-problems, be it for a group of coupling constraints that link together all variables. If the group of coupling constraints is dualized, the k sub-problems can be solved independently, and they are often easy to solve because they are smaller in size and "purer" in structure than the initial problem. Therefore dedicated operations research algorithms can be used to solve them. This is a typical example of the **decomposition** scheme. For example, (Gendron and Crainic, 1994) dualize the flow conservation constraints in a network design problem. This leads to continuous knapsack sub-problems which are easily solved by sorting the knapsack cost coefficients.

4. Constraint Programming and Mixed Integer Programming

Combining constraint programming and linear programming or mixed integer programming is the most studied combination in hybrid optimization.

The **multiple search** scheme is not often used, probably because constraint programming really works better than mixed integer programming on some applications, while mixed integer programming really works better than constraint programming on the others, thereby leading to the selection of one of these paradigms to be the master. Yet some exceptions might occur. For example, if we consider the above-mentioned network dimensioning problem ((Le Pape et al., 2002)), it appeared in the course of the study that on some instances constraint programming provided in 5 minutes solutions much better than those found by mixed integer programming in 10 minutes, and vice-versa. In such a case, running the two algorithms, each for 5 minutes, would be beneficial. Needless to say, more complex multiple search strategies can also be considered. For example, constraint programming can be used to quickly construct good solutions and mixed integer programming to get lower bounds and optimal solutions — using the upper bounds from constraint programming to prune the search tree. Branches that are cut off by one of the two algorithms can also be transformed into additional constraints for the other algorithm, preventing a given set of decisions to be made on the same branch, and thereby preventing the exploration of a useless branch.

The **decomposition** scheme with constraint programming as the master is the most often used, because it allows the hybrid algorithm to benefit both from the efficiency of the linear solver and from the robustness of constraint programming to the addition of side constraints. The most classical **decomposition** consists in passing a linear sub-problem, *i.e.*, a set of linear constraints and possibly a linear objective to minimize, to a linear programming solver. The linear solver determines whether the given constraints can be simultaneously satisfied and provides a lower bound on the objective. It can also adjust variable domains when inequalities are implied. CLP(R) (Jaffar and Lassez, 1987), CHIP (Hentenryck, 1989), Prolog III (Colmerauer, 1990), are early examples of constraint programming languages embedding a linear solver. Such an integration can be viewed as a particular case of global constraint propagation: the linear solver operates on a global constraint expressed as a conjunction of linear constraints, *i.e.*, on the continuous relaxation of a sub-problem. Note that beside domain adjustments, the linear solver also provides the optimal continuous solution of the sub-problem. When the sub-problem is "close" to the complete problem, this solution can also be used by the constraint programming master as a heuristic guide towards a good solution of the complete problem. A particular example in the field of dynamic scheduling is given in (Sakkout and Wallace, 2000). In this example, the linear solver includes only temporal constraints (some of which have been added to the initial problem in order to ensure the satisfaction of resource sharing constraints) and the definition of the optimization criterion as the total deviation of start times of activities from the start times of the same activities in a reference schedule. An interesting characteristic of this model is that the optimal continuous solution of the linear sub-problem is guaranteed to be integral; hence, either this solution satisfies all the resource constraints and it is optimal, or it violates some resource constraint which can be used to branch on the order of two conflicting activities.

Refinements of this **decomposition** scheme are presented in (Milano et al., 2002). For example, when strong bounds on the objective are known, reduced costs resulting from linear solving can also be used to adjust variable domains (*cf.* Chapter 1). (Milano and van Hoeve, 2002) present a different use of reduced costs to guide constraint programming. First a linear relaxation of the original problem is solved. The resulting reduced costs are used to rank values inside each variable domain, which are accordingly divided into a *good* set and a *bad* set of values. The branching strategy at each node is to choose one branching variable and reduce its domain to its *good* set in one child node and to its *bad* set in the other child node. The global constraint programming search tree is then explored with limited discrepancy search (Harvey and Ginsberg, 1995), first exploring the most promising problems, *i.e.*, those where most variables range on their *good* set. Results on pure TSP problems and some variants show that the proposed ranking based on reduced costs is extremely accurate: in almost all cases, even if the relaxation is loose, the optimal solution is found in the sub-problem where all variables range on their *good* set. This sub-problem is explored first, which decreases the computational time needed to reach the optimal solution. However, the tightness of the lower bound is instrumental for completing rapidly the proof of optimality.

In some cases, the linear model can also be tightened as the constraint programming engine progresses towards a solution. (Refalo, 1999) discusses the case of piecewise linear functions. Given is a constraint $y = f(x)$, where f is a piecewise linear function. The convex hull of this constraint is maintained at all nodes in the search tree: cuts are added to the linear formulation when the domain of x is reduced. For small transportation problems with costs y depending on shipped quantities x, such a refinement enabled to save between one and two orders of magnitude in CPU time.

Transformations of non-linear constraints, in particular disjunctive constraints, into linear constraints can also be considered (Rodosek et al., 1999; Refalo, 2000). In general, this leads to the introduction of additional binary variables in the linear model. Hoist scheduling problem results provided in (Rodosek and Wallace, 1998) show that the resulting cooperative algorithm is more robust than pure constraint programming or mixed integer programming.

Other forms of **decomposition** also appear in the literature.

The first example is branch-and-price, where the sub-problem is solved with constraint programming (*cf.* Chapter 7 for an in-depth discussion of constraint programming based branch-and-price). This allows to combine the power of branch-and-price to optimize very big problems and the flexibility of constraint programming to take into account side constraints. (Rousseau et al., 2002) applies this cooperation technique to the vehicle routing problem with time windows. (Yunes et al., 2000) applies it to crew scheduling and crew rostering problems for a brazilian urban transit bus company: the hybrid method is shown on multiple and various real world instances to perform better than both integer programming and constraint programming approaches used alone.

The second example is Lagrangean relaxation (*cf.* Section 3 for an introduction). A typical example of this **decomposition** scheme is (Benoist et al.,

2001). The Traveling Tournament Problem is a variant of the sports league tournament: one has to organize a set of matches between all teams of a sports league in a given number of weeks, such that each team plays each other team twice, once at home and once away, the sequence of meetings for each team abiding some constraints (for example no more than 3 consecutive home games are allowed). The objective is to minimize the sum of the distance travelled by all teams. The coupling constraints in this problem are that team A must be at home when team B visits it and that no more than one visitor stays at each team's home simultaneously. Once these coupling constraints have been dualized, each sub-problem dealing with the sequence of games of one team can be solved independently with a dedicated TSP constraint-based solver. The computation of this Lagrangean lower bound is embedded in a global constraint that is used in the main constraint programming model driving the tree search.

(Sellmann and Fahle, 2003) give a general framework to solve linear problems such that the constraints can be divided in two families A and B, for which an efficient propagation algorithm exists: respectively $Prop(A)$ and $Prop(B)$. However, combining the two efficient propagation algorithms in a straightforward manner may not be efficient because it is likely that tight bounds on the objective are not obtained by taking only a subset of the restrictions into account. Therefore, Sellman and Fahle propose the following **decomposition** scheme: dualize the A constraints family and use $Prop(B)$ for propagation in each corresponding Lagrangean sub-problem, then apply $Prop(A)$ on the Lagrangean sub-problem obtained when dualizing the B constraints family. This scheme can be generalized to more than two families of constraints. It is applied to the Automatic Recording Problem that contains a knapsack constraint and a maximum weighted stable set constraint on an interval graph. The numerical results show the coupling method significantly decreases computation time and the number of choice points. In both examples (Benoist et al., 2001; Sellmann and Fahle, 2003), the tight bounds on the objective allow to remove some values from the variables domains. Hence Lagrangean relaxation enables constraint programming to be applied more successfully to optimization problems.

The third example is Benders decomposition (Benders, 1962) which is often understood as the "dual" of column generation: instead of generating columns iteratively, the set of columns is given but the rows are generated iteratively. Whereas Lagrangean relaxation is particularly helpful in presence of coupling constraints, Benders decomposition is particularly helpful in presence of coupling variables. The principle is to solve a relaxed master problem to fix the values of the coupling variables. The rest of the variables are then assigned in k sub-problems (that can now be solved independently because the coupling variables are fixed), and each solution to the sub-problems generates a Benders cut that is added to the relaxed master problem. Both the master problem and the sub-problems are usually solved with linear programming, but each can be solved with constraint programming. For example, (Thorsteinsson, 2001) presents Branch-and-Check, a framework that encompasses Benders decomposition and allows for the sub-problems to be modeled and solved with constraint programming. (Benoist et al., 2002) presents a Benders decomposition applied to a workforce scheduling problem where the sub-problem is solved as usual with linear programming, but where the master problem is solved with constraint programming based on a flow global constraint. This cooperation allows to handle a wide variety of additional constraints thanks to the flexibility of constraint programming. (Eremin and Wallace, 2001) presents an implementation of hybrid Benders decomposition in ECLiPSe, where the master problem is also solved by constraint programming. With this implementation, the programmer need only specify which variables belong to which sub-problems, the dual form of any sub-problem is automatically derived, and the Benders decomposition is extracted automatically. Such implementations are very helpful to promote hybrid algorithms because they permit to build prototypes more quickly and they allow specialists of only one component of the cooperative solver to design hybrid algorithms.

5. Operations Research Algorithms and Local Search

As in the case of constraint programming and mixed integer programming, efficient operations research algorithms can be used as slaves of local search masters. In a local search context, a specific algorithm can help in two ways: either to explore a given neighborhood or to select the neighborhood to be explored.

For example, Adams, Balas, and Zawack (Adams et al., 1988) applied a procedure referred to as "shifting bottleneck" (later generalized as "shuffle" (Applegate and Cook, 1991) and "large neighborhood search" (Shaw, 1998)) to the job-shop scheduling problem. The basic idea behind large neighborhood search is straightforward. Given a solution, a neighborhood is constructed by relaxing some, but not all, of the decisions in the solution. For example, one can keep the values of some of the variables and relax the others, *i.e.*, allow these variables to take new values, provided that all the constraints of the problem are satisfied. A new solution is then searched for in this neighborhood, using whatever optimization technique is the most appropriate. The decisions to be relaxed at each step can be chosen randomly or according to some given patterns. In the particular case of the shifting bottleneck procedure, a resource (the current "bottleneck") is chosen. All the ordering decisions determining the sequencing of the activities on the resources are kept, except those that concern the bottleneck. The resulting neighborhood is explored with the help of a specific sequencing algorithm described in (Carlier, 1982), which is not polynomial in the worst case, yet appears to be efficient in almost all cases.

The shifting bottleneck procedure is no longer state-of-the-art on the job-shop scheduling problem, but constraint programming forms of large neighborhood search are still used extensively in this domain, as reported for example in (Caseau and Laburthe, 1995; Nuijten and Le Pape, 1998; Caseau et al., 2001).

In the best case, the use of an operations research algorithm can also lead to the adoption of a simpler representation in the local search space, dealing only with some of the problem variables, while the operations research algorithm is able to optimally fill in the details and compute the exact cost of a solution. For example, many applications of local search to scheduling problems assume that, given a set of sequencing decisions, a longest path algorithm will be used to compute the exact start and end times of each activity. The local search algorithm explores the space of possible sequences, while the longest path algorithm provides the optimal start and end times for a given sequence, thereby reducing an apparently large neighborhood to its locally optimal point.

Operations research algorithms can also be used to select composite moves. For example, for a vehicle routing problem, large neighborhood search consists in iterating composite moves: at each step, several customers are moved from their positions in their current routes to other positions in the same route or to different routes. (Xu and Kelly, 1996) model ejections and insertions of customers from routes as a network flow problem. The objective function considers both an approximation of the distance changes implied by the moves and a penalty mechanism to heuristically ensure the feasibility of moves, for example with respect to the capacity constraints. The constraints represent the flow balance constraints: if a customer is removed from its current route, it must be assigned to another route. This network flow model can be solved with a linear programming solver or even better with network algorithms in order to select a feasible and cost-improving composite move at each step of the large neighborhood search.

6. Mixed Integer Programming and Local Search

For mixed integer programs, two aspects of the problem are important: tightening the continuous relaxation by adding cuts in order to increase the lower bound (when minimizing), finding good integer solutions in order to decrease the upper bound. The traditional way of obtaining integer solutions is to explore a tree: at each node, a variable that is fractional in the current relaxation is chosen and several child nodes are created by adding to each a new constraint to partition the search space while cutting off the current relaxation. For example, if integer variable x has fractional value f in the relaxation, two child nodes are created: one with the new constraint $x \leq \lfloor f \rfloor$, the other with the new constraint $x \geq \lceil f \rceil$. Thus, as the depth in the tree increases, the number of variables with fractional values is expected to decrease, and it is hoped that at some node, the relaxation will have an integer solution. However, if the relaxation is a very poor approximation of the convex hull of the integer solutions, and if the branching decisions add variable bounds that fail to improve rapidly this approximation, and if, as a consequence, the relaxation is not a good guide to decide on which part of the tree to concentrate, it is very difficult to find nodes where the relaxation solution is spontaneously integer. Using additional algorithms such as local search to find integer solutions might then be beneficial. This is why heuristics are often combined with branch-and-bound to find quickly integer solutions. There are few domain independent heuristics and there mainly exist domain specific heuristics that take advantage of the high-level problem structure. Two kinds of heuristics exist:

- Rounding heuristics start from an integer-infeasible solution, typically the solution of the continuous relaxation, and try to infer from it an integer-feasible solution. In other words, they perform local search around the continuous relaxation. The most studied heuristic of this kind is the Pivot & Complement heuristic (Balas and Martin, 1980); other heuristics such as those presented in (Lokketangen and Woodruff, 2001; Aboudi and Jörnsten, 1994) use Pivot & Complement as a subroutine inside a metaheuristic. Another common strategy is what is described in (Bixby et al., 2000; Fischetti and Lodi, 2002) as hard variable fixing or diving: fix some variables to integer values, infer some new bounds on the remaining variables due to these fixings, solve the LP relaxation (or not, for a less expensive variant of the heuristic), and repeat until all variables have been fixed. Randomized rounding (Raghavan and Thompson, 1987) is another rounding heuristic: consider a pure 0-1 integer program; if v_i is the value for variable x_i in the continuous relaxation, then x_i is rounded up to 1 with probability v_i and rounded down to 0 with probability $1 - v_i$. Thus, for each constraint, the expected value of the right hand side is equal to its value for the relaxation, which is compatible with the left hand side, and the expected value of the rounded solution is the value of the continuous relaxation. Of course, it is only on some models (for example, set covering) that randomized rounding leads to feasible solutions, where all constraints are valid simultaneously and not only in expectation. On some models, the rounded solution can be theoretically proved to be feasible and to have a cost within a certain range of the optimal solution. When the rounded solution is not feasible, repair heuristics can be used: (Walser, 1998) transforms the MIP problem into an over-constrained problem and uses an algorithm similar to Walksat to reduce the number of violated constraints. An example of domain-specific rounding heuristic appears in (Cavalcante et al., 2001): the relaxation is used to order jobs in a scheduling problem, which produces approximation algorithms with performance guarantees.

- Local improvement heuristics start from one or several integer solutions and try to find a better integer solution. For example, (Chabrier et al., 2002) uses a specific heuristic for the vehicle routing problem with time windows based on large neighborhood search to improve on the integer solutions found by mixed integer programming (in fact a column generation approach). In this case, cooperation is even tighter: the local search heuristic not only provides integer solutions, but it also yields new columns for the column generation master problem, because it is not restricted to combining existing columns to generate integer solutions.

Such cooperation between local search and mixed integer programming clearly corresponds to the **multiple search** cooperative scheme described in the introduction. Local search algorithms succeed in finding quickly integer solutions, which obviously helps the MIP solver. But the MIP solver acts in turn as a good diversification scheme for the local search algorithm. One of the problems of local search algorithms is indeed that they tend to get stuck in local optima, so a good diversification scheme is needed to explore a different part of the solution space. Often, degrading the objective value is allowed during diversification. In this cooperation scheme however, the MIP solver diversifies the local search and always finds a new solution with better cost than the local optima, because its upper bound is always updated with the cost of the best current integer solution. Besides, this cooperative scheme combines the power of an incomplete search method (local search) for finding solutions, and the power of a complete method (mixed integer programming) for proving the optimality of those solutions.

Two recent heuristics obey to an entirely different paradigm: a **decomposition** scheme where the initial master is the MIP solver, but where local search is called regularly and acts as a secondary master, defining neighborhoods around solutions found by the MIP solver, which are explored by a recursive MIP slave. This paradigm is based on assumptions on the solution space, on which local search is also based and which can be summarized as the following "locality principle": there exist good solutions in the neighborhood of a good solution. Local branching (Fischetti and Lodi, 2002) (see also (Vasquez and Hao, 2001) for a similar approach applied to the multidimensional knapsack problem) explores the neighborhood of the current integer solution, defining the recursive MIP by adding a constraint to the global MIP stating that all subsequent solutions in the recurvive MIP must be within Hamming distance k of the current integer solution, that is there will be at most k variables that have different values in the current solution and in each subsequent solution. Relaxation Induced Neighborhood Search (RINS) (Danna et al., 2003) explores both the neighborhood of the current integer solution and the neighborhood of the continuous relaxation, defining the recursive MIP on the variables that have different values in the current integer solution and in the continuous relaxation, fixing the other variables to their common value. (Danna et al., 2003) presents an in-depth comparison of local branching and RINS on multiple hard difficult MIP models. It appears computationally that RINS performs consistently better than local branching.

7. Constraint Programming and Local Search

As mixed integer programming, constraint programming can be used in a **decomposition** scheme as a slave to a local search algorithm. The local search master can benefit from constraint programming for at least two reasons:

- Constraint propagation can be used to determine characteristics that are shared by all the solutions or, similarly, by all the solutions that respect a given set of decisions and are better than (or close to) the best solution found so far. The solution space to be explored by local search can then be restricted to the solutions that meet these characteristics — provided this does not lead to a dramatic disconnection of the solution space. As mentioned in (Vasquez, 2002), this has two advantages: the time needed to explore a neighborhood is reduced, and neighborhoods with no good solutions are not explored. Constraint programming can be used as a preprocessing to local search, as in (Vasquez, 2002), or as a technique to explore the neighborhood, as in (Pesant and Gendreau, 1996). In (Vasquez, 2002), a pure tabu search algorithm and a combination of constraint propagation and tabu search are applied to thirty instances of a frequency assignment problem. Results show that the pure tabu search algorithm provided most of the best results, but also some of the worst. The combination was altogether more robust. In (Pesant and Gendreau, 1996), constraint programming is used to find the best non-tabu neighbor of a given solution to a traveling salesman problem with time windows. In the reported results, the average ratio between the number of backtracks and the size of the neighborhood is 9% (29% for the worst instance), showing that many neighbors are indeed discarded by constraint programming.

- Using constraint programming techniques as part of a local search algorithm may also enable one to keep as much as possible of the local search efficiency enjoyed on the "pure" problems and gain in flexibility by allowing side constraints to be taken into account. For example, Kilby, Prosser, and Shaw report results obtained on vehicle routing problems with side constraints, showing that the use of constraint programming to construct initial solutions and explore neighborhoods enhances robustness with respect to the side constraints (Kilby et al., 2000).

Globally, such approaches are promising whenever local search operators, possibly large neighborhood search operators, provide a good basis for the exploration of the search space, and either side constraints or effective constraint propagation algorithms can be used to prune the search space.

The **multiple search** scheme can also be used to combine constraint programming and local search, taken as alternative ways to explore the search space. For example, (Caseau and Laburthe, 1995) describe an algorithm for the job-shop scheduling problem which combines constraint programming and local search. The overall algorithm finds an approximate solution to start with, makes local changes and repairs on it to quickly decrease the makespan (*i.e.*, the time at which all activities are completed) and, finally, performs a constraint-based exhaustive search for decreasing makespans. Given a schedule S, a critical path is defined as a sequence of activities A_1, ..., A_n such that for all i in $\{1 \ldots n-1\}$, A_i precedes A_{i+1} and Σ *duration*(A_i) = *makespan*(S). Two types of local moves are considered:

- "Repair" moves swap two activities scheduled on the same machine to shrink or reduce the number of critical paths.
- "Shuffle" moves (Applegate and Cook, 1991) keep part of the solution and search through the rest of the solution space to complete it. Each shuffle move is implemented as a constraint-based search algorithm with a limited number of backtracks (typically 10, progressively increased to 100 or 1000), under the constraint that the makespan of the solution must be improved (with an improvement step of a given δ, typically 1% of the makespan, progressively decreased to one time unit).

Hence, in this example, both generic schemes are applied: local search is followed by exhaustive constraint programming tree search, and part of the local search relies on constraint programming to limit the exploration of the neighborhood to relevant problem solutions. Excellent computational results have been obtained with this approach (Caseau and Laburthe, 1995; Caseau et al., 2001) as well as with other constraint-based implementations of shuffle moves, as reported in (Baptiste et al., 1995; Nuijten and Le Pape, 1998).

In the same spirit, the best algorithm we are aware of for the preemptive job-shop scheduling problem (Le Pape and Baptiste, 1999) relies on the combination of:

- a strong constraint propagation algorithm (edge-finding);
- a local optimization operator called "Jackson derivation". Given a schedule S, its Jackson derivation $J(S)$ is obtained by (i) defining the due-date of an activity A in a job J as the start time in S of the successor of A along job J and (ii) rescheduling all activities by giving the highest priority to

activities with the smallest due-dates. This operator and the symmetric operator based on the end time of the predecessor of each activity A are systematically applied each time a new solution is found.

- limited discrepancy search (Harvey and Ginsberg, 1995) around the best schedule found so far. Limited discrepancy search is an alternative to depth-first search, which relies on the intuition that heuristics make few mistakes through the search tree. Thus, considering the path from the root node of the tree to the first solution found by a depth-first search algorithm, there should be few "wrong turns" (*i.e.*, few nodes which were not immediately selected by the heuristic). The basic idea is to restrict the search to paths that do not diverge more than w times from the choices recommended by the heuristic. Each time this limited search fails to improve on the best current schedule, w is incremented and the process is iterated, until either a better solution is found or it is proven that there is no better solution. It is easy to prove that when w gets large enough, limited discrepancy search is complete. Yet it can be seen as a form of local search around the recommendation of the heuristic.

On ten well-known problem instances, each with 100 activities, experimental results show that each of the three techniques mentioned above brings improvements in efficiency, the average deviation to optimal solutions after 10 minutes of CPU time falling from 13.72% when none of these techniques is used to 0.23% when they are all employed.

(Focacci and Shaw, 2002) provide an example in which local search is used as a slave of a constraint programming master, as an alternative to dominance rules to prune the constraint programming search tree. For a minimization problem, dominance rules state that if a partial solution s can be extended to a complete solution whose cost is less than the optimal extension of partial solution s', then all extensions of s' can be pruned. This enables to drastically reduce the search space. However, pruning s' could be counter-productive if the sub-tree rooted in s has not been explored yet. Indeed, by exploring the sub-tree rooted in s' immediately, one may obtain quickly an extension of s' better than the best solution known so far and immediately tighten the remainder of the search. As a result, the systematic use of dominance rules may make it difficult to find good solutions in limited CPU time. Therefore, Focacci and Shaw propose to prune only the partial solutions that cannot be extended to solutions better than the best known solution. In this case, s' is handled as a no-good, *i.e.*, a partial solution that cannot be extended to a complete solution, because it would not be legal with respect to the current upper bound on cost. Any algorithm can be used to prove that a partial solution is dominated: an incomplete local search algorithm is used in (Focacci and Shaw, 2002) (a nearest neighbor heuristic and a relocate operator for the traveling salesman problem). When this algorithm discovers that the sub-problem defined by the partial solution under consideration is a relaxation of the sub-problem defined by one of the stored no-goods, the partial solution can be pruned. On numerous symmetric and asymmetric TSP instances, the hybrid method highly reduces the CPU time and the number of fails, compared to a pure constraint programming approach and to the standard application of dominance rules.

Conclusion

In this chapter we gave an overview of hybrid algorithms. We tried to abstract from the examples two generic cooperation schemes: **decomposition** and **multiple search**. It appears however that there is still a lot of work to be done to create a comprehensive taxonomy of cooperation techniques and to understand on which kind of problems which kind of cooperation has more chances to be efficient and robust.

The reader is referred to the next chapters for a more extensive discussion of some of the cooperation techniques we mentioned: Chapter 3 for a global framework to unify constraint programming and integer programming; Chapter 4 for the integration of operations research algorithms in constraint programming to develop global constraints and filtering algorithms; Chapter 5 for integrating constraint programming and mixed integer programming; Chapter 6 for constraint programming based branch-and-price; Chapter 7 for more details on integrating constraint programming and local search; and Chapter 8 for randomized algorithms and algorithm portfolio design — another example of **multiple search**.

References

Aboudi, R. and Jörnsten, K. (1994). Tabu search for general zero-one integer programs using the pivot and complement heuristic. *ORSA Journal on Computing*, 6(1):82–93.

Adams, J., Balas, E., and Zawack, D. (1988). The shifting bottleneck procedure for job-shop scheduling. *Management Science*, 34(3):391–401.

Applegate, D. and Cook, W. (1991). A computational study of the job-shop scheduling problem. *ORSA Journal on Computing*, 3(2):149–156.

Balas, E. and Martin, C. (1980). Pivot and complement — a heuristic for 0-1 programming. *Management Science*, 26(1):224–234.

Baptiste, P. and Le Pape, C. (1995). A theoretical and experimental comparison of constraint propagation techniques for disjunctive scheduling. In *Proceedings of the Fourteenth International Joint Conference on Artificial Intelligence*.

Baptiste, P., Le Pape, C., and Nuijten, W. (1995). Incorporating efficient operations research algorithms in constraint-based scheduling. In *Proceedings of the First International Joint Workshop on Artificial Intelligence and Operations Research*.

Baptiste, P., Le Pape, C., and Nuijten, W. (1999). Satisfiability tests and time-bound adjustments for cumulative scheduling problems. *Annals of Operations Research*, 92:305–333.

Baptiste, P., Le Pape, C., and Péridy, L. (1998). Global constraints for partial CSPs: A case study of resource and due-date constraints. In *Proceedings of the Fourth International Conference on Principles and Practice of Constraint Programming*.

Barnhart, C., Johnson, E. L., Nemhauser, G. L., Savelsbergh, M. W., and Vance, P. H. (1998). Branch-and-price: Column generation for solving huge integer programs. *Operations Research*, 46:316–329.

Benders, J. (1962). Partitioning procedures for solving mixed-variables programming problems. *Numerische Mathematik*, 4:238–252.

Benoist, T., Gaudin, E., and Rottembourg, B. (2002). Constraint programming contribution to Benders decomposition: A case study. In *Proceedings of the*

Eigth International Conference on Principles and Practice of Constraint Programming (CP'2002), pages 603–617.

Benoist, T., Laburthe, F., and Rottembourg, B. (2001). Lagrange relaxation and constraint programming collaborative schemes for traveling tournament problems. In *Proceedings of the Third International Workshop on Integration of AI and OR Techniques in Constraint Programming for Combinatorial Optimisation Problems (CP-AI-OR'01)*, pages 15–26.

Bixby, R. E., Fenelon, M., Gu, Z., Rothberg, E., and Wunderling, R. (2000). *MIP: Theory and Practice — Closing the Gap*, pages 19–49. Kluwer Academic Publishers.

Carlier, J. (1982). The one-machine sequencing problem. *European Journal of Operational Research*, 11(1):42–47.

Carlier, J. and Pinson, E. (1989). An algorithm for solving the job-shop problem. *Management Science*, 35(2):164–176.

Carlier, J. and Pinson, E. (1990). A practical use of Jackson's preemptive schedule for solving the job-shop problem. *Annals of Operations Research*, 26:269–287.

Caseau, Y. and Laburthe, F. (1994). Improved CLP scheduling with task intervals. In *Proceedings of the Eleventh International Conference on Logic Programming*.

Caseau, Y. and Laburthe, F. (1995). Disjunctive scheduling with task intervals. Technical report, École Normale Supérieure.

Caseau, Y. and Laburthe, F. (1996). Cumulative scheduling with task intervals. In *Proceedings of the Joint International Conference and Symposium on Logic Programming*.

Caseau, Y. and Laburthe, F. (2000). Solving various weighted matching problems with constraints. *Constraints*, 5(2):141–160.

Caseau, Y., Laburthe, F., Le Pape, C., and t Rottembourg, B. (2001). Combining local and global search in a constraint programming environment. *Knowledge Engineering Review*, 16(1):41–68.

Cavalcante, C., de Souza, C. C., Savelsbergh, M., Wang, Y., and Wolsey, L. (2001). Scheduling projects with labor constraints. *Discrete Applied Mathematics*, 112:27–52.

Chabrier, A., Danna, E., and Le Pape, C. (2002). Coopération entre génération de colonnes avec tournées sans cycle et recherche locale appliquée au routage de véhicules. In *Huitièmes Journées Nationales sur la résolution de Problèmes NP-Complets (JNPC'2002)*.

Colmerauer, A. (1990). An introduction to PROLOG III. *Communications of the ACM*, 33(7):69–90.

Danna, E., Rothberg, E., and Le Pape, C. (2003). Exploring relaxation induced neighborhoods to improve MIP solutions. Technical report, ILOG.

Desrochers, M. (1988). An algorithm for the shortest path problem with resource constraints. Technical Report G-88-27, Les cahiers du GERAD.

Eremin, A. and Wallace, M. (2001). Hybrid Benders decomposition algorithms in constraint logic programming. In *Proceedings of the Seventh International Conference on Principles and Practice of Constraint Programming (CP'2001)*.

Feillet, D., Dejax, P., Gendreau, M., and Gueguen, C. (2001). An exact algorithm for the elementary shortest path problem with resource constraints: Application to some vehicle routing problems. submitted to Networks.

Fischetti, M. and Lodi, A. (2002). Local branching. In *Proceedings of the Integer Programming Conference in honor of Egon Balas*.

Focacci, F., Laborie, P., and Nuijten, W. (2000). Solving scheduling problems with setup times and alternative resources. In *Proceedings of the Fifth International Conference on Artificial Intelligence Planning and Scheduling*.

Focacci, F., Lodi, A., and Milano, M. (1999). Cost-based domain filtering. In *Proceedings of the Fifth International Conference on Principles and Practice of Constraint Programming*.

Focacci, F. and Shaw, P. (2002). Pruning sub-optimal search branches using local search. In *Proceedings of the Fourth International Workshop on Integration of AI and OR Techniques in Constraint Programming for Combinatorial Optimisation Problems (CP-AI-OR'02)*, pages 181–189.

Gendron, B. and Crainic, T. G. (1994). Relaxations for multicommodity capacitated network design problems. Technical Report CRT-965, Centre de recherche sur les transports, Université de Montréal.

Harvey, W. D. and Ginsberg, M. L. (1995). Limited discrepancy search. In *Proceedings of the Fourteenth International Joint Conference on Artificial Intelligence*.

Hentenryck, P. V. (1989). *Constraint Satisfaction in Logic Programming*. MIT Press.

Jaffar, J. and Lassez, J.-L. (1987). Constraint logic programming. In *Proceedings of the Fourteenth Annual ACM Symposium on Principles of Programming Languages*.

Kilby, P., Prosser, P., and Shaw, P. (2000). A comparison of traditional and constraint-based heuristic methods on vehicle routing problems with side constraints. *Constraints*, 5(4):389–414.

Lokketangen, A. L. and Woodruff, D. L. (2001). Integrating pivot based search with branch and bound for binary MIP's. *Control and Cybernetics, Special issue on Tabu Search*, 29(3):741–760.

Le Pape, C. and Baptiste, P. (1999). Heuristic control of a constraint-based algorithm for the preemptive job-shop scheduling problem. *Journal of Heuristics*, 5(3):305–325.

Le Pape, C., Perron, L., Régin, J.-C., and Shaw, P. (2002). Robust and parallel solving of a network design problem. In *Proceedings of the Eighth International Conference on Principles and Practice of Constraint Programming (CP'2002)*.

Milano, M., Ottosson, G., Refalo, P., and Thorsteinsson, E. S. (2002). The role of integer programming techniques in constraint programming's global constraints. *INFORMS Journal on Computing, Special Issue on* The Merging of Mathematical Programming and Constraint Programming, *14(4), 2002.*.

Milano, M. and van Hoeve, W. J. (2002). Reduced cost-based ranking for generating promising subproblems. In *Proceedings of the Eighth International Conference on Principles and Practice of Constraint Programming (CP'2002)*, pages 1–16.

Nuijten, W. and Le Pape, C. (1998). Constraint-based job-shop scheduling with ILOG Scheduler. *Journal of Heuristics*, 3(4):271–286.

Nuijten, W. P. M. (1994). *Time and Resource Constrained Scheduling: A Constraint Satisfaction Approach*. PhD thesis, Eindhoven University of Technology.

Nuijten, W. P. M. and Aarts, E. H. L. (1994). Constraint satisfaction for multiple capacitated job-shop scheduling. In *Proceedings of the Eleventh European Conference on Artificial Intelligence*.

Pesant, G. and Gendreau, M. (1996). A view of local search in constraint programming. In *Proceedings of the Second International Conference on Principles and Practice of Constraint Programming.*

Pinson, E. (1988). *Le problème de job-shop*. PhD thesis, University Paris VI.

Raghavan, P. and Thompson, C. (1987). Randomized rounding : A technique for provably good algorithms and algorithmic proofs. *Combinatorica*, 7(4):365–374.

Refalo, P. (1999). Tight cooperation and its application in piecewise linear optimization. In *Proceedings of the Fifth International Conference on Principles and Practice of Constraint Programming.*

Refalo, P. (2000). Linear formulation of constraint programming models and hybrid solvers. In *Proceedings of the Sixth International Conference on Principles and Practice of Constraint Programming*, pages 369–383, Singapore. Springer-Verlag.

Régin, J.-C. (1994). A filtering algorithm for constraints of difference in CSPs. In *Proceedings of the Twelfth National Conference on Artificial Intelligence.*

Rodosek, R. and Wallace, M. (1998). A generic model and hybrid algorithm for hoist scheduling problems. In *Proceedings of the Fourth International Conference on Principles and Practice of Constraint Programming.*

Rodosek, R., Wallace, M., and Hajian, M. T. (1999). A new approach to integrating mixed integer programming and constraint logic programming. *Annals of Operations Research*, 86:63–87.

Rousseau, L.-M., Gendreau, M., and Pesant, G. (2002). Solving small VRPTWs with constraint programming based column generation. In *Proceedings of the Fourth International Workshop on Integration of AI and OR Techniques in Constraint Programming for Combinatorial Optimisation Problems (CP-AI-OR'02)*, pages 333–344.

Sakkout, H. E. and Wallace, M. (2000). Probe backtrack search for minimal perturbation in dynamic scheduling. *Constraints*, 5(4):359–388.

Sellmann, M. and Fahle, T. (2003). Constraint programming based lagrangian relaxation for the automatic recording problem. *Annals of Operations Research*, 118:17–33.

Shaw, P. (1998). Using constraint programming and local search methods to solve vehicle routing problems. In *Proceedings of the Fourth International Conference on Principles and Practice of Constraint Programming (CP'98)*, pages 417–431.

Thorsteinsson, E. S. (2001). Branch-and-check: A hybrid framework integrating mixed integer programming and constraint logic programming. In *Proceedings of the Seventh International Conference on Principles and Practice of Constraint Programming (CP'2001).*

Vasquez, M. (2002). Arc-consistency and tabu search for the frequency assignment problem with polarization. In *Proceedings of the Fourth International Workshop on Integration of AI and OR Techniques in Constraint Programming for Combinatorial Optimisation Problems (CP-AI-OR'02)*, pages 359–372.

Vasquez, M. and Hao, J.-K. (2001). A hybrid approach for the 0-1 multidimensional knapsack problem. In *Proceedings of the 17th International Joint Conference on Artificial Intelligence (IJCAI'2001)*, pages 328–333.

Walser, J. P. (1998). *Domain-Independent Local Search for Linear Integer Optimization.* PhD thesis, Technischen Fakultät der Universität des Saarlandes.

Xu, J. and Kelly, J. P. (1996). A network flow-based tabu search heuristic for the vehicle routing problem. *Transportation Science*, 30(4):379–393.

Yunes, T., Moura, A., and de Souza, C. (2000). A hybrid approach for solving large scale crew scheduling problems. In *Practical Aspects of Declarative Languages*, pages 293–307.

Chapter 3

BRANCH-AND-INFER: A FRAMEWORK FOR COMBINING CP AND IP

Alexander Bockmayr
Université Henri Poincaré, LORIA, B.P. 239
54506 Vandœuvre-lès-Nancy, France.
bockmayr@loria.fr

Thomas Kasper
GBU Supply Chain Management, IBD3 Optimization
SAP AG, Neurottstr. 16, 69190 Walldorf, Germany.
thomas.kasper@sap.com

Abstract We develop a unifying view of integer linear programming and finite domain constraint programming. We present the two modeling and solution approaches in a uniform framework, branch-and-infer. The goal of this framework is to clarify the relationship between the two techniques, and to indicate possible ways towards their integration. We illustrate the different concepts by examples from discrete tomography and supply chain optimization.

Keywords: Integer linear programming, Constraint Programming, Symbolic Constraints, Discrete Tomography, Supply Chain Optimization

1. Introduction

Integer linear programming (IP) and finite domain constraint programming (CP) are two general approaches for solving constraint satisfaction and constrained optimization problems. The purpose of this chapter is to develop a unifying view of the two methodologies, and to present them in a common framework, branch-and-infer.

Practical problem solving usually involves two steps: model building and model solving. In order to compare IP and CP, we ask for each of them the following questions:

- How expressive is the language that we can use to build a model ?
- How efficient are the algorithms that support this language when the model is solved ?

The branch-and-infer framework helps to clarify the relationship between IP and CP. For example, CP has a much richer constraint language, while IP has more efficient algorithms to solve linear arithmetic constraints. Branch-and-infer also indicates different ways of combining IP and CP. For example, it shows how the concept of symbolic constraints, which plays a key role in CP, can be transferred to IP.

Before we begin our discussion, we recall some basic definitions and fix our notation. For additional details, we refer to Chapter 1. Problems in IP and CP have two main ingredients:

- variables $x_1, \ldots, x_n$, and
- constraints $c_1, \ldots, c_m$.

Each variable x_j takes its values in a finite set D_j, which is called the *domain* of x_j. Each constraint c defines a relation $R_c \subset D_1 \times \cdots \times D_n$. An assignment $(v_1, \ldots, v_n)$ to the variables $(x_1, \ldots, x_n)$ *satisfies* the constraint c if $(v_1, \ldots, v_n) \in R_c$. In a *constraint satisfaction problem*, we are given a set of variables $V = \{x_1, \ldots, x_n\}$, with corresponding domains $D = \{D_1, \ldots, D_n\}$, and a set of constraints $C = \{c_1, \ldots, c_m\}$. We are looking for variable assignments $(v_1, \ldots, v_n)$ that satisfy simultaneously all the constraints $c_1, \ldots, c_m$. We call such $(v_1, \ldots, v_n)$ a *feasible* solution and denote by $\mathrm{Sol}(C)$ the set of all feasible solutions of C. In a *constrained optimization problem*, we are given in addition an *objective function* $f : D_1 \times \cdots \times D_n \to \mathcal{R}$, and we are searching for *optimal solutions*, i.e., feasible solutions of the constraint satisfaction problem C with a maximum or minimum value of the objective function.

2. Modeling in CP and IP

For a given practical problem, there may exist many ways to express it as a constraint satisfaction or constrained optimization problem. There may be various choices for introducing the variables, and once the variables have been defined, many different constraints may be stated. Integer linear programming and finite domain constraint programming have in common that they both involve variables and constraints. However, the types of the variables and constraints that are used, and the way the constraints are solved, are different in the two approaches.

Integer linear programming relies completely on linear equations and inequalities in integer variables, i.e., there are two only types of constraints

- Linear arithmetic: Linear equations or inequalities

$$a_1x_1 + \cdots + a_nx_n \left\{ \begin{array}{c} \leq \\ = \\ \geq \end{array} \right. \beta,$$

  with rational numbers $a_i, \beta \in \mathcal{Q}$.

- Integrality: The constraint

$$\texttt{integer}(x_1, \ldots, x_n)$$

  stating that the variables $x_1, \ldots, x_n$ have to take their values in the integer numbers.

Integer linear programming problems can always be written in the form

$$Ax \leq b, x \in \mathcal{Z}^n,$$

with a matrix $A \in \mathcal{Q}^{m \times n}$ and a column vector $b \in \mathcal{Q}^m$. Note that a linear equation $a_1x_1 + \cdots + a_nx_n = \beta$ is equivalent to the pair of inequalities $a_1x_1 + \cdots + a_nx_n \leq \beta$ and $-a_1x_1 - \cdots - a_nx_n \leq -\beta$. Very often, the variables are restricted to take only the values 0 or 1, which can be expressed by adding inequalities of the form $0 \leq x \leq 1$.

While a large variety of problems can be modeled naturally with linear equations and inequalities, there exist natural properties that cannot be expressed so easily in this framework. Consider for example n variables $x_1, \ldots, x_n$ with common domain $\{1, \ldots, m\}$, and suppose we want to express that these variables should take pairwise different values. We may start by using disequalities $x_i \neq x_j, 1 \leq i < j \leq n$, which may be transformed into disjunctions of linear inequalities:

$$\begin{array}{lcrclcrcl} x_i \neq x_j & \iff & x_i & < & x_j & \vee & x_i & > & x_j \\ & \iff & x_i & \leq & x_j - 1 & \vee & x_i & \geq & x_j + 1 \end{array}$$

Since in IP we can handle only conjunctions of linear inequalities, the disjunctions have to be eliminated. One way to achieve this might be to introduce for each disjunction an indicator variable $y_{ij} \in \{0, 1\}$, telling us which member of the disjunction is true. This leads to the constraints

$$\begin{array}{rcll} x_i - x_j + 1 & \leq & m\, y_{ij}, & \text{for all } 1 \leq i < j \leq n, \\ x_j - x_i + 1 & \leq & m\,(1 - y_{ij}), & \text{for all } 1 \leq i < j \leq n. \end{array}$$

Note that if $y_{ij} = 0$, the first inequality becomes $x_i \leq x_j - 1$, while the second inequality is trivially satisfied. Similarly, if $y_{ij} = 1$, the first inequality is trivial and the second one becomes $x_j + 1 \leq x_i$.

Alternatively, one might use $m \cdot n$ variables $z_{ik} \in \{0, 1\}$, where $z_{ik} = 1$ means $z_i = k$. The corresponding linear constraints are

$$\begin{aligned} z_{i1} + \cdots + z_{im} &= 1, \text{ for all } i = 1, \ldots, n, \\ z_{1k} + \cdots + z_{nk} &\leq 1, \text{ for all } k = 1, \ldots, m. \end{aligned}$$

In finite domain constraint programming, there is a much richer constraint language. In addition to linear equations and inequalities, there are various other constraints:

- Disequalities $x_i \neq x_j$ and more generally $a_1 x_1 + \cdots + a_n x_n \neq \beta$,
- Symbolic constraints, like `alldifferent`(...) or `cumulative`(...).

In principle, *symbolic constraints*, also called *global constraints*, might define arbitrary relations on the domains of the variables. However, in order to be practically useful, they should express key properties arising in a variety of problems, and they should allow for efficient filtering algorithms (see Chapter 1 and 4).

For example, the constraint `alldifferent`$(x_1, \ldots, x_n)$ states that the variables $x_1, \ldots, x_n$ should take pairwise different values. From a declarative point of view, this is equivalent to the system of disequalities $x_i \neq x_j$, for all $1 \leq i < j \leq n$. Grouping together these constraints in a new so-called global constraint allows making more powerful inferences during resolution. For example, in the system $x_1 \neq x_2, x_2 \neq x_3, x_1 \neq x_3$, with 0-1 variables x_1, x_2, x_3, each constraint can be satisfied individually or locally, while the global constraint `alldifferent`(x_1, x_2, x_3) is infeasible.

The `cumulative` constraint has been introduced to model scheduling problems (Aggoun and Beldiceanu, 1993; Baptiste et al., 2001). Suppose there are n tasks. Task j has starting time s_j, duration d_j and needs r_j units of a given resource. The constraint

$$\texttt{cumulative}([\texttt{s}_1, \ldots, \texttt{s}_\texttt{n}], [\texttt{d}_1, \ldots, \texttt{d}_\texttt{n}], [\texttt{r}_1, \ldots, \texttt{r}_\texttt{n}], \texttt{l}, \texttt{e})$$

states that the tasks have to be executed in such a way that the global resource limit l is never exceeded, and e is the end of the schedule.

Symbolic constraints are one of the key achievements of constraint programming. On the modeling level, they increase the expressive power of the constraint language by allowing one to state complex relations in a compact way. On the algorithmic level, they provide a way to integrate specialized solution methods into a general solver.

While there exist powerful methods to handle systems of linear arithmetic constraints, in particular the Simplex algorithm, there is no general method to solve arbitrary sets of symbolic constraints. Therefore, in constraint programming, symbolic constraints are handled individually or locally. Each constraint

	v + h convex	*v convex*	*h convex*	*no restriction*
connected	P	NP-complete	NP-complete	NP-complete
no restriction	NP-complete	NP-complete	NP-complete	P

Table 3.1. Complexity of the polyomino reconstruction problem

comes with a number of filtering and constraint propagation techniques that allow one to remove values from the domains of the variables that cannot belong to a solution.

3. An illustrating example: discrete tomography

To illustrate the variety of models that may exist in integer and constraint programming, we consider the reconstruction of pictures in discrete tomography (Bockmayr et al., 1998). A *two-dimensional binary picture* is given by a binary matrix $X \in \{0,1\}^{m \times n}$. Intuitively, a pixel is black iff the corresponding matrix element is 1. A binary picture X is

- *horizontally convex*, if the set of 1's in each row is convex, i.e., $x_{ij_1} = x_{ij_2} = 1$ implies $x_{ij} = 1$, for all $1 \leq i \leq m, 1 \leq j_1 < j < j_2 \leq n$.
- *vertically convex*, if the set of 1's in each column is convex, i.e., $x_{i_1 j} = x_{i_2 j} = 1$ implies $x_{ij} = 1$, for all $1 \leq i_1 < i < i_2 \leq m, 1 \leq j \leq n$.
- *connected* or a *polyomino*, if the set of 1's in the matrix is connected with respect to the adjacency relation where each matrix element is adjacent to its two vertical and horizontal neighbors.

Given two vectors $h = (h_1 \ldots, h_m) \in \mathcal{N}^m, v = (v_1, \ldots, v_n) \in \mathcal{N}^n$, the *reconstruction problem* of a binary picture from *orthogonal projections* consists in finding $X \in \{0,1\}^{m \times n}$ such that

- $\sum_{j=1}^{n} x_{ij} = h_i$, for $i = 1, \ldots, m$ (horizontal projections)
- $\sum_{i=1}^{n} x_{ij} = v_j$, for $j = 1, \ldots, n$ (vertical projections)

The complexity of the reconstruction problem depends on the additional properties that are required for the picture (Woeginger, 2001), see Table 3.1.

3.1 Integer programming models

The above properties may be modeled in many different ways. In integer linear programming, one typically uses 0-1 variables x_{ij}. The binary picture $X \subseteq \{0,1\}^{m \times n}$ with horizontal and vertical projections $h \in \mathcal{N}^m, v \in \mathcal{N}^n$ is horizontally convex iff the following set of linear inequalities is satisfied:

$$h_i \cdot x_{ik} + \sum_{j=k+h_i}^{n} x_{ij} \leq h_i, \text{ for all } 1 \leq i \leq m, 1 \leq k \leq n.$$

X is vertically convex iff

$$v_j \cdot x_{kj} + \sum_{i=k+v_j}^{m} x_{ij} \leq v_j, \text{ for all } 1 \leq k \leq m, 1 \leq j \leq n.$$

The connectivity of a horizontally convex picture can be expressed as follows:

$$\sum_{j=k}^{k+h_i-1} x_{ij} - \sum_{j=k}^{k+h_i-1} x_{i+1j} \leq h_i - 1, \text{ for all } 1 \leq i \leq m-1, 1 \leq k \leq n-h_i+1.$$

This leads to $O(mn)$ variables and constraints.

3.2 Constraint programming models

In finite domain constraint programming, 0-1 variables are usually avoided. For each row resp. column in the given $m \times n$-matrix, we introduce a finite domain variable

- $x_i \in \{1, \ldots, n\}$, for all $i = 1, \ldots, m$, resp.
- $y_j \in \{1, \ldots, m\}$, for all $j = 1, \ldots, n$.

If $h = (h_1, \ldots, h_n)$ and $v = (v_1, \ldots, v_m)$ are the horizontal and vertical projections, then $x_i = j$ says that the block of h_i 1's for row i starts at column j. Analogously, $y_j = i$ expresses that the block of v_j 1's for column j starts in row i.

Conditional propagation. To ensure that the values of the variables x_i and y_j are compatible with each other, we impose the constraints

$$x_i \leq j < x_i + h_i \Longleftrightarrow y_j \leq i < y_j + v_j, \forall i = 1, \ldots, m, \forall j = 1, \ldots, n.$$

Such constraints may be realized by *conditional propagation* rules of the form `if` C `then` P, saying that, as soon as the remaining values for the variables satisfy the condition C, the constraints P become active. This models horizontal/vertical projections and convexity. To ensure connectivity, we have to forbid that the block in row $i+1$ ends left of the block in row i or that the block in row $i+1$ starts right of the block in row i. Negating this disjunction yields the linear inequalities

$$x_i \leq x_{i+1} + h_{i+1} - 1 \text{ and } x_{i+1} \leq x_i + h_i, \forall i = 1, \ldots, m-1.$$

The above constraints are sufficient to model the reconstruction problem. However, we may try to improve propagation by adding further constraints, which are redundant from the declarative point of view, but provide additional filtering techniques on the operational side. *Adding redundant constraints* is an important technique in constraint programming, which may help to speed up the solution process, similar to cutting planes in integer programming. Like in IP, there is a problem-dependent tradeoff between the cost of the filtering algorithm and the domain reductions that are obtained.

Cumulative. For example, we may use the `cumulative` constraint. We identify each horizontal block in the image with a task $(x_i, h_i, 1)$, which starts at time x_i, has duration h_i, and requires 1 resource unit. For each column j, we introduce an additional task $(j, 1, m - v_j + 1)$, which starts at time j, has duration 1, and uses $m - v_j + 1$ resource units. These complementary tasks model vertical projections numbers. The capacity of the resource is $m + 1$ and all the tasks end before time $n + 1$. Thus, the constraint

$$\texttt{cumulative(}\begin{array}{lllllll} [& x_1, & \ldots, & x_m, & 1, & \ldots, & n &], \\ [& h_1, & \ldots, & h_m, & 1, & \ldots, & 1 &], \\ [& 1, & \ldots, & 1, & m - v_1 + 1, & \ldots, & m - v_n + 1 &], \\ & m+1 & , & & n+1 & &) & \end{array}$$

models horizontal/vertical projection numbers, and horizontal convexity, see Fig. 3.1.

Diffn. Another possibility is to use the `diffn` constraint. Here, we look at polyomino reconstruction as packing of two-dimensional rectangles. We model the problem by an extended version of the `diffn` constraint (Beldiceanu and Contejean, 1994), which expresses that a given set of rectangles in n-dimensional should not overlap. In the first argument, we define the rectangles. For each black horizontal block in the picture, we introduce a rectangle

$$R_i = [x_i, i, h_i, 1],$$

with origins (x_i, i) and lengths $(h_i, 1), i = 1, \ldots, m$. To model vertical convexity, we introduce $2n$ additional rectangles

$$S_{1,j} = [j, 0, 1, l_{j,1}], \quad S_{2,j} = [j, m + 1 - l_{j,2}, 1, l_{j,2}],$$

which correspond to two white blocks in each column. The variables l_{jk} define the height of these rectangles. To ensure that each white block has a nonzero surface, we introduce two additional rows 0 and $m + 1$, see Fig. 3.2 for an illustration.

The second argument of the `diffn` constraint says that the total number of rows and columns is $m + 2$ resp. n. In the third argument, we express that the

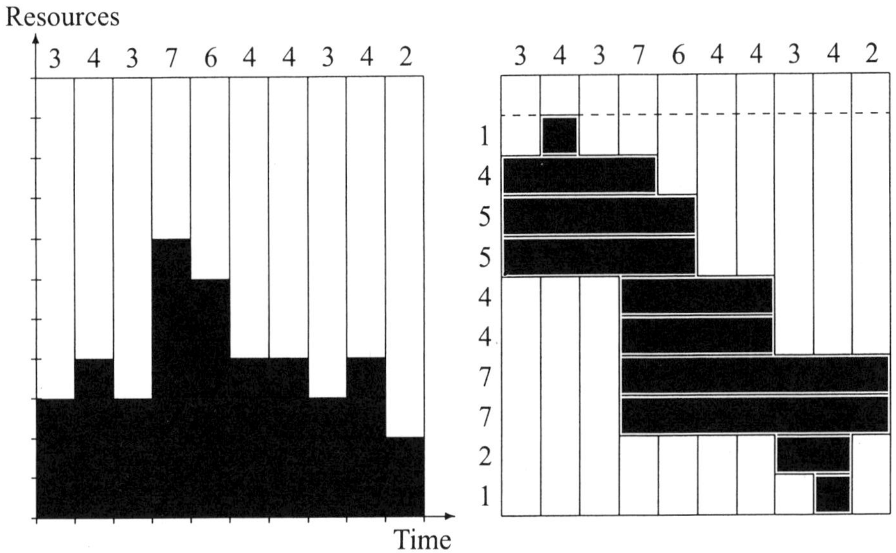

Figure 3.1. `Cumulative` constraint in discrete tomography

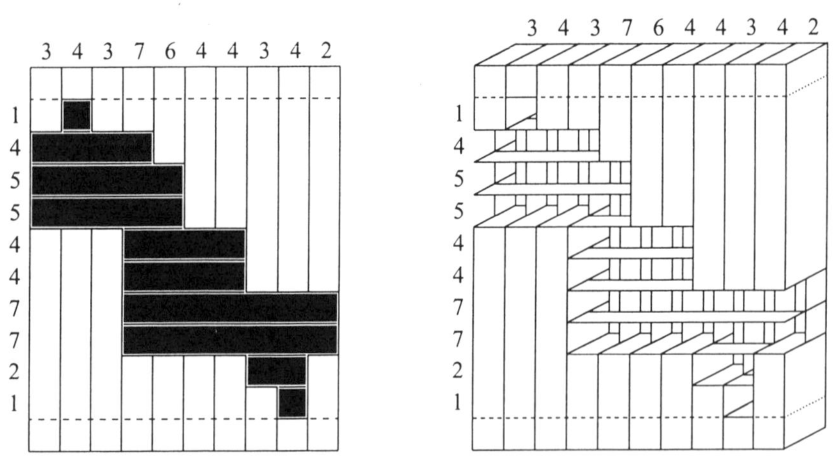

Figure 3.2. Two- and three-dimensional `diffn` constraint in discrete tomography

distance between the two white rectangles in column j has to be equal to v_j. To model connectivity, we state in the fourth argument that each pair of successive rectangles has a contact in at least one position. This is represented by the list $[[1, 2, c_1], ..., [m-1, m, c_{m-1}]]$, with domain variables $c_i \geq 1$. Thus, the whole reconstruction problem can be modeled by a single `diffn` constraint:

$$\begin{aligned}\texttt{diffn(}\; & [R_1, \ldots, R_m, S_{1,1}, \ldots, S_{1,n}, S_{2,1}, \ldots, S_{2,n}], \\ & [n, m+2], \\ & [[m+1, m+n+1, v_1], \ldots, [m+n, m+2*n, v_n]], \\ & [[1, 2, c_1], \ldots, [m-1, m, c_{m-1}]]\,)\end{aligned}$$

Note that this model involves only the row variables x_i, not the column variables y_j. It is also possible to use row and column variables simultaneously. This leads to another model based on a single `diffn` constraint in 3 dimensions, see Fig. 3.2. Here, the third dimension is used to ensure that row and column variables define the same picture.

4. Branch and Infer

After illustrating IP and CP modeling on an example, we now develop a unifying framework, branch-and-infer, to clarify the relationship between the two approaches.

4.1 Primitive and non-primitive constraints

In both IP and CP, we can notice that there is a class of constraints that can be solved easily, which we will call *primitive*, and a class of constraints that are difficult to solve, which we will call *non-primitive*. The intuitive notion of easy and hard constraints can be made more precise by saying that there exist polynomial algorithms to solve conjunctions of primitive constraints, while adding non-primitive constraints makes a problem NP-hard.

The primitive constraints in IP are linear equations and inequalities as long as they are solved over the rational numbers. What makes IP NP-hard, is the integrality constraint $\texttt{integer}(x_1, \ldots, x_n)$.

In CP, the primitive constraints are domain constraints $a \leq x, x \leq b, x \neq c$, with integers a, b, c, and equations between two variables $x = y$. Systems of such constraints can be solved over the integer numbers in polynomial time. This means that $\texttt{integer}(x_1, \ldots, x_n)$ may be seen as a primitive constraint in CP. Non-primitive constraints in CP are disequalities between two variables, $x \neq y$, general linear constraints $a^T x \diamond \beta, \diamond \in \{=, \leq, \geq, <, >, \neq\}$, and all kinds of symbolic constraints. Solving conjunctions of primitive constraints and such non-primitive constraints is NP-hard. For example, coloring problems can be formulated by the integrality constraint and disequations of the form $x \neq y$.

	Primitive	*Non-Primitive*
IP	$\sum a_i x_i \diamond b,\ \diamond \in \{=, \leq, \geq\}$	`integer`
CP	$x \leq u, x \geq l, x \neq v, x = y$ `integer`	$\sum a_i x_i \diamond b,\ \diamond \in \{=, \leq, \geq, <, >, \neq\}$ `alldifferent, cumulative, ...`

Table 3.2. Primitive and non-primitive constraints

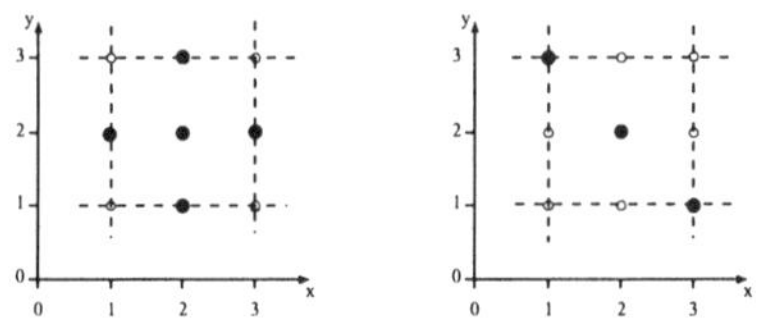

Figure 3.3. Primitive constraints in CP

Integer linear programming problems can be expressed by linear inequalities and the integrality constraint.

Table 3.2 summarizes the primitive and non-primitive constraints in IP and CP.

4.2 Inferring primitive from non-primitive constraints

If a model contains only primitive constraints, there is an efficient way to solve it, both in IP and CP. The difficult problems are therefore those that contain both primitive and non-primitive constraints, and the question is how to handle the non-primitive constraints. Since we know how to solve primitive constraints, a natural approach to deal with the non-primitive constraints might be trying to reduce them to primitive constraints. To what extent is this possible? More formally, given a non-primitive constraint c, is there a set of primitive constraints $c_1, \ldots, c_m$ that has the same set of solutions? The answer depends on whether we are using IP or CP.

In CP, it is in general not possible to replace a non-primitive constraint by an equivalent set of primitive constraints. For example, the point sets given in Fig. 3.3, which might be the solution sets of some non-primitive constraint, cannot be defined by a conjunction of primitive constraints.

In IP the situation is different. The convex hull $\text{conv}(S)$ of any bounded set $S \subset \mathcal{Z}^n$ of integer points forms a polyhedron $P \subset \mathcal{R}^n$ that can be defined by a system of primitive constraints, see Figure 3.4. In other words, there exists a system of linear inequalities $Ax \leq b$ such that

$$P = \text{conv}(S) = \{x \in \mathcal{Q}^n \mid Ax \leq b\}.$$

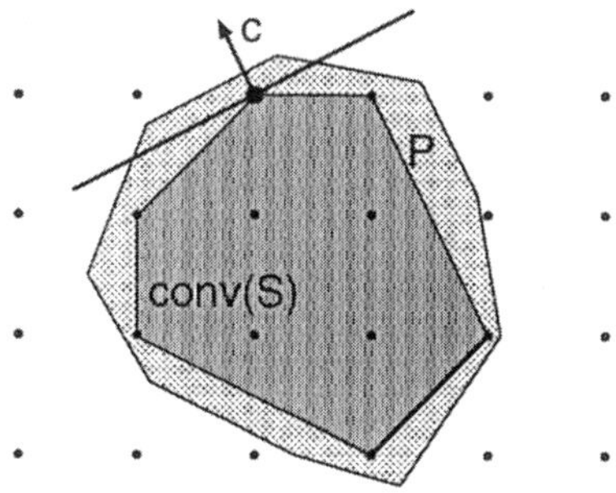

Figure 3.4. Primitive constraints in IP

Furthermore, if $c^T x$ is a linear objective function, then

$$\max\{c^T x \mid x \in S \subset \mathcal{Z}^n\} = \max\{c^T x \mid Ax \leq b, x \in \mathcal{Q}^n\}.$$

In this sense, it is possible in IP to replace a non-primitive constraint by an equivalent set of primitive constraints. The problem, however, is that this might be computationally prohibitive. Note that linear optimization over the integers is NP-hard, while linear optimization over the rationals is polynomial. Therefore, unless P = NP, we cannot hope for a polynomial size description of the polyhedron P. Indeed, results in polyhedral combinatorics show that for NP-hard integer programming problems, the convex hull description requires exponentially many inequalities.

In summary, we may conclude that in both IP and CP there is no efficient way to replace an arbitrary non-primitive constraint by an equivalent set of primitive constraints. However, what we can still do, is to *infer* some primitive constraints that are implied by the given non-primitive constraint. In other words, we may try to obtain a partial description of the non-primitive constraint in terms of primitive constraints. This is precisely what happens in both IP and CP.

In IP, we can derive cutting planes that tighten the linear inequality description of the integer solution set (cf. Chapter 1). Cutting planes have interesting logical properties, see e.g. (Bockmayr and Eisenbrand, 2000). There exist pure cutting plane algorithms that in principle can solve any integer linear programming problem. However, exponentially many cutting planes may be needed, and the practical performance of these algorithms is poor.

In CP, filtering algorithms reduce the domains of the variables. From a logical point of view, this corresponds to the generation of new bound constraints $a \leq x, x \leq b$ (bound reasoning) or disequalities $x \neq c$ (domain reasoning). Note that all these constraints are primitive for CP. More powerful constraint reasoning techniques have been studied in the literature, see e.g. (Debruyne and Bessière, 2001), which allow removing tuples of values (instead of values)

	CP	IP
Inferences	$x \geq l, x \leq u$ Bound reasoning $x \neq v$ Domain reasoning	$\sum a_i x_i \leq b$ Cutting planes
Relaxation	Discrete – $\mathcal{Z}$	Continuous – $\mathcal{Q}$
Algorithms	Filtering Propagation	Separation Linear programming

Table 3.3. Inference in IP and CP

from the Cartesian product $D_1 \times \cdots \times D_n$, over which a constraint is defined. However, the corresponding filtering algorithms become rather expensive, so that their use in practice is very limited.

In our framework, inferred primitive constraints are logical consequences of a given non-primitive constraint, i.e., feasible solutions are preserved. Recently, so-called *optimization constraints* have been proposed in the literature, which incorporate inference techniques that may remove feasible solutions that are not optimal (Focacci et al., 1999; Sellmann, 2002).

Cooperative solving. The idea of constraint solving by inferring primitive from non-primitive constraints leads to a computational architecture which is well-known in constraint programming, see Fig. 3.5. The primitive constraints are collected in a constraint store. They define a *relaxation* of the problem C, i.e., a constraint set $relax(C)$ logically implied by C. There exist efficient solution methods for solving the relaxation. The non-primitive constraints are handled locally by an *inference agent* that derives from a given non-primitive constraint and the current relaxation new primitive constraints that tighten this relaxation. The only interaction that is possible between non-primitive constraints is via the constraint store. This implies that the expressive power of the primitive constraints determines what kind of information can be shared by the non-primitive constraints, or in the terminology of constraint programming, the amount of propagation that is possible between different non-primitive constraints.

We close this section with a small example that illustrates the different inference mechanisms in IP and CP, see Figure 3.6. Consider the system of linear

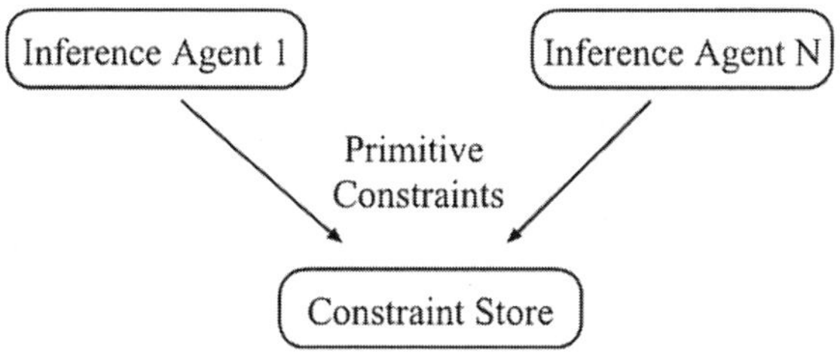

Figure 3.5. Architecture for constraint-based solving

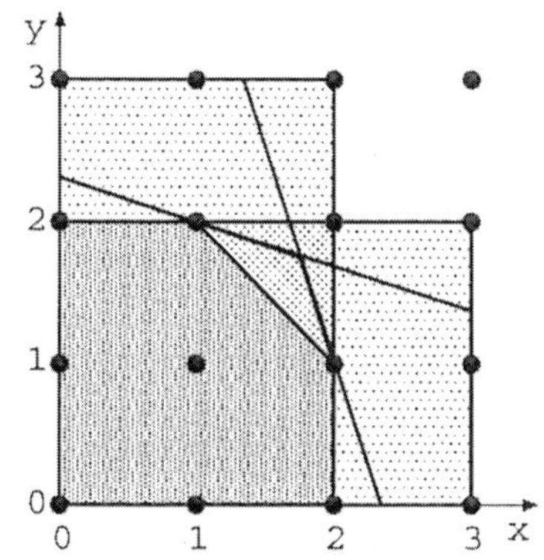

Figure 3.6. Inference in IP and CP

arithmetic constraints

$$\begin{aligned} 3x + y &\leq 7, \\ 3y + x &\leq 7, \\ x + y &= z, \\ x, y \in \{0, \ldots, 3\}, &\ \ z \in \{0, \ldots, 6\}. \end{aligned}$$

In CP, each of the linear constraints is non-primitive, and using domain filtering, we may infer that $x, y \leq 2$, and $z \leq 4$. In IP, we can reason globally on the constraint system. Using linear programming, we may infer $z \leq 3.5$, and after integer rounding, which can be seen as cutting plane generation, we obtain $z \leq 3$.

4.3 Search

Inferring and communicating primitive constraints is a powerful principle to solve non-primitive constraints. However, as we have seen before, this alone is not sufficient. In CP, the inference process may get stuck because no more primitive constraints can be inferred, i.e., the domains of the variables cannot be further reduced. In IP, cutting planes tend to become less effective when many of them are generated. This means that the relaxation defined by the constraint store is not sufficiently tightened anymore.

In both cases, we need an alternative solution principle, common to IP and CP, which is *branching*. If the inference process on a constraint problem C gets stuck, we may do a case analysis by considering a disjunction of primitive constraints $c_1 \vee \cdots \vee c_k$ that together cover all possible cases, i.e., C and $C \wedge (c_1 \vee \cdots \vee c_k)$ have the same set of solutions. Then we replace C by k new subproblems $c_1 \wedge C, \ldots, c_k \wedge C$. Since c_i is primitive, the constraint store of C gets tightened, and we may restart the inference process. Repeating this branching operation in a systematic way leads to a search tree. There are many ways to choose the branching constraint $c_1 \vee \cdots \vee c_k$. Often one uses disjunctions of the form $x \leq a \vee x \geq a+1$, where x is a variable and a a value from its domain.

The two solution principles that we have developed so far, *branch* and *infer*, can be formalized by transition rules of the form

$$- : \frac{\langle P, S\rangle}{\langle P', S'\rangle} \quad \text{if } \ \mathit{Cond}$$

saying that from a computation state $\langle P, S\rangle$ we may proceed to a computation state $\langle P', S'\rangle$ if the conditions in Cond are satisfied. Here $P = \{C_1, \ldots, C_m\}$ (resp. P') denotes a set of constraint (sub-)problems $C_1, \ldots, C_m$, which logically corresponds to the disjunction $C_1 \vee \cdots \vee C_m$. The set S (resp. S') denotes a set of feasible solutions. For a set T and an element t, we will write $t.T$ instead of $\{t\} \cup T$. When solving a constraint problem C, the initial state is $\langle\{C\}, \emptyset\rangle$, and the final state is $\langle\emptyset, \{S\}\rangle$. If $S = \emptyset$, then the problem is infeasible. $\mathrm{Prim}(C)$ denotes the set of primitive constraints in a constraint problem C, and $\rightarrow$ is logical implication. The following two transitions rules form the core of the *branch-and-infer* framework.

$$\texttt{bi\_infer}: \ \frac{\langle (c \wedge C) \,.\, P, S\rangle}{\langle (p \wedge (c \wedge C)) \,.\, P, S\rangle} \quad \text{if} \ \left\{ \begin{array}{c} c \text{ is non-primitive,} \\ p \text{ is primitive,} \\ \mathrm{Prim}(C) \wedge c \rightarrow p, \\ \mathrm{Prim}(C) \not\rightarrow p. \end{array} \right.$$

$$\texttt{bi\_branch}: \ \frac{\langle C \,.\, P, S\rangle}{\langle (c_1 \wedge C) \,.\, \ldots \,.\, (c_k \wedge C) \,.\, P, S\rangle} \quad \text{if} \ \left\{ \begin{array}{l} C \equiv C \wedge (\bigvee_{i=1}^{k} c_i) \\ c_i \text{ primitive} \\ \mathrm{Prim}(C) \not\rightarrow c_i \end{array} \right.$$

Note that these rules are the same for IP and CP and thus provide a unifying framework for the two approaches (Bockmayr and Kasper, 1998; Kasper, 1998). However, the notion of primitive and non-primitive constraints in the two cases is different.

4.4 Pruning the search tree

In principle, branching alone is sufficient to solve an arbitrary constraint satisfaction problem. By repeated application of the branching rule, we may enumerate all possible variable assignments and thus determine all the solutions. From a practical point of view, however, this will work only for very small problems. Since the size of the search tree is exponential, it is necessary to prune it as much as possible. The key to pruning is the inference process done in every node of the search tree.

In CP, filtering algorithms narrow the domains of the variables, which reduces the number of successors of the current node in the tree.

In IP, cutting plane generation tightens the current linear programming relaxation, which leads to better bounds on the objective function. In the context of a *branch-and-bound* algorithm, this can be used to prune the search tree. Consider the problem $\max\{f(x) \mid x \in \mathrm{Sol}(C)\}$ of maximizing an objective function f over the set of solutions of a constraint problem C. Any feasible solution s^* of C yields a lower bound $f(s^*)$ for the maximum value of f. In order to obtain an upper bound for the optimal value, we may maximize f over the primitive constraints in C. These define a relaxation of C, which may be easier to solve. Comparing lower and upper bounds may allow us to prune the search tree. If for a subproblem C corresponding to one node in the search tree, the upper bound obtained by solving the relaxed problem $\max\{f(x) \mid x \in \mathrm{Sol}(\mathrm{Prim}(C))\}$ is smaller than the best known lower bound, then the subtree rooted at the node C cannot contain a better solution. Therefore, it needs not be further explored.

The different pruning rules used in IP and CP can be formalized in the following way (Bockmayr and Kasper, 1998; Kasper, 1998).

- Unsatisfiability of the relaxation

$$\texttt{bi\_clash}: \frac{\langle C\,.\,P, S\rangle}{\langle P, S\rangle} \quad \text{if } \mathrm{Prim}(C) \rightarrow \bot$$

- Lower bounds

$$\texttt{bi\_climb}: \frac{\langle C\,.\,P, \{s\}\rangle}{\langle (c \wedge C)\,.\,(c \wedge P), \{s^*\}\rangle} \quad \text{if } \begin{cases} s^* = \texttt{extract}(\mathrm{Prim}(C)) \\ f(s^*) > f(s) \\ c \equiv [f(x) \geq f(s^*) + 1]. \end{cases}$$

  Here `extract` is a function that tries to compute a feasible solution from the relaxation.

- Lower and upper bounds

$$\texttt{bi\_bound}: \frac{\langle C\,.\,P, \{s\}\rangle}{\langle P, \{s\}\rangle} \quad \text{if } \max\{f(x) \mid x \in \mathrm{Sol}(\mathrm{Prim}(C))\} \leq f(s)$$

- Feasible solutions from the relaxation

$$\texttt{bi\_opt}: \frac{\langle C\,.\,P, \{s\}\rangle}{\langle P, \{s^*\}\rangle} \text{ if } \begin{cases} \max\{f(x) \mid x \in \text{Sol}(\text{Prim}(C))\} = f(s^*) \\ s^* \in \text{Sol}(C) \\ f(s^*) > f(s) \end{cases}$$

To solve constrained optimization problems, CP uses the rules `bi_infer`, `bi_branch`, `bi_clash`, `bi_climb`. The branch-and-cut approach in IP is captured by the rules `bi_infer`, `bi_branch`, `bi_clash`, `bi_bound`, and `bi_opt`.

5. Symbolic constraints in IP

Symbolic constraints are a key idea of CP and one of the main reasons for its success. It seems therefore natural to introduce this concept also in IP. The branch-and-infer framework tells us how this might be done.

5.1 Symbolic constraint abstractions

In the branch-and-infer framework, we distinguish primitive and non-primitive constraints. The set of primitive constraints is solved globally in the constraint store. Each non-primitive constraint is handled locally by its corresponding inference agent. The declarative semantics of a non-primitive constraint specifies under which conditions the constraint is satisfied. The operational semantics describes how the corresponding inference agent works.

Separating declarative and operational semantics allows us to model with non-primitive constraints by looking only at their declarative meaning, without knowing the operational behavior. This means that we can *abstract* from specifying an equivalent set of primitive constraints (if there exists one) and a corresponding inference procedure. The benefits of this abstraction can be summarized as follows:

Expressiveness. Using non-primitive constraints, we can specify a solution set $S \subseteq \mathcal{Q}^n$ that cannot be defined by the available primitive constraints, i.e., the expressiveness of the constraint language is increased.

Efficiency. While primitive constraints are very general, one can build problem-specific non-primitive constraints. Since they are handled locally, the inference agent can exploit the problem-specific properties captured by the non-primitive constraint. This may improve the efficiency of constraint solving.

Cooperative solving. The architecture of branch-and-infer supports cooperation between different non-primitive constraints in a natural way. The *local* treatment of non-primitive constraints allows us to develop inference agents that work independently. The *global* structure of the constraint store to which the inference agents are connected, serves as a communication medium between the different agents. Using the primitive constraints in the store for their reasoning, *different constraint-specific inference algorithms may cooperate* in order to find a solution.

This view of non-primitive constraints is similar to the idea of procedural or data abstraction in modern programming languages. Due to this similarity, we will also use the term *symbolic constraint abstraction*. Finding good constraint abstractions is a non-trivial task because there is a tradeoff between expressive power and efficiency. On the one hand, a constraint abstraction should be generic enough in order to apply to many problem situations. On the other hand, there must be sufficiently common domain-specific knowledge that can be exploited during the inference process.

A library of symbolic constraints for IP is currently being developed in the SCIL project (Althaus et al., 2002).

5.2 Extending expressiveness

Constraint abstractions enable us to extend the expressiveness of a constraint language. This can be done in three different ways.

First, we may use a symbolic constraint to characterize a set $S \subseteq \mathcal{Q}^n$, for which an equivalent set of primitive constraints does not exist. For instance, by introducing a symbolic constraint for a non-linear inequality in general integer variables, it becomes possible to describe *non-convex* solution sets S.

Second, we may use a symbolic constraint in cases where a description in terms of primitive constraints exists, but is not available at modeling time. Take, for example, a problem that can be modeled naturally by non-linear 0-1 inequalities. If we have to use primitive constraints, the problem must be stated in terms of linear inequalities. This is always possible, but depending on the non-linear constraint, finding an equivalent set of linear inequalities may be quite complicated. In particular, exponentially many inequalities may be needed.

Finally, even if a complete set of primitive constraints is known at modeling time, it may be computationally prohibitive to add all these constraints to the model. A typical example is the traveling salesman problem (TSP) (Jünger et al., 1995) with its exponential number of subtour elimination constraints. To model this, we may introduce a symbolic constraint

```
tsp(Adjacencies, Weights)
```

- `Adjacencies`: A list of 0-1 variables $[x_{12}, \ldots, x_{(n-1)n}]$.
- `Weights`: A list of nonnegative rational numbers $[w_{12}, \ldots, w_{(n-1)n}]$.

The adjacency of two nodes i and j in the graph is represented by a variable $x_{ij}, i < j$, which has value 1, if the edge is used in a tour, and 0 otherwise. The weights w_{ij} represent the cost imposed by using the edge between nodes i and j in a tour.

Hiding the set of TSP-constraints inside a new global constraint `tsp` offers clear and concise modeling. The inference agent implementing `tsp` can use a cutting plane procedure that will add only those problem-defining inequalities that cut off the current LP solution.

5.3 Improving efficiency

Constraint abstractions contribute to automating the difficult task of obtaining an efficient solution procedure from a model, since nontrivial problem structures can be made explicit using symbolic constraints. Since each symbolic constraint is handled locally by the corresponding inference agent, the problem structure can be exploited by the inference algorithms. This opens the door to apply results from polyhedral combinatorics, i.e., *problem-specific* cutting planes, inside a *general-purpose* IP solver.

For example, in the TSP case, instead of implementing only separation procedures for the problem-defining degree and subtour elimination constraints, we can add additional separation routines for further facet-defining valid inequalities like *comb inequalities* (Jünger et al., 1995), which may considerably enhance the solution process.

The concept of constraint abstractions is complementary to work in computational integer programming that aims at automatically extracting and generating canonical structures from a given integer linear programming program formulation (Cordier et al., 1999). In the simplest case, these structures consist of one inequality of the model (e.g. knapsack inequality) that is treated individually to infer strong valid inequalities of the convex hull of solutions. Although this idea of improving the efficiency of constraint solving is similar to the way it is done by branch-and-infer, the approach taken in (Cordier et al., 1999) has some limitations in comparison to symbolic constraints.

- The technique is limited to problems that can be modeled completely by (in-)equalities and whose (in-)equality set is completely known at modeling time. In particular *all* inequalities of the model have to be given to the solver. Thus, problems like TSP are not within the scope of this technique.

- Only rather simple structures can be exploited since in general it is difficult to automatically extract the semantics of a set of inequalities. As a consequence, the inferred cutting planes can have only limited strength compared to those derivable from symbolic constraints that exploit the knowledge available to the modeler.

Another aspect of constraint abstractions is that one can provide several symbolic constraints with the same declarative, but different operational semantics. This can be useful if several inference algorithms for a symbolic constraint are available that apply different heuristics. In such a case, one can offer several symbolic constraints which are declaratively equivalent. However, each of the corresponding inference agents uses a specific heuristic. Thus, it becomes possible to select on the modeling level a combination of heuristics that achieves the best results.

5.4 Compositionality

Symbolic constraints are not designed to compete with special-purpose algorithms for the problem that they define. The aim is rather to exploit problem-specific algorithms inside a general-purpose solver. A key point is that different symbolic constraints can be *used together* in one and the same model. This is important, because complex real-world problems often contain well-understood subproblems, whose structure should be exploited during the solution process.

From the declarative point of view, constraint abstractions allow us to create compact descriptions of the problem properties, which help us to better understand the model. In the ideal case, substructures of the problem have direct counterparts in the constraint language, so that the different symbolic constraints can serve as building blocks in the model construction. This makes it easier to build a model and to reason about it than when modeling with inequalities alone. In particular, substructures of the problem can be made explicit on a conceptual level, which makes the model much more readable.

On the operational side, each symbolic constraint can communicate with other symbolic constraints via the constraint store. If an inference agent adds a new primitive constraint to the constraint store, this constraint becomes visible to all other non-primitive constraints and may be used in their inference process.

6. Example: Symbolic constraints for supply chain planning

Developing an extended constraint language for IP containing generic symbolic constraints is a non-trivial task. In addition to the right balance between expressive power and solver efficiency, symbolic constraints should be designed in such a way that they can serve as building blocks in a model that fit together well. In the following, we illustrate the use of symbolic constraints by a small constraint language for supply chain management (Simchi-Levi et al., 2003), which is a major application area for IP, CP, and their combination.

A *supply chain* typically involves a number of organisations that are linked by material, information and financial flows. These organizations may be suppliers of raw material or components, producers of intermediate or final products, logistic service providers, or customers, see Figure 3.7 for a simple example.

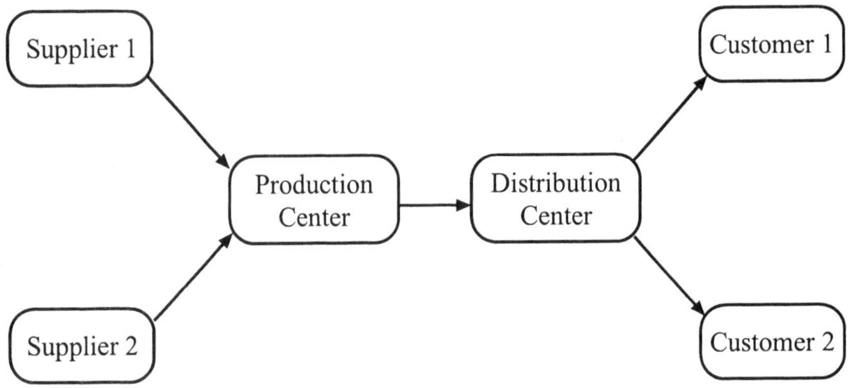

Figure 3.7. Supply Chain Example

Supply chain management can be understood as the task of integrating organizational units along the supply chain and coordinating the material, information and financial flow in order to fulfill customer demands, with the aim of improving the competitiveness of the supply chain as a whole (Stadtler, 2000).

Along the supply chain, a huge number of decisions have to be made to meet these goals (e.g. how to produce and transport goods in order to satisfy the customer demands in time). In order to prepare these decisions, *planning* is used. According to the decisions to be made, planning tasks are usually classified into three different planning levels (Fleischmann et al., 2000):

- *Long-term planning:* Decisions on this level are called strategic decisions and usually concern the design and the structure of the supply chain. Example: facility location problems

- *Mid-term planning:* This planning level involves outlining rough quantities and time windows for the resources and flows in the given supply chain. The planning horizon ranges from several weeks up to 2 years. Example: production planning problems

- *Short-term planning:* On the most detailed level, activities and instructions are specified for immediate execution. Example: machine scheduling problems

In order to deal with the complexity of the planning problem, the three levels are often combined in a hierarchical planning scheme (Schneeweiss, 1999). An increasing (resp. decreasing) level of detail is achieved by disaggregating (resp.

aggregating) data or results, while going down (resp. up) in the hierarchy, for example:

- Products are aggregated into products groups.
- Resources are aggregated into resource families.
- Time is aggregated into periods.

In this paper, we focus on mid-term planning problems (Rohde and Wagner, 2000). This means that we want to synchronize the product flow along the supply chain, taking into account capacities for production and transportation. As a result, we wish to obtain for each time period information on what has to be procured, produced, transported or delivered, and the corresponding quantities.

To model and solve mid-term planning problems, we can use mixed integer linear programming. The underlying mathematical model corresponds to a generalized multi-commodity network flow problem (Ahuja et al., 1993) with side constraints. While the transportation planning part can naturally be viewed as a network flow problem, the production can be described by a generalized network with gains and losses. This allows one to describe how many units of raw products are required to produce a certain final product. Since the planning horizon consists of several periods, in each period, it has to be decided how much to produce, to transport etc. The combined transportation and production network in each period is linked via the stock-level of the different products.

Modeling mid-term planning problems with symbolic constraints leads to a decomposition of the model into different submodels such that (in the ideal case) each symbolic constraint represents one submodel. There are several possibilities to perform this decomposition. Since the complete model is a multi-commodity generalized network flow problem with side constraints, a natural decomposition is as follows:

- encapsulate generalized network flow constraints in one symbolic constraint.
- encapsulate all constraints restricting the amount of flow in the network into one constraint.
- encapsulate all side constraints into one symbolic constraint.

In the following, we present three symbolic constraints that are based on this decomposition.

6.1 Stock-resource constraint

The `StockResource` constraint describes the stock-level for one product over the planning horizon. We will use the index set $T = \{1, \ldots, n\}$ to denote planning periods. The stock-level for a product changes depending on production, procurement or transport. In practice, one may also want to consider safety stocks, back logs, target days of supply, or shelf-life features. We present here a simplified version, which is close to lot-sizing problems, see e.g. (Wolsey and Belvaux, 2001). The constraint has the following *declarative semantics*

`StockResource(InitialStock, StockLevel, FlowIn, FlowOut)`

- `InitialStock:` A variable or a rational number s_0.
- `StockLevel:` A list $[s_1, \ldots, s_n]$ of variables or rational numbers
- `FlowIn:` A list $[I_1, \ldots, I_n]$, where $I_t = [[p_{1t}, x_{1t}, y_{1t}], \ldots, [p_{mt}, x_{mt}, y_{mt}]], t \in T$, is a list of lists $[p_{it}, x_{it}, y_{it}]$ consisting of 3 elements. p_{it} is a rational number, x_{it} is a rational variable, and y_{it} is a 0-1 variable.
- `FlowOut:` A list $[O_1, \ldots, O_n]$ where $O_t = [[r_{1t}, v_{1t}], \ldots, [r_{ht}, v_{ht}]]$, with $t \in T$, is a list of lists $[r_{jt}, v_{jt}]$ consisting of 2 elements. r_{jt} is a rational number and v_{jt} is a rational variable.

The constraint is satisfied if

$$s_t = s_{t-1} + \sum_{i=1}^{m} p_{it}x_{it} - \sum_{j=1}^{h} r_{jt}v_{jt}, \quad t \in T \tag{3.1}$$

$$x_{it} \leq My_{it}, \quad y_{it} \in \{0,1\}, \quad i \in \{1, \ldots, m\}, \quad t \in T \tag{3.2}$$

Constraint (3.1) represents the stock balance equality. It says that the stock-level in period t is equal to the stock-level from the previous period plus all receipts minus all demands. The coefficients p_{jt} and r_{jt} allow modeling the bill of materials. Constraint (3.2) can be used to express setup. The constant M corresponds to an upper bound for the variable x_{jt}. Concerning the *operational semantics* of this symbolic constraint, we may exploit results about the facial structure of the lot-sizing polytope (Pochet and Wolsey, 1995; Wolsey, 2002).

6.2 Finite capacity resource

The second symbolic constraint controls the maximum flow in the network. For example, there may be maximum production or transport capacities depending on the available machines or vehicles. To express such constraints, we introduce a new symbolic constraint `FiniteCapacityResource`. In practical applications, there may be further requirements like extending the available capacity to model extra shifts, or requiring a minimum resource utilization. To keep things easy, we consider here only simple capacity constraints. We express the capacity via the upper bound of a variable, which allows measuring resource utilization. The constraint can be seen as a generalization of the multiple knapsack constraint and has the following *declarative semantics*:

`FiniteCapacityResource(Capacities, Activities)`

- `Capacities:` A list of $[y_1, \ldots, y_n]$ of non-negative rational variables or values
- `Activities:` A list $[R_1, \ldots, R_n], R_t = [[r_{1t}, x_{1t}], \ldots, [r_{mt}, x_{mt}]], t \in T$, of nonnegative rational numbers r_{it} and rational variables x_{it}.

The constraint is satisfied if

$$\sum_{i=1}^{m} r_{it} x_{it} \leq y_t, \quad t \in T \tag{3.3}$$

The coefficients r_{it} denote the resource requirement of activity i in period t. The variables x_{it} represent the activity levels. These are limited by the available capacity, which is represented by the upper bound of the variable y. Concerning the *operational semantics*, we can exploit knowledge about the (generalized) knapsack polytope, (e.g. (Ceria et al., 1998; Marchand and Wolsey, 1999; Trick, 2002; Fahle and Sellmann, 2002)).

6.3 State resource

In many practical situations, a machine has to be setup before production can start. This can already be modeled with the `StockResource` constraint. If the setup capacity consumption is small compared to the period capacity, we may neglect that in reality a setup state is usually conserved across several periods. Only a small fraction of the available capacity will be consumed by redundant setups in case that the product is produced over several contiguous periods. If, however, the setup consumption is large compared to the period capacity (e.g. half of it), then the model is not sufficiently precise. In this case, it would be more accurate to conserve the setup state to the next period. If the same product is made without interruption, this would allow saving redundant setups in subsequent periods. In order to model this behavior, we introduce a new symbolic constraint `StateResource`:

`StateResource(States, L, U)`

- `States:` A list $[S_1, \ldots, S_n]$ where S_t is a list of two lists and two variables $[[x_{1t}, \ldots, x_{mt}], [y_{1t}, \ldots, y_{mt}], min_t, max_t]$ and $t \in T$. The variables x_{it} and y_{jt} are 0-1 variables or values.
- `L:` a nonnegative integer variable or value
- `U:` a nonnegative integer variable or value

The 0-1 variable x_{it} expresses that the setup state for i is created in period t, while the 0-1 variable y_{it} indicates whether or not setup state i is the last setup state in period t. The constraint has the following *declarative semantics*:

$$\sum_{i=1}^{m} y_{it} \le 1, \qquad t \in T \tag{3.4}$$

$$y_{it} \le x_{it} + y_{it-1}, \qquad i \in \{1,\ldots,m\}, t \in T \tag{3.5}$$

$$y_{it-1} \le 1 - x_{it}, \qquad i \in \{1,\ldots,m\}, t \in T \tag{3.6}$$

$$y_{it-1} + y_{it} \le 2 - x_{jt}, \qquad i \ne j \in \{1,\ldots,m\}, t \in T \tag{3.7}$$

$$min_t \le \sum_{i=1}^{m} x_{it} \le max_t, \qquad t \in T \tag{3.8}$$

$$L \le \sum_{i=1}^{m} \sum_{t \in T} x_{it} \le U, \tag{3.9}$$

$$x_{it}, y_{it} \in \{0,1\}, \qquad i \in \{1,\ldots,m\}, t \in T \tag{3.10}$$

Inequality (3.4) expresses that at most one setup state can be the last one in a period. Constraint (3.5) models that setup state i can only be the last setup state in period t if setup state i is created in period t or if it was the last one (and consequently conserved) from the previous period. Inequality (3.6) represents that if setup state i is the last setup state in period $t-1$, then it cannot be created in period t (since it is conserved across the border). Constraint (3.7) models that if setup state i is conserved from period $t-1$ to t, and from t to $t+1$, then there can be no different setup in period t. The inequalities (3.8) express lower and upper bounds for the number of new setups per period while constraint (3.9) models lower and upper bounds on the number of new setups on the complete planning horizon.

For the *operational semantics*, we can exploit some recent work on the capacitated lot-sizing problem with linked lot-sizes (Sürie and Stadtler, 2002).

6.4 Combined example

In order to see how the three symbolic constraints are typically used in a planning problem, consider the following scenario: We have to satisfy the demand for two final products A and B over a planning horizon of 6 weeks. One period corresponds to one week. In order to produce A and B, a raw product R is required. For sake of simplicity, one unit of R is needed in order to produce one unit of A resp. B. The products A and B are produced on the same machine, which has a limited capacity of 40 hours per week. In order to produce one unit of A (resp. B) 2 (resp. 3) hours are needed. In particular, for producing A (resp. B), the machine has to be setup. This causes setup costs and consumes machine capacity of 15 hours for A and 18 hours for B. The production takes place at a plant location, while the raw product R has to be transported from a supplier, where an initial stock level of 1000 units is available. As a consequence, we have product R at the supplier location and at the plant location. We use the following index sets:

- $T = \{1, \ldots, 6\}$ denotes the number of periods.
- $P = \{A, B, R\}$ denotes the set of products.

The model uses the following non-negative rational variables

- s_t^p: stock-level of product $p \in P$ in period $t \in T$ at the plant location
- $\tilde{s}_t^p$: stock level of product p in period t at the supplier location
- x_t^p: number of produced lots of product $p \in P$ in period $t \in T$
- $\tilde{x}_t^p$: number of transported lots of product p in period t from the supplier location to the plant location

the 0-1 variables

- y_t^p: indicates whether or not one has to do a setup for product $p \in P$ in period $t \in T$.

and, the parameters:

- d_t^p: demand for product $p \in P$ in period $t \in T$
- f_t^p: fixed cost for setting up the machine for product $p \in P$ in period $t \in T$

Now the problem can be modeled by the following symbolic constraints:

$$\begin{array}{lll}
\min & \sum_{p \in P} \sum_{t \in T} s_t^p \;+\; \sum_{p \in P} \sum_{t \in T} x_t^p \;+\; \sum_{p \in P} \sum_{t \in T} f_t^p y_t^p & \\
\text{s.t.} & \texttt{StockResource(} & 0, [s_1^A, \ldots, s_6^A],\ [[1, x_1^A, y_1^A], \ldots, [1, x_6^A, y_6^A]], \\
& & [[d_1^A, 1], \ldots, [d_6^A, 1]]), \\
& \texttt{StockResource(} & 0, [s_1^B, \ldots, s_6^B],\ [[1, x_1^B, y_1^B], \ldots, [1, x_6^B, y_6^B]], \\
& & [[d_1^B, 1], \ldots, [d_6^B, 1]]), \\
& \texttt{StockResource(} & 0, [s_1^R, \ldots, s_6^R],\ [[1, \tilde{x}_1^R, 1], \ldots, [1, \tilde{x}_6^R, 1]], \\
& & [[1, x_1^A], \ldots, [1, x_6^A], [1, x_1^B], \ldots, [1, x_6^B]]), \\
& \texttt{StockResource(} & 1000, [\tilde{s}_1^R, \ldots, \tilde{s}_6^R],\ [[0, 0, 0], \ldots, [0, 0, 0]], \\
& & [[1, \tilde{x}_1^R], \ldots, [1, \tilde{x}_6^R]], \\
& \multicolumn{2}{l}{\texttt{FiniteCapacityResource(}[40, \ldots, 40],} \\
& & [[[2, x_1^A], [15, y_1^A], [3, x_1^B], [18, y_1^B]], \ldots, \\
& & [[2, x_6^A], [15, y_6^A], [3, x_6^B], [18, y_6^B]]])
\end{array} \tag{3.11}$$

In model (3.11), the required setup times take nearly half of the week capacity. Therefore, it has to be taken into account that once the machine has been setup, its status holds across the time periods until the setup state is changed for the next product. We use the following variables:

- The 0-1 variable z_t^p denotes whether a setup has be to performed for product $p \in P$ in period $t \in T$.
- The 0-1 variable w_t^p denotes whether setup for product $p \in P$ is the last setup state in period $t \in T$.

Then, we may extend the model using the `StateResource` constraint.

$$\begin{array}{ll} \texttt{StateResource}([[z_1^A, z_1^B, 0, 2], \ldots, [z_6^A, z_6^B, 0, 2]], & \\ \qquad\qquad [[w_1^A, w_1^B], \ldots, [w_6^A, w_6^B]], 0, 12) & \\ y_t^A \leq z_t^A + w_t^A, \quad t \in T & \\ y_t^B \leq z_t^B + w_t^A, \quad t \in T & \end{array} \tag{3.12}$$

The inequalities in (3.12) connect the state variables with the setup state of the `StockResource` constraint in (3.11).

7. Summary

We have introduced a unifying framework, branch-and-infer, to describe and compare integer linear programming and finite domain constraint programming, both from the viewpoint of model building and model solving. Finite domain constraint programming offers a variety of arithmetic and symbolic constraints that allows one to model and solve combinatorial problems in many different ways. Integer linear programming admits only linear equations and inequalities, but has developed very efficient methods to handle them.

Acknowledgments

We thank an anonymous referee and the editor for their detailed comments on a previous version of this chapter.

References

Aggoun, A. and Beldiceanu, N. (1993). Extending CHIP in order to solve complex scheduling and placement problems. *Mathl. Comput. Modelling*, 17(7):57 – 73.

Ahuja, R. K., Magnanti, T. L., and Orlin, J. B. (1993). *Network flows : theory, algorithms and applications*. Prentice Hall.

Althaus, E., Bockmayr, A., Elf, M., Kasper, T., Jünger, M., and Mehlhorn, K. (2002). SCIL - Symbolic Constraints in Integer Linear Programming. In *10th European Symposium on Algorithms, ESA'02, Rome*, pages 75–87. Springer, LNCS 2461.

Baptiste, P., Pape, C. L., and Nuijten, W. (2001). *Constraint-Based Scheduling*, volume 39 of *International Series in Operations Research and Management Science*. Kluwer.

Beldiceanu, N. and Contejean, E. (1994). Introducing global constraints in CHIP. *Mathl. Comput. Modelling*, 20(12):97 – 123.

Bockmayr, A. and Eisenbrand, F. (2000). Combining logic and optimization in cutting plane theory. In *Frontiers of Combining Systems, FroCoS'2000, Nancy*, pages 1 – 17. Springer, LNCS 1794.

Bockmayr, A. and Kasper, T. (1998). Branch-and-infer: A unifying framework for integer and finite domain constraint programming. *INFORMS J. Computing*, 10(3):287 – 300.

Bockmayr, A., Kasper, T., and Zajac, T. (1998). Reconstructing binary pictures in discrete tomography. In *16th European Conference on Operational Research, EURO XVI, Bruxelles*.

Ceria, S., Cordier, C., Marchand, H., and Wolsey, L. (1998). Cutting planes for integer programs with general integer variables. *Mathematical Programming*, 81:201–214.

Cordier, C., Marchand, H., Laundry, R., and Wolsey, L. (1999). bc-opt: A branch-and-cut code for mixed integer programs. *Mathematical Programming*, 86:335–354.

Debruyne, R. and Bessière, C. (2001). Domain filtering consistencies. *Journal of Artificial Intelligence Research*, 14:205 – 230.

Fahle, T. and Sellmann, M. (2002). Cost-based filtering for the constrained knapsack problem. *Annals of Operations Research*, 115:73–93.

Fleischmann, B., Meyr, H., and Wagner, M. (2000). Advanced planning. In Stadtler, H. and Kilger, C., editors, *Supply Chain Management and Advanced Planning: Concepts, Models, Software and Case Studies*, pages 57–71. Springer.

Focacci, F., Lodi, A., and Milano, M. (1999). Cost-based domain filtering. In *Principles and Practice of Constraint Programming, CP'99, Alexandria, VA*, pages 189–203. Springer, LNCS 1713.

Jünger, M., Reinelt, G., and Rinaldi, G. (1995). The traveling salesman problem. In Ball, M., Magnanti, T., Monma, C., and Nemhauser, G., editors, *Network models*, volume 7 of *Handbooks in operations research and management science*, chapter 4, pages 225–330. North Holland.

Kasper, T. (1998). *A unifying logical framework for integer linear programming and finite domain constraint programming*. PhD thesis, Technische Fakultät, Univ. des Saarlandes.

Marchand, H. and Wolsey, L. (1999). The 0-1 knapsack problem with a single continuous variable. *Mathematical Programming*, 85:15–33.

Pochet, Y. and Wolsey, L. (1995). Algorithms and reformulation for lot sizing problems. In Cook, W., Lovasz, L., and Seymour, P., editors, *Combinatorial Optimization: Papers from the DIMACS Special Year*, volume 20, pages 245 – 293. DIMACS Series in Discrete Mathematics and Theoretical Computer Science.

Rohde, J. and Wagner, M. (2000). Master planning. In Stadtler, H. and Kilger, C., editors, *Supply Chain Management and Advanced Planning: Concepts, Models, Software and Case Studies*, pages 117–133. Springer.

Schneeweiss, C. (1999). *Hierarchies in distributed decision making*. Springer.

Sellmann, M. (2002). *Reduction techniques in constraint programming and combinatorial optimization*. PhD thesis, Univ. Paderborn.

Simchi-Levi, D., Kaminsky, P., and Simchi-Levi, E. (2003). *Designing and Managing the Supply Chain: Concepts, Strategies, and Cases*. McGraw-Hill, 2nd edition.

Stadtler, H. (2000). Supply chain planning - An overview. In Stadtler, H. and Kilger, C., editors, *Supply Chain Management and Advanced Planning: Concepts, Models, Software and Case Studies*, pages 7–28. Springer.

Sürie, C. and Stadtler, H. (2002). The capacitated lot-sizing problem with linked lot-sizes. Schriften zur Quantitativen Betriebswirtschaftslehre 2, Technische Universität Darmstadt. To appear in Management Science.

Trick, M. (2002). A dynamic programming approach for consistency and propagation for knapsack constraints. In *3rd International Workshop on Integration of AI and OR techniques in Constraint Programming for Combinatorial Optimization Problems, CP-AI-OR'01, Ashford, UK*, pages 113 – 124.

Woeginger, G. J. (2001). The reconstruction of polyominoes from their orthogonal projections. *Information Processing Letters*, 77(5-6):225–229.

Wolsey, L. (2002). Solving multi-item lot-sizing problems with an MIP solver using classification and reformulation. Discussion Paper No. 2002/12, CORE.

Wolsey, L. and Belvaux, G. (2001). Modelling practical lot-sizing problems as mixed integer programs. *Management Science*, 47:993–1007.

Chapter 4

GLOBAL CONSTRAINTS AND FILTERING ALGORITHMS

Jean-Charles Régin
ILOG
1681, route des Dolines,
Sophia Antipolis, 06560 Valbonne, France
jcregin@ilog.fr

Abstract

Constraint programming (CP) is mainly based on filtering algorithms; their association with global constraints is one of the main strengths of CP. This chapter is an overview of these two techniques. Some of the most frequently used global constraints are presented. In addition, the filtering algorithms establishing arc consistency for two useful constraints, the alldifferent and the global cardinality constraints, are fully detailed. Filtering algorithms are also considered from a theoretical point of view: three different ways to design filtering algorithms are described and the quality of the filtering algorithms studied so far is discussed. A categorization is then proposed. Over-constrained problems are also mentioned and global soft constraints are introduced.

Keywords: Global constraint, filtering algorithm, arc consistency, alldifferent, global cardinality constraint, over-constrained problems, global soft constraint, graph theory, matching.

1. Introduction

A constraint network (CN) consists of a set of variables, domains of possible values associated with each of these variables, and a set of constraints that link up the variables and define the set of combinations of values that are allowed. The search for an instantiation of all variables that satisfies all the constraints is called a Constraint Satisfaction Problem (CSP), and such an instantiation is called a solution of a CSP.

A lot of problems can be easily coded in terms of CSP. For instance, CSP has already been used to solve problems of scene analysis, placement, resource allocation, crew scheduling, time tabling, scheduling, frequency allocation, car sequencing, and so on. An interesting paper of (Simonis, 1996) presents a survey on industrial studies and applications developed over the last ten years.

Unfortunately, a CSP is an NP-Complete problem. Thus, much work has been carried out in order to try to reduce the time needed to solve a CSP. Constraint Programming (CP) is one of these techniques.

Constraint Programming proposes to solve CSPs by associating with each constraint a filtering algorithm that removes some values of variables that cannot belong to any solution of the CSP. These filtering algorithms are repeatedly called until no new deduction can be made. This process is called the *propagation mechanism*. Then, CP uses a search procedure (like a backtracking algorithm) where filtering algorithms are systematically applied when the domain of a variable is modified. Therefore, with respect to the current domains of the variables, using filtering algorithms, CP removes once and for all certain inconsistencies that would have been discovered several times otherwise. Thus, if the cost of the calls of the filtering algorithms at each node is less than the time required by the search procedure to discover many times the same inconsistency, then the resolution will be speeded up.

One of the most interesting properties of a filtering algorithm is arc consistency. We say that a filtering algorithm associated with a constraint establishes arc consistency if it removes all the values of the variables involved in the constraint that are not consistent with the constraint. For instance, consider the constraint $x + 3 = y$ with the domain of x equals to $D(x) = \{1, 3, 4, 5\}$ and the domain of y equal to $D(y) = \{4, 5, 8\}$. Then the establishing of arc consistency will lead to $D(x) = \{1, 5\}$ and $D(y) = \{4, 8\}$.

Since constraint programming is based on filtering algorithms, it is quite important to design efficient and powerful algorithms. Therefore, this topic caught the attention of many researchers, who then discovered a large number of algorithms. Nevertheless, many studies on arc consistency have been limited to binary constraints that are defined in extension, in other words by the list of allowed combinations of values. This limitation was justified by the fact that any constraint can always be defined in extension and by the fact that any non-binary constraint network can be translated into an equivalent binary one with additional variables (Rossi et al., 1990). However, in practice, this approach has several drawbacks:

- it is often inconceivable to translate a non-binary constraint into an equivalent set of binary ones because of the underlying computational and memory costs (particularly for non-representable ones, (Montanari, 1974)).

- the structure of the constraint is not used at all. This prevents us from developing more efficient filtering algorithm dedicated to this constraint. Moreover, some non-binary constraints lose much of their structure when encoded into a set of binary constraints. This leads, for example, to a much less efficient pruning behavior for arc consistency algorithms handling them.

The advantage of using the structure of a constraint can be emphasized on a simple example: the constraint $x \leq y$. Let $min(D)$ and $max(D)$ be respectively the minimum and the maximum value of a domain. It is straightforward to establish that all the values of y in the range $[min(D(x)), max(D(y)]$ are consistent with the constraint and all the values of x in the range

$$[min(D(x)), max(D(y)]$$

are consistent with the constraint. This means that arc consistency can be efficiently and easily established by removing the values that are not in the above ranges. Moreover, the use of the structure is often the only way to avoid memory consumption problems when dealing with non-binary constraints. In fact, this approach prevents one from explicitly representing all the combinations of values allowed by the constraint.

Thus, researchers interested in the resolution of real life applications with constraint programming, and notably those developing languages that encapsulate CP (like PROLOG), designed specific filtering algorithms for the most common simple constraints (like $=, \neq, <, \leq$, ...) and also general frameworks to exploit efficiently some knowledge about binary constraints (like AC-5, (Van Hentenryck et al., 1992)). However, they have been confronted with two new problems: the lack of expressiveness of these simple constraints and the weakness of domain reduction of the filtering algorithms associated with these simple constraints. It is, indeed, quite convenient when modeling a problem in CP to have at one's disposal some constraints corresponding to a set of constraints. Moreover, these new constraints can be associated with more powerful filtering algorithms because they can take into account the simultaneous presence of simple constraints to further reduce the domains of the variables. These constraints encapsulating a set of other constraints are called **global constraints**.

One of the most well known examples is the alldifferent constraint, especially because the filtering algorithm associated with this constraint is able to establish arc consistency in a very efficient way.

An alldifferent constraint defined on X, a set of variables, states that the values taken by variables must be all different. This constraint can be represented by a set of binary constraints. In this case, a binary constraint of difference is built for each pair of variables belonging to the same constraint of difference. But the pruning effect of arc consistency for these constraints is limited. In fact, for a binary alldifferent constraint between two variables i and j, arc-consistency removes a value from domain of i only when the domain of j is reduced to a single value. Let us suppose we have a CSP with

	Mo	Tu	We	Th	...
peter	D	N	O	M	
paul	D	B	M	N	
mary	N	O	D	D	
...					

A = {M,D,N,B,O}, P = {peter, paul, mary, ...}
W = {Mo, Tu, We, Th, ...}
M: morning, D: day, N: night B: backup, O: day-off

Figure 4.1. An Assignment Timetable.

3 variables x_1, x_2, x_3 and an alldifferent constraint involving these variables with $D(x_1) = \{a, b\}$, $D(x_2) = \{a, b\}$ and $D(x_3) = \{a, b, c\}$. Establishing arc consistency for this alldifferent constraint removes the values a and b from the domain of x_3, while arc-consistency for the alldifferent represented by binary constraints of difference does not delete any value.

We can further emphasize the advantage of global constraints on a more realistic example that involves global cardinality constraints (GCC).

A GCC is specified in terms of a set of variables $X = \{x_1, ..., x_p\}$ which take their values in a subset of $V = \{v_1, ..., v_d\}$. It constrains the number of times a value $v_i \in V$ is assigned to a variable in X to be in an interval $[l_i, u_i]$. GCCs arise in many real life problems. For instance, consider the example derived from a real problem and given in (Caseau et al., 1993) (cf. Figure 4.1). The task is to schedule managers for a directory-assistance center, with 5 activities (set A), 7 persons (set P) over 7 days (set W). Each day, a person can perform an activity from the set A. The goal is to produce an assignment matrix that satisfies the following global and local constraints:

- **general constraints** restrict the assignments. First, for each day we may have a minimum and maximum number for each activity. Second, for each week, a person may have a minimum and maximum number for each activity. Thus, for each row and each column of the assignment matrix, there is a global cardinality constraint.

- **local constraints** mainly indicate incompatibilities between two consecutive days. For instance, a morning schedule cannot be assigned after a night schedule.

Each general constraint can be represented by as many min/max constraints as the number of involved activities. Now, these min/max constraints can be easily handled with, for instance, the **atmost/atleast** operators proposed in (Van Hentenryck and Deville, 1991). Such operators are implemented using local propagation. But as is noted in (Caseau et al., 1993): "The problem is that efficient resolution of a timetable problem requires a global computation on the

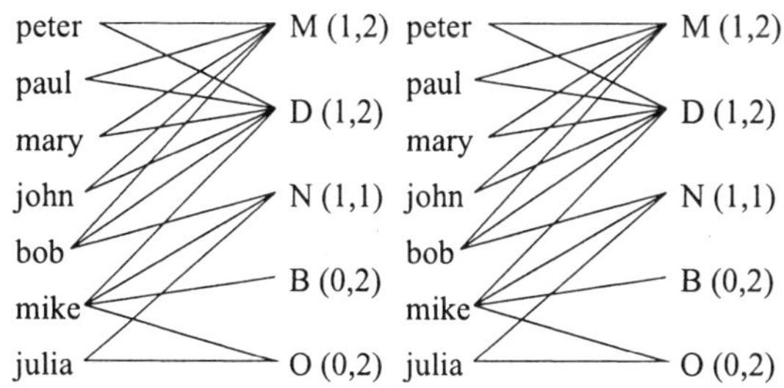

Figure 4.2. An example of a global cardinality constraint.

set of min/max constraints, and not the efficient implementation of each of them separately." Hence, this way is not satisfactory. Therefore global cardinality constraints associated with efficient filtering algorithms (like ones establishing arc consistency) are needed.

In order to show the difference in global and local filtering, consider a GCC associated with a day (cf figure 4.2). The constraint can be represented by a bipartite graph called a value graph (left graph in Figure 4.2). The left set corresponds to the person set, the right set to the activity set. There exists an edge between a person and an activity when the person can perform the activity. For each activity, the numbers between parentheses express the minimum and the maximum number of times the activity has to be assigned. For instance, John can work the morning or the day but not the night; one manager is required to work the morning, and at most two managers work the morning. We recall that each person has to be associated with exactly one activity.

Encoding the problem with a set of atmost/atleast constraints leads to no deletion. Now, we can carefully study this constraint. Peter, Paul, Mary, and John can work only in the morning and during the day. Moreover, morning and day can be assigned together to at most 4 persons. Thus, no other persons (i.e. Bob, Mike, nor Julia) can perform activities M and D. So we can delete the edges between Bob, Mike, Julia and D, M. Now only one possibility remains for Bob: N, which can be assigned at most once. Therefore, we can delete the edges {mike,N} and {julia,N}. This reasoning leads to the right graph in Figure 4.2. It corresponds to the establishing of arc consistency for the global constraint.

This chapter is organized as follows. First, some preliminaries are reviewed and the definition and the significance of global constraints are discussed. Some of the most frequently used global constraints are then presented. Section 3 deals with the possible types of filtering algorithms (FA). Three types of filtering algorithm are presented. In section 4, the filtering algorithms establishing arc consistency for the alldifferent and the global cardinality constraint are detailed. Section 5 deals with over-constrained problems and shows the advantages of modeling the Maximal Constraint Satisfaction problem by a global constraint Max-Sat. This section also introduces the global soft constraints and two general definitions of violation costs associated with global constraints. The soft alldifferent constraint is taken as example. In Section 6 the quality of filtering algorithms is discussed and a classification proposed. Some miscellaneous considerations about filtering algorithms are mentioned in Section 7. Finally, we conclude.

2. Global Constraints

2.1 Preliminaries

A finite **constraint network** $\mathcal{N}$ is defined as a set of n **variables** $X = \{x_1, \ldots, x_n\}$, a set of current **domains** $\mathcal{D} = \{D(x_1), \ldots, D(x_n)\}$ where $D(x_i)$ is the finite set of possible **values** for variable x_i, and a set $\mathcal{C}$ of **constraints** between variables. We introduce the particular notation

$$\mathcal{D}_0 = \{D_0(x_1), \ldots, D_0(x_n)\}$$

to represent the set of initial domains of $\mathcal{N}$. Indeed, we consider that any constraint network $\mathcal{N}$ can be associated with an initial domain $\mathcal{D}_0$ (containing $\mathcal{D}$), on which constraint definitions were stated.

A **constraint** C on the ordered set of variables $X(C) = (x_{i_1}, \ldots, x_{i_r})$ is a subset $T(C)$ of the Cartesian product $D_0(x_{i_1}) \times \cdots \times D_0(x_{i_r})$ that specifies the **allowed** combinations of values for the variables $x_1, \ldots, x_r$. An element of $D_0(x_1) \times \cdots \times D_0(x_r)$ is called a **tuple on** $X(C)$. $|X(C)|$ is the **arity** of C.

A value a for a variable x is often denoted by (x, a). $var(C, i)$ represents the ith variable of $X(C)$, while $index(C, x)$ is the position of variable x in $X(C)$. $\tau[k]$ denotes the kth value of the tuple τ. $\tau[index(C, x)]$ will be denoted by $\tau[x]$ when no confusion is possible. $D(X)$ denotes the union of domains of variables of X (i.e. $D(X) = \cup_{x_i \in X} D(x_i)$). $\#(a, \tau)$ is the number of occurrences of the value a in the tuple τ.

Let C be a constraint. A tuple τ on $X(C)$ is **valid** if $\forall (x, a) \in \tau, a \in D(x)$. C is **consistent** iff there exists a tuple τ of $T(C)$ which is valid. A value $a \in D(x)$ is **consistent with** C iff $x \notin X(C)$ or there exists a valid tuple τ of $T(C)$ with $a = \tau[index(C, x)]$. (τ is the called a **support** for (x, a) on C.) A constraint is **arc consistent** iff $\forall x_i \in X(C), D(x_i) \neq \emptyset$ and $\forall a \in D(x_i)$, a is consistent with C.

A **filtering algorithm** associated with a constraint C is an algorithm which may remove some values that are inconsistent with C, and that does not remove any consistent values. If the filtering algorithm removes all the values inconsistent with C we say that it establishes the arc consistency of C.

The **propagation** is the mechanism that consists of calling the filtering algorithm associated with the constraints involving a variable x each time the domain of this variable is modified. Note that if the domains of the variables are finite, then this process terminates because a domain can be modified only a finite number of times.

2.2 Definition and Advantages

Two kinds of constraints can be identified: **non-decomposable constraints** and **global constraints**.

Non-decomposable constraints are constraints that cannot be expressed by a set of other constraints, whereas global constraints are constraints that are equal to a set of other constraints (non-decomposable or global).

As example of non-decomposable constraints is: the arithmetic constraints like $x < y$, $x = y$, $x \neq y$, $x + y = z$, $x * y = z$...; and the constraints given in extension.

Formally, the global constraints can be defined as follows:

Definition 1 *Let $\mathcal{C} = \{C_1, C_2, .., C_n\}$ be a set of constraints. The constraint C_G equals to the conjunction of all the constraints of $\mathcal{C}$: $C_G = \wedge\{C_1, C_2, ..C_n\}$ is a* **global constraint**.
The set of tuples of $\mathcal{C}$ is equal to the set of solutions of the constraint network $(\cup_{C \in \mathcal{C}} X(C), \mathcal{D}_{X(\mathcal{C})}, \{C_1, C_2, .., C_n\})$.

Global constraints are often defined from a set of variables and some prototypes of non-decomposable constraints. For instance, an alldifferent constraint is just defined by: alldifferent(X) which means that it corresponds to all the constraints $\neq$ stated for each pair of variables of X.

Global constraints have three main advantages:

- Expressiveness: it is more convenient to define one constraint corresponding to a set of constraints than to define independently each constraint of this set.
- Since a global constraint corresponds to a set of constraints it is possible to deduce some information from the simultaneous presence of constraints.
- Powerful filtering algorithms can be designed because the set of constraints can be taken into account as a whole.

Specific filtering algorithms make it possible to use Operations Research techniques or graph theory.

The last point is emphasized by the following property:

Property 1 *The establishing of arc consistency on $\mathcal{C} = \wedge\{C_1, C_2, .., C_n\}$ is stronger (that is, cannot remove fewer values) than the establishing of arc consistency of the network $(\cup_{C \in \mathcal{C}} X(C), \mathcal{D}_{X(\mathcal{C})}, \{C_1, C_2, .., C_n\})$.*

proof: By Definition 1 the set of tuples of $\mathcal{C} = \wedge\{C_1, C_2, .., C_n\}$ corresponds to the set of solution of $(\cup_{C \in \mathcal{C}} X(C), \mathcal{D}_{X(\mathcal{C})}, \{C_1, C_2, .., C_n\})$. Therefore, the establishing of arc consistency of $\wedge\{C_1, C_2, .., C_n\}$ removes all the values that do not belong to a solution of $(\cup_{C \in \mathcal{C}} X(C), \mathcal{D}_{X(\mathcal{C})}, \{C_1, C_2, .., C_n\})$ which is stronger than the arc consistency of the previous network.

Therefore, arc consistency on global constraints is a strong property. The following proposition is an example of the gap between arc consistency for a global constraint and arc consistency for the network corresponding to this global constraint

Property 2 *Arc Consistency for C =alldifferent(X) corresponds to the arc consistency of a Constraint Network with an exponential number of constraints defined by:*
$\forall A \subseteq X$: $|D(A)| = |A| \Rightarrow D(X - A)$ *is reduced to* $D(X) - D(A)$

proof: see (Régin, 1995).

2.3 Examples

The purpose of this section is not to be exhaustive, but to present some of the global constraints that are useful in pratice. We will give a short review of:

- cumulative
- diff-n
- cycle
- sort
- alldifferent and permutation
- symmetric alldifferent
- global cardinality
- global cardinality with costs
- sum and scalar product of alldifferent variables
- sum and binary inequalities
- sequence
- stretch
- minimum global distance
- k-diff
- number of distinct values

Cumulative Constraint. Here is the definition of the constraint cumulative(S, D, H, u) from (Beldiceanu and Contejean, 1994): "The cumulative constraint matches directly the single resource scheduling problem, where the S variables correspond to the start of the tasks, the D variables to the duration of the tasks, and the H variable to the heights of the resources that is the amounts of resource used by each task. The natural number u is the total amount of available resource that must be shared at any instant by the different tasks. The cumulative constraint states that, at any instant i of the schedule, the summation of the amount of resource of the tasks that overlap i does not exceed the upper limit u."

Definition 2 *Consider A a set of activities where each activity i is associated with 3 variables: s_i the start variable representing the start time of the activity, d_i the duration variable representing the duration of the activity, and h_i the consumption representing the amount of resource which is needed by the activity.*
A **cumulative constraint** *is a constraint C associated with a positive integer u and A a set of activities, such that:*

$$T(C) = \{ \quad \tau \text{ s.t. } \tau \text{ is a tuple of } X(C) \text{ and } \forall i = [1, |A|] : \tau[s_i] = a \Leftrightarrow \sum_{j/\tau[s_j] \leq a \leq \tau[s_j]+\tau[d_j]-1} \tau[h_j] \leq u \}$$

A filtering algorithm is detailed in (Beldiceanu and Carlsson, 2002). Its complexity is $O(mn \log n + mnp)$ where m is the number of resources, n the number of tasks, and p the number of tasks that are not totally fixed. Other algorithms have been proposed for disjunctive scheduling problems. In this case, each resource can execute at most one activity at a time. For instance, the reader can consult (Baptiste et al., 1998), or (Carlier and Pinson, 1994) for a presentation of the edge-finder algorithm with the lowest worst case complexity so far.

Diff-n Constraint. We present here only the diff-n/1 constraint. We quote (Beldiceanu and Contejean, 1994): "The diff-n constraint was introduced in CHIP in order to handle multi-dimensional placement problems that occur in scheduling, cutting or geometrical placement problems. The intuitive idea is to extend the alldifferent constraint which works on a set of domain variables to a nonoverlapping constraint between a set of objects defined in a n-dimensional space."

Definition 3 *Consider R a set of multidirectional rectangles. Each multidirectional rectangle i is associated with 2 set of variables $O_i = \{o_{i1}, .., o_{in}\}$ and $L_i = \{l_{i1}, .., l_{in}\}$. The variables of O_i represent the origin of the rectangle for every dimension, for instance the variable o_{ij} corresponds to the origin of the rectangle for the jth dimension. The variables of L_i represent the length of the rectangle for every dimension, for instance the variable l_{ij} represents the length of the rectangle for the jth dimension.*
A **diff-n constraint** *is a constraint C associated with a set R of multidirectional rectangles, such that:*

$$T(C) = \{ \quad \tau \text{ s.t. } \tau \text{ is a tuple of } X(C) \text{ and } \forall i \in [1, m], \forall j \in [1, m], j \neq i, \exists k \in [1, n] \text{ s.t. } \tau[o_{ik}] \geq \tau[o_{jk}] + \tau[l_{jk}] \text{ or } \tau[o_{jk}] \geq \tau[o_{ik}] + \tau[l_{ik}] \}$$

This constraint is mainly used for packing problems. In (Beldiceanu et al., 2001), an $O(d)$ filtering algorithm for the non-overlapping constraint between two d-dimensional boxes and a filtering algorithm for the non-overlapping constraint between two convex polygons are presented.

Cycle Constraint. We present here only the cycle/2 constraint. Here is the idea of this constraint (Beldiceanu and Contejean, 1994): "The cycle constraint was introduced in CHIP to tackle complex vehicle routing problems. The cycle/2 constraint can be seen as the problem of finding N distinct circuits in a directed graph in such a way that each node is visited exactly once. Initially, each domain variable x_i corresponds to the possible successors of the ith node of the graph."

Definition 4 *A* **cycle constraint** *is a constraint C associated with a positive integer n and defined on a set X of variables, such that:*

$T(C) = \{$ *τ s.t. τ is a tuple of $X(C)$*
and the graph defined from the arcs $(k, \tau[k])$
has n connected components
and every connected component is a cycle$\}$

This constraint is mentioned in the literature but no filtering algorithm is explicitly given. It is mainly used for vehicle routing problems or crew scheduling problems.

Sort Constraint. This constraint has been proposed by (Bleuzen-Guernalec and Colmerauer, 1997): "A sortedness constraint expresses that an n−tuple $(y_1, ..., y_n)$ is equal to the n−tuple obtained by sorting in increasing order the terms of another n−tuple $(x_1, ..., x_n)$".

Definition 5 *A* **sort constraint** *is a constraint C defined on two sets of variables $X = \{x_1, ..., x_n\}$ and $Y = \{y_1, ..., y_n\}$ such that*

$T(C) =$ *$\{\tau$ s.t. τ is a tuple on $X(C)$ and $\exists f$ a permutation of $[1..n]$ s.t.*
$\forall i \in [1..n]\ \tau[x_{f(i)}] = \tau[y_i]\}$

The best filtering algorithm establishing bound consistency has been proposed by (Melhorn and Thiel, 2000). Its running time is $O(n)$ plus the time required to sort the interval endpoints of the variables of X. If the interval endpoints are from an integer range of size $O(n^k)$ for some constant k the algorithm runs in linear time, because this sort becomes linear.

A sort constraint involving 3 sets of variables has also been proposed by (Zhou, 1996; Zhou, 1997). The n added variables are used for making explicit a permutation linking the variables of X and those of Y. Well known difficult job shop scheduling problems have been solved thanks to this constraint.

Alldifferent and Permutation Constraints. The alldifferent constraint constrains the values taken by a set of variables to be pairwise different. The permutation constraint is an alldifferent constraint in which $|D(X(C))| = |X(C)|$.

Definition 6 *An* **alldifferent constraint** *is a constraint C such that*
$T(C) = \{\tau \text{ s.t. } \tau \text{ is a tuple on } X(C) \text{ and } \forall a_i \in D(X(C)) : \#(a_i, \tau) \leq 1\}$

This constraint is used in many real world problems like rostering or resource allocation. It is quite useful to express that two things cannot be at the same place at the same moment.
A filtering algorithm establishing arc consistency for the alldifferent is given in this chapter and also in (Régin, 1994). Its complexity is in $O(m)$ with $m = \sum_{x \in X} |D(x)|$, after the computation of the consistency of the constraint which requires $O(\sqrt{n}m)$. When the domain of the variables are intervals, (Melhorn and Thiel, 2000) proposed a filtering algorithm establishing bound consistency with a complexity which is asymptotically the same as for sorting the internal endpoints. If the interval endpoints are from an integer range of size $O(n^k)$ for some constant k the algorithm runs in linear time. Therefore, Melhorn's algorithm is linear for a permutation constraint.

On the other hand, (Leconte, 1996) has proposed an algorithm which considers that the domains are intervals, but which can create "holes" in the domain. His filtering algorithm is in $O(n^2 d)$.

Symmetric Alldifferent Constraint. The symmetric alldifferent constraint constrains some entities to be grouped by pairs. It is a particular case of the alldifferent constraint, a case in which variables and values are defined from the same set S. That is, every variable represents an element e of S and its values represent the elements of S that are compatible with e. This constraint requires that all the values taken by the variables are different (similar to the classical alldifferent constraint) and that if the variable representing the element i is assigned to the value representing the element j, then the variable representing the element j is assigned to the value representing the element i.

Definition 7 *Let X be a set of variables and σ be a one-to-one mapping from $X \cup D(X)$ to $X \cup D(X)$ such that*
$\forall x \in X$: $\sigma(x) \in D(X)$; $\forall a \in D(X)$: $\sigma(a) \in X$ *and* $\sigma(x) = a \Leftrightarrow x = \sigma(a)$.
A **symmetric alldifferent constraint** *defined on X is a constraint C associated with σ such that:*

$$\begin{aligned} T(C) = \{ & \tau \text{ s.t. } \tau \text{ is a tuple on } X \\ & \text{and } \forall a \in D(X) : \#(a, \tau) = 1 \\ & \text{and } a = \tau[index(C, x)] \Leftrightarrow \sigma(x) = \tau[index(C, \sigma(a))]\} \end{aligned}$$

This constraint has been proposed by (Régin, 1999b). It is useful to be able to express certain items that should be grouped as pairs, for example in the problems of sports scheduling or rostering. Arc consistency can be established in $O(nm)$ after computing the consistency of the constraint which is equivalent to the search for a maximum matching in a non-bipartite graph, which can be performed in $O(\sqrt{n}m)$ by using the complex algorithm of (Micali and Vazirani, 1980).

In (Régin, 1999b), another filtering algorithm is proposed. It is difficult to characterize it but its complexity is $O(m)$ per deletion. In this paper, it is also shown how the classical alldifferent constraint plus some additional constraints can be useful to solve the original problem. The comparison between this approach, the symmetric alldifferent constraint, and the alldifferent constraint has been carried out by (Henz et al., 2003).

Global Cardinality Constraint. A global cardinality constraint (GCC) constrains the number of times every value can be taken by a set of variables. This is certainly one of the most useful constraints in practice. Note that the alldifferent constraint corresponds to a GCC in which every value can be taken at most once.

Definition 8 *A* **global cardinality constraint** *is a constraint C in which each value $a_i \in D(X(C))$ is associated with two positive integers l_i and u_i with $l_i \leq u_i$ and*
$T(C) = \{ \tau$ *s.t.* τ *is a tuple on* $X(C)$
and $\forall a_i \in D(X(C)) : l_i \leq \#(a_i, \tau) \leq u_i \}$
It is denoted by $gcc(X, l, u)$.

This constraint is present in almost all rostering or car-sequencing problems. A filtering algorithm establishing arc consistency for this constraint is described in (Régin, 1996) and is detailed in section 4.3. The consistency of the constraint can be checked in $O(nm)$ and the arc consistency can be computed in $O(m)$ providing that a maximum flow has been defined.

Global Cardinality Constraint with Costs. A global cardinality constraint with costs (costGCC) is the conjunction of a GCC constraint and a sum constraint:

Definition 9 *A* **cost function on a variable set** *X is a function which associates with each value (x, a), $x \in X$ and $a \in D(x)$ an integer denoted by $cost(x, a)$.*

Definition 10 *A* **global cardinality constraint with costs** *is a constraint C associated with* `cost` *a cost function on $X(C)$, an integer H and in which each value $a_i \in D(X(C))$ is associated with two positive integers l_i and u_i*
$T(C) = \{ \tau$ *s.t.* τ *is a tuple on* $X(C)$
and $\forall a_i \in D(X(C)) : l_i \leq \#(a_i, \tau) \leq u_i$
and $\Sigma_{i=1}^{|X(C)|} cost(var(C, i), \tau[i]) \leq H \}$
It is denoted by $costgcc(X, l, u, cost, H)$.

This constraint is used to model some preferences between assignments in resource allocation problems. Note that there is no assumption made on the sign of costs.

The integration of costs within a constraint is quite important, especially to solve optimization problems, because it improves back-propagation, which is due to the modification of the objective variable. In other words, the domain of the variables can be reduced when the objective variable is modified. (Caseau and Laburthe, 1997) have used an alldifferent constraint with costs, but only the consistency of the constraint has been checked, and no specific filtering has been used. The first proposed filtering algorithm comes from (Focacci et al., 1999a) and (Focacci et al., 1999b), and is based on reduced cost. A filtering algorithm establishing arc consistency has been proposed by (Régin, 1999a) and (Régin, 2002). The consistency of this constraint can be checked by searching for a minimum cost flow and arc consistency can be established in $O(|\Delta|S(m, n+d, \gamma))$ where $|\Delta|$ is the number of values that are taken by a variable in a tuple, and where $S(m, n+d, \gamma))$ is the complexity of the search for shortest paths from a node to every node in a graph with m arcs and n nodes with a maximal cost γ.

Sum and Scalar product of alldifferent Variables. An interesting example of costGCC is the constraint on the sum of all different variables. More precisely, for a given set of variable X, this constraint is the conjunction of the constraint $\sum_{x_i \in X} x_i \leq H$ and alldifferent(X). Similarly, we can define the constraint which is the conjunction of the constraint $\sum_{x_i \in X} \alpha_i x_i \leq H$ and alldifferent(X).

Definition 11 *A* **scalar product of alldifferent variables constraint** *is a constraint C associated with α a set of coefficients, one for each variable, an integer H, such that:*

$$T(C) = \{ \tau \text{ s.t. } \tau \text{ is a tuple on } X(C) \\ \text{and } \forall a_i \in D(X(C)) : \#(a_i, \tau) \leq 1 \\ \text{and } \Sigma_{i=1}^{|X(C)|} \alpha_i \tau[i] \leq H \}$$

The following model is used to compute arc consistency for this constraint (Régin, 1999a).

Let us define the boundaries and cost function as follows:

- For each value $a_i \in D(X)$ we define $l_i = 0$ and $u_i = 1$
- For each variable $x \in X$ and for each value $a \in D(x)$, $cost(x, a) = \alpha_i a$

Then, it is easy to prove that the costGCC constraint $costgcc(X, l, u, cost, H)$ represents the conjunction of the constraint $\sum_{x_i \in X} \alpha_i x_i \leq H$ and alldifferent(X).

Therefore, establishing arc consistency for this constraint is equivalent to establish arc consistency to the costGCC constraint defined as above.

Note that we could generalize this constraint to deal with a global cardinality constraint defined on the variables instead of an alldifferent constraint.

This constraint is used, for instance, to solve the Golomb ruler problem.

Sum and Binary Inequalities Constraint. This constraint is the conjunction of a sum constraint and a set of distance constraints, that is constraints of the form $x_j - x_i \leq c$.

Definition 12 *Let $SUM(X, y)$ be a sum constraint, and $\mathcal{I}_{neq}$ be a set of binary inequalities defined on X. The* **sum and binary inequalities constraint** *is a constraint C associated with $SUM(X, y)$ and $\mathcal{I}_{neq}$ such that:*

$T(C) = \{$ *τ s.t. τ is a tuple of $X \cup y$*
and $(\sum_{i=1}^{|X|} \tau[i]) = \tau[y]$
and the values of τ satisfy $\mathcal{I}_{neq}$ $\}$

This constraint has been proposed by (Régin and Rueher, 2000). It is used to minimize the delays in scheduling applications.
Bound consistency can be computed in $O(n(m + n \log n))$, where m is the number of inequalities and n the number of variables. It is also instructive to remark that the bound consistency filtering algorithm still works when $y = \Sigma_{i=1}^{i=n} \alpha_i x_i$ where α is non-negative real number.

Sequence Constraint. A global sequencing constraint C is specified in terms of a ordered set of variables $X(C) = \{x_1, ..., x_p\}$ which take their values in $D(C) = \{v_1, ..., v_d\}$, some integers q, min and max and a given subset V of $D(C)$. On one hand, a gsc constrains the number of variables in $X(C)$ instantiated to a value $v_i \in D(C)$ to be in an interval $[l_i, u_i]$. On the other hand, a gsc constrains for each sequence S_i of q consecutive variables of $X(C)$, that at least min and at most max variables of S_i are instantiated to a value of V.

Definition 13 *A* **global sequencing constraint** *is a constraint C associated with three positive integers min, max, q and a subset of values $V \subseteq D(C)$ in which each value $v_i \in D(C)$ is associated with two positive integers l_i and u_i and*

$T(C) = \{$ *t such that t is a tuple of $X(C)$*
and $\forall v_i \in D(C) : l_i \leq \#(v_i, t) \leq u_i$
and for each sequence S of q consecutive
variables: $min \leq \sum_{v_i \in V} \#(v_i, t, S) \leq max$ $\}$

This constraint arises in car sequencing or in rostering problems. A filtering algorithm is described in (Régin and Puget, 1997). Some problems of the CSP-Lib have been closed using this constraint.

Stretch Constraint. This constraint has been proposed by (Pesant, 2001). This constraint can be seen as the opposite of the sequence constraint. The stretch constraint aims to group the values by sequence of consecutive values, whereas the sequence is often used to obtain a homogenous repartition of values.

A stretch constraint C is specified in terms of an ordered set of variables $X(C) = \{x_1, ..., x_p\}$ which take their values in $D(C) = \{v_1, ..., v_d\}$, and two sets of integers $l = \{l_1, ..., l_d\}$ and $u = \{u_1, ..., u_d\}$, where every value v_i of $D(C)$ is associated with l_i the ith integer of L and u_i the ith integer of U. A stretch constraint states that if $x_j = v_i$ then x_j must belong to a sequence of consecutive variables that also take value v_i and the length of this sequence (the span of the stretch) must belong to the interval $[l_i, u_i]$.

Definition 14 *A* **stretch constraint** *is a constraint C associated with a subset of values $V \subseteq D(C)$ in which each value $v_i \in D(C)$ is associated with two positive integers l_i and u_i and*

$$T(C) = \{\ t \text{ s.t. } t \text{ is a tuple of } X(C) \text{ and } \forall x_j \in [1..|X(C)|], (x_j = v_i \text{ and } v_i \in D(C)) \Leftrightarrow \exists p, q \text{ with } q \geq p, q - p + 1 \in [l_i, u_i] \text{ s.t. } j \in [p, q] \text{ and } \forall k \in [p, q]\ x_k = v_i\}$$

This constraint is used in rostering or in car sequencing problems (especially in the paint shop part).

A filtering algorithm has been proposed by (Pesant, 2001). The case of cyclic sequence (that is, the successor of the last variable is the first one) is also taken into account by this algorithm. Its complexity is in $O(m^2 max(u) max(l))$. G. Pesant also described filtering algorithms for some variations of this constraint, notably one that deals with patterns and constrains the successions of patterns (that is some patterns cannot immediately follow some other patterns).

Global Minimum Distance Constraint. This constraint has been proposed by (Régin, 1997) and is mentioned in (ILOG, 1999). A global minimum distance constraint defined on X, a set of variables, states that for any pair of variable x and y of X the constraint $|x - y| \geq k$ must be satisfied.

Definition 15 *A* **global minimum distance constraint** *is a constraint C associated with an integer k such that*

$$T(C) = \{\ \tau \text{ s.t. } \tau \text{ is a tuple of } X(C) \text{ and } \forall a_i, aj \in \tau : |a_i - a_j| \geq k\}$$

This constraint can be used to model frequency allocation problems.
A filtering algorithm has been proposed for this constraint by (Régin, 1997). Note that there is a strong relation between this constraint and the sequence constraint. A $1/q$ sequence constraint constrained two variables assigned to the same value to be separated by at least $q - 1$ variables, in regard to the variable ordering. Here we want to select the values taken by a set of variables such that are all pairs of values are at least k units apart.

k-diff Constraint. The k-diff constraint constrains the number of variables that are different to be greater than or equal to k.

Definition 16 *A* **k-diff constraint** *is a constraint C associated with an integer k such that*

$$T(C) = \{\tau \text{ s.t. } \tau \text{ is a tuple on } X(C) \text{ and } |\{a_i \in D(X(C)) \text{ s.t. } \#(a_i, \tau) \leq 1\}| \geq k\}$$

This constraint has been proposed by (Régin, 1995). It is useful to model some parts of over-constrained problems where it corresponds to a relaxation of the alldifferent constraint.

A filtering algorithm establishing arc-consistency is detailed in (Régin, 1995). Its complexity is the same as for the alldifferent constraint, because the filtering algorithm of the alldifferent constraint is used when the cardinality of the maximum matching is equal to k. When this cardinality is strictly greater than k, we can prove that the constraint is arc consistent (see (Régin, 1995)).

Number of Distinct Values Constraint. The number of distinct values constraint constrains the number of distinct values taken by a set of variables to be equal to another variable.

Definition 17 *An* **number of distinct values constraint** *is a constraint C defined on a variable y and a set of variables X such that*

$$T(C) = \{\tau \text{ s.t. } \tau \text{ is a tuple on } X(C) \text{ and } |\{a_i \in D(X(C)) \text{ s.t. } \#(a_i, \tau) \leq 1\}| = \tau[y]\}$$

This constraint is quite useful for modeling some complex parts of problems.

A filtering algorithm based on the search of a lower bound of the dominating set problem (Damaschke et al., 1990) has been proposed by (Beldiceanu, 2001). When all the domains of the variables are intervals this leads to an $O(n)$ algorithm, if the intervals are already sorted.

3. Filtering Algorithms

There are several ways to design a filtering algorithm associated with a constraint. However, for global constraints we can see at least three different and important types of filtering algorithms:

- the filtering algorithms based on constraints addition. That is, from the simultaneous presence of constraints the filtering algorithm consists of adding some new constraints.
- the filtering algorithms using the general filtering algorithm (GAC-Schema) establishing arc consistency. In this case, there is no new algorithm to write provided that an algorithm checking the consistency of the constraint is given.
- the dedicated filtering algorithms. That is, a custom-written filtering algorithm is designed in order to take into account and to exploit the structure of the constraint.

For convenience, we introduce the notion of pertinent filtering algorithm for a global constraint:

Definition 18 *A filtering algorithm associated with* $\mathcal{C} = \wedge\{C_1, C_2, .., C_n\}$ *is* **pertinent** *if it can remove more values than the propagation mechanism called on the network* $(\cup_{C \in \mathcal{C}} X(C), \mathcal{D}_{X(\mathcal{C})}, \{C_1, C_2, .., C_n\})$.

3.1 Algorithms Based on Constraints Addition

A simple way to obtain a pertinent filtering algorithm is to deduce from the simultaneous presence of constraints, some new constraints. In this case, the global constraint is replaced by a set of constraints that is a superset of the one defining the global constraint. That is, no new filtering algorithm is designed.

For instance, consider a set of 5 variables: $X = \{x_1, x_2, x_3, x_4, x_5\}$ with domains containing the integer values from 0 to 4; and four constraints $atleast(X, 1, 1), atleast(X, 1, 2), atleast(X, 1, 3)$, and $atleast(X, 1, 4)$ which mean that each value of $\{1, 2, 3, 4\}$ has to be taken at least one time by a variable of X in every solution.

An $atleast(X, \#time, val)$ constraint is a local constraint. If such a constraint is considered individually then the value val cannot be removed while it belongs to more than one domain of a variable of X. A filtering algorithm establishing arc consistency for this constraint consists of assigning a variable x to val if and only if x is the only one variable whose domain contains val.

Thus, after the assignments $x_1 = 0$, $x_2 = 0$, and $x_3 = 0$, no failure is detected. The domains of x_4 and x_5, indeed, remain the same because every value of $\{1, 2, 3, 4\}$ belongs to these two domains. Yet, there is obviously no solution including the previous assignments, because 4 values must be taken at least 1 time and only 2 variables can take them.

For this example we can deduce another constraint by applying the following property: If 4 values must be taken at least 1 time by 5 variables, then the other values can be taken at most $5 - 4 = 1$, that is we have $atmost(x, 1, 0)$.

This idea can be generalized for a $gcc(X, l, u)$. Let $card(a_i)$ be a variable associated with each value a_i of $D(X)$ which counts the number of domains of X that contain a_i. We have $l_i \leq card(a_i) \leq u_i$. Then, we can simply deduce the constraint $\sum_{a_i \in D(X)} card(a_i) = |X|$; and each time the minimum or the maximum value of $card(a_i)$ is modified, the values of l_i and u_i are accordingly modified and the GCC is modified.

This method is usually worthwhile because it is easy to implement. However, the difficulty is to find the constraints that can be deduced from the simultaneous presence of other constraints.

3.2 General Arc Consistency Filtering Algorithm

The second way to easily define a powerful filtering algorithm, but which may be time consuming, is to use the general arc consistency algorithm (Bessière and Régin, 1997).

In constraint programming, to solve a problem, we begin by designing a model using predefined constraints, such as sum, alldifferent, and so on. Next, we define other constraints specific to the problem. Then we call a procedure to search for a solution.

Often when we are solving a real problem, say $\mathcal{P}$, the various simple models that we come up with cannot be solved within a reasonable period of time. In such a case, we may consider a sub-problem of the original problem, say $\mathcal{R}$. We then try to improve the resolution of $\mathcal{R}$ with the hope of thus eventually solving $\mathcal{P}$. That is, we try to identify sub-problems of $\mathcal{P}$ where the resolution can be improved by defining a particular constraint for each of these sub-problems along with a filtering algorithm associated with these constraints.

More precisely, for each possible relevant sub-problem of $\mathcal{P}$, we construct a *global* constraint that is the conjunction of the constraints involved in the sub-problem. Suppose that we then apply arc consistency to these new constraints and that this improves the resolution of $\mathcal{P}$ (i.e., the number of backtracks markedly decreases). In this case, we know that it is worthwhile to write another algorithm dedicated to solving the sub-problem $\mathcal{R}$ under consideration. In contrast, if the number of backtracks decreases only slightly, then we know that the resolution of $\mathcal{R}$ has only a modest effect on the resolution of $\mathcal{P}$. By proceeding in this way, we can improve the resolution of $\mathcal{P}$ much faster. Therefore, a general algorithm can be really useful in practice.

3.2.1 Preliminaries. Suppose that you are provided with a function, denoted by EXISTSOLUTION(P), which is able to know whether a particular problem $P = (X, \mathcal{C}, \mathcal{D})$ has a solution or not. In this section, we present two general filtering algorithms establishing arc consistency for the constraint corresponding to the problem, that is the global constraint $C(P) = \wedge\mathcal{C}$

These filtering algorithms correspond to particular instantiations of a more general algorithm: GAC-Schema (Bessière and Régin, 1997).

For convenience, we will denote by $P_{x=a}$ the problem P in which it is imposed that $x = a$, in other words $P_{x=a} = (X, \mathcal{C} \cup \{x = a\}, \mathcal{D})$.

Establishing arc consistency on $C(P)$ is done by looking for supports for the values of the variables in X. A support for a value (y, b) on $C(P)$ can be searched by using any search procedure since a support for (y, b) is a solution of problem $P_{y=b}$.

3.2.2 A First Algorithm. A simple algorithm consists of calling the function ExistSolution with $P_{x=a}$ as a parameter for every value a of every variable x involved in P, and then to remove the value a of x when ExistSolution$(P_{x=a})$ has no solution. Algorithm 4.1 is a possible implementation.

```
SimpleGeneralFilteringAlgorithm(C(P): constraint; deletionSet: list): Bool
for each a ∈ X do
  for each a ∈ D(x) do
    if ¬ ExistSolution(P_{x=a}) then
      remove a from D(x)
      if D(x) = ∅ then return False
      add (y, b) to deletionSet
return True
```

Algorithm 4.1. Simple general filtering algorithm establishing arc consistency

This algorithm is quite simple but it is not efficient because each time a value will be removed, the existence of a solution for all the possible assignments needs to be recomputed.

If $O(P)$ is the complexity of function ExistSolution(P) then we can recapitulate the complexity of this algorithms as follows:

	Consistency checking		Establishing Arc consistency	
	best	worst	best	worst
From scratch	$\Omega(P)$	$O(P)$	$nd \times \Omega(P)$	$nd \times O(P)$
After k modifications	$k \times \Omega(P)$	$k \times O(P)$	$knd \times \Omega(P)$	$knd \times O(P)$

3.2.3 A better general algorithm. This section shows how a better general algorithm establishing arc consistency can be designed provided that function ExistSolution(P) returns a solution when there is one instead of being Boolean.

First, consider that a value (x, a) has been removed from $D(x)$. We must study the consequences of this deletion. So, for all the values (y, b) that were supported by a tuple containing (x, a) another support must be found. The list of the tuples containing (x, a) and supporting a value is the list $\mathcal{S}_C(x, a)$; and the values supported by a tuple τ is given by $S(\tau)$.

Therefore, Line 1 of Algorithm 4.2 enumerates all the tuples in the $\mathcal{S}_C$ list and Line 2 enumerates all the values supported by a tuple. Then, the algorithm tries to find a new support for these values either by "inferring" new ones (Line 3) or by explicitly calling function ExistSolution (Line 4).

Here is an example of this algorithm:
Consider $X = \{x_1, x_2, x_3\}$ and $\forall x \in X, D(x) = \{a, b\}$;

```
GENERALFILTERINGALGORITHM(C(P): constraint; x: variable; a: value, deletionSet: list): Bool
1  for each τ ∈ S_C(x, a) do
       for each (z, c) ∈ τ do remove τ from S_C(z, c)
2      for each (y, b) ∈ S(τ) do
           remove (y, b) from S(τ)
           if b ∈ D(y) then
3              σ ← SEEKINFERABLESUPPORT(y, b)
               if σ ≠ nil then add (y, b) to S(σ)
               else
4                  σ ← EXISTSOLUTION(P_{y=b})
                   if σ ≠ nil then
                       add (y, b) to S(σ)
                       for k = 1 to |X(C)| do add σ to S_C(var(C(P), k), σ[k])
                   else
                       remove b from D(y)
                       if D(y) = ∅ then return False
                       add (y, b) to deletionSet
   return True
```

Algorithm 4.2. function GENERALFILTERINGALGORITHM

and $T(C(P)) = \{(a,a,a), (a,b,b), (b,b,a), (b,b,b)\}$ (i.e. these are the possible solutions of P.)
First, a support for (x_1, a) is sought: (a,a,a) is computed and (a,a,a) is added to $S_C(x_2, a)$ and $S_C(x_3, a)$, (x_1, a) in (a,a,a) is added to $S((a,a,a))$.
Second, a support for (x_2, a) is sought: (a,a,a) is in $S_C(x_2, a)$ and it is valid, therefore it is a support. There is no need to compute another solution.
Then a support is searched for all the other values.

Now, suppose that value a is removed from x_2, then all the tuples in $S_C(x_2, a)$ are no longer valid: (a,a,a) for instance. The validity of the values supported by this tuple must be reconsidered, that is the ones belonging to $S((a,a,a))$, so a new support for (x_1, a) must be searched for and so on.

The program which aims to establish arc consistency for $C(P)$ must create and initialize the data structures (S_C, S), and call the function GENERALFILTERINGALGORITHM$(C(P), x, a, deletionSet)$ (see Algorithm 4.2) each time a value a is removed from a variable x involved in $C(P)$, in order to propagate the consequences of this deletion. $deletionSet$ is updated to contain the deleted values not yet propagated. S_C and S must be initialized in a way such that:

- $S_C(x, a)$ contains all the allowed tuples τ that are the current support for some value, and such that $\tau[index(C(P), x)] = a$.

- $S(\tau)$ contains all values for which τ is the current support.

Function SEEKINFERABLESUPPORT of algorithm 4.2 "infers" an already checked allowed tuple as support for (y, b) if possible, in order to ensure that it never looks for a support for a value when a tuple supporting this value has already been checked. The idea is to exploit the property: "If (y, b) belongs to a tuple supporting another value, then this tuple also supports (y, b)". Therefore, elements in $S_C(y, b)$ are good candidates to be a new support for (y, b). Algorithm 4.3 is a possible implementation of this function.

```
SEEKINFERABLESUPPORT(y: variable, b: value): tuple
while S_C(y,b) ≠ ∅ do
  | σ ← first(S_C(y,b))
  | if σ is valid then return σ    /* σ is a support */
  | else remove σ from S_C(y,b)
return nil
```

Algorithm 4.3. function SEEKINFERABLESUPPORT

The complexity of the GENERALFILTERINGALGORITHM is given in the following table:

	Consistency checking		Establishing Arc consistency	
	best	worst	best	worst
From scratch	$\Omega(P)$	$O(P)$	$nd \times \Omega(P)$	$nd \times O(P)$
After k modifications	$\Omega(1)$	$k \times O(P)$	$nd \times \Omega(P)$	$knd \times O(P)$

Moreover, the space complexity of this algorithm is $O(n^2d)$, where d is the size of the largest domain and n is the number of variables involved in the constraint. This space complexity depends on the number of tuples needed to support all the values. Since there are nd values and only one tuple is required per value, we obtain the above complexity.

3.2.4 Discussion and Example. Algorithm 4.2 can be efficiently improved, if the search for a solution of P can be made according to a predefined ordering of the tuple. In this case, a more complex algorithm can be designed. Moreover, it is also possible to use the solver in itself to search for a solution in P. All these algorithms are fully detailed in (Bessière and Régin, 1997) and (Bessière and Régin, 1999). These papers also detail how Algorithm 4.2 can be adapted to constraints that are given by the list of tuples that satisfy the constraint (in this case the resolution of P corresponds to the search for a valid tuple in that list) or by the list of forbidden combinations of value for the constraint (i.e. the complement of the previous list).

(Bessière and Régin, 1999) have proposed to study a configuration problem as an example of the application of the general filtering algorithm establishing arc consistency. The general formulation is: given a supply of components and bins of given types, determine all assignments of components to bins satisfying specified assignment constraints subject to an optimization criterion.

In the example we will consider that there are 5 types of components: {glass, plastic, steel, wood, copper}. There are three types of bins: {red, blue, green} whose capacity constraints are: red has capacity 5, blue has capacity 5, green has capacity 6.

The containment constraints are:

- red can contain glass, copper, wood
- blue can contain glass, steel, copper
- green can contain plastic, copper, wood

The requirement constraints are (for all bin types): wood requires plastic. Certain component types cannot coexist: glass excludes copper

Certain bin types have capacity constraints for certain components:

- red contains at most 1 of wood
- green contains at most 2 of wood
- for all the bins there is either no plastic or at least 2 plastic.

Given an initial supply of 12 of glass, 10 of plastic, 8 of steel, 12 of wood, and 8 of copper, what is the minimum total number of bins required to contain the components?

A description of a possible implementation of a similar problem is given in (ILOG, 1999). We will call it the "standard model".

Almost all the constraints between types of bins and components are local. The filtering algorithm associated with them leads to few domain reductions. Therefore, they can be grouped inside a single global constraint. That is, problem P is formed by all these constraints and P is solved by using another CP solver.

Here are the results we obtained:

	# Backtracks	time (s)
standard model	1,361,709	430
new algorithm	12,659	11

These results clearly show the advantages of global constraints and prove that a general filtering algorithm establishing arc consistency may be useful in practice to solve some real life problems. However, in practice, when the problems become big the complexity of the GAC-Schema often prevents its use, and specific filtering algorithm establishing arc consistency have to be used. In (Bessière and Régin, 1999) some other examples show by using GAC-Schema that sometimes arc consistency is useless. Even in this case the search for a good model is improved because wrong models can be identified more quickly.

3.3 Dedicated Filtering Algorithms

The third method to design a pertinent filtering algorithm is to use the structure of the constraint in order to define some properties identifying that some values are not consistent with the global constraint.

The use of the structure of a constraint has four main advantages:

- The search for a support can be speeded up.
- Some inconsistent values can be identified without explicitly checking for every value whether it has a support or not.
- The call of the filtering algorithm can be limited to some events that can be clearly identified.
- Advantages of (possible) incrementality.

For instance, consider the constraint $(x < y)$. Then:

- The search for a support for a value a of $D(x)$ is immediate because any value b of $D(y)$ such that $b > a$ is a support, so a is consistent with the constraint if $a < max(D(y))$.
- We can immediately state that $max(D(x)) < max(D(y))$ and $min(D(y)) > min(D(x))$ which mean that all values of $D(x)$ greater than or equal to $max(D(y))$ and all values of $D(y)$ less than or equal to $min(D(x))$ can be removed.
- Since the deletions of values of $D(y)$ depends only on $max(D(y))$ and the deletions of values of $D(x)$ depends only on $min(D(x))$, the filtering algorithm must be called only when $max(D(y))$ or $min(D(x))$ are modified. It is useless to call it for the other modifications.

We propose an original contribution for a well-known problem: the n-queens problem.

The n-queens problem involves placing n queens on a chess board in such a way that none of them can capture any other using the conventional moves allowed by a queen. In other words, the problem is to select n squares on a chessboard so that any pair of selected squares is never aligned vertically, horizontally, nor diagonally.

This problem is usually modeled by using one variable per queen; the value of this variable represents the column in which the queen is set. If x_i represents the variable corresponding to queen i (that is the queen in row i) the constraints can be stated in the following way. For every pair (i, j), with $i \neq j$, $x_i \neq x_j$ guarantees that the columns are distinct; and $x_i + i \neq x_j + j$ and $x_i - i \neq x_j - j$ together guarantee that the diagonals are distinct.
These relations are equivalent to defining an alldifferent constraint on the variables x_i, an alldifferent constraint on the variables $x_i + i$, and an alldifferent

queen					
i	x		x		x
$i+1$					
$i+2$			X		

queen					
i	x			x	
$i+1$					
$i+2$					
$i+3$	X			X	

Figure 4.3. Rules of the ad-hoc filtering algorithm for the n-queens problem.

constraint on the variables $x_i - i$.

We propose to use a specific constraint that is defined on x_i and try to take into account the simultaneous presence of three alldifferent constraints. Consider a queen q: if there are more than three values in its domain, this queen cannot lead to the deletion of one value of another queen, because three directions are constrained (the column and the two diagonals) and so at least one value of queen q does not belong to one of these directions. Therefore, a first rule can be stated:

- While a queen has more than three values in its domain, it is useless to study the consequence of the deletion of one of its values.

From a careful study of the problem we can deduce some rules (see Figure 4.3):

- If a queen i has 3 values $\{a, b, c\}$, with $a < b < c$ in its domain then the value b of queens $i-k$ and the value b of queen $i+k$ can be deleted if $b = a+k$ and $c = b + k$;
- If $D(x_i) = \{a, b\}$ with $a < b$, then the values a and b of queens $i-(b-a)$ and of queens $i + (b - a)$ can be deleted.
- If $D(x_i) = \{a\}$, then the value $a + j$ for all queens $i + j$, and the value $a - j$ for all queens $i - j$ can be deleted.
- While a queen has more than 3 values in its domain, then this constraint cannot deduce anything.

Therefore, a careful study of a constraint can lead to efficient filtering algorithms. This method is certainly the most promising way. However, it implies a lot of work. In (Bessière and Régin, 1999), it is proposed to try to use first the general arc consistency algorithm in order to study if the development of a powerful filtering algorithm could be worthwhile for the considered problem.

4. Two Successful Filtering Algorithms

In this section, the filtering algorithms associated with two of the most frequently used constraints in practice - the alldifferent and the global cardinality constraint - are presented. The advantages of these filtering algorithms is that they clearly show how Operational Research algorithms can be integrated into Constraint Programming.

4.1 Preliminaries

The definitions about graph theory are from (Tarjan, 1983). The definitions, theorems and algorithms about flow are based on books of (Berge, 1970; Lawler, 1976; Tarjan, 1983; Ahuja et al., 1993).

A **directed graph** or **digraph** $G = (X, U)$ consists of a **node set** X and an **arc set** U, where every arc (u, v) is an ordered pair of distinct nodes. We will denote by $X(G)$ the node set of G and by $U(G)$ the arc set of G.

A **path** from node v_1 to node v_k in G is a list of nodes $[v_1, ..., v_k]$ such that (v_i, v_{i+1}) is an arc for $i \in [1..k-1]$. The path **contains** node v_i for $i \in [1..k]$ and arc (v_i, v_{i+1}) for $i \in [1..k-1]$. The path is **simple** if all its nodes are distinct. The path is a **cycle** if $k > 1$ and $v_1 = v_k$.

If $\{u, v\}$ is an edge of a graph, then we say that u and v are the **ends** or the **extremities** of the edge. A **matching** M on a graph is a set of edges no two of which have a common node. The **size** $|M|$ of M is the number of edges it contains. The **maximum matching problem** is that of finding a matching of maximum size. M **covers** X when every node of X is an endpoint of some edge in M.

Let M be a matching. An edge in M is a **matching edge**; every edge not in M is **free**. A node is **matched** if it is incident to a matching edge and **free** otherwise.

An **alternating path** or **cycle** is a simple path or cycle whose edges are alternately matching and free. The **length** of an alternating path or cycle is the number of edges it contains.

Let G be a graph for which each arc (i, j) is associated with two integers l_{ij} and u_{ij}, respectively called the **lower bound capacity** and the **upper bound capacity** of the arc.

A **flow** in G is a function f satisfying the following two conditions:

- For any arc (i, j), f_{ij} represents the amount of some commodity that can "flow" through the arc. Such a flow is permitted only in the indicated direction of the arc, i.e., from i to j. For convenience, we assume $f_{ij} = 0$ if $(i, j) \notin U(G)$.
- A **conservation law** is observed at each node: $\forall j \in X(G) : \sum_i f_{ij} = \sum_k f_{jk}$.

We will consider two problems of flow theory:

- **the feasible flow problem**: Does there exist a flow in G that satisfies the **capacity constraints**? That is, find f such that $\forall (i, j) \in U(G)\ l_{ij} \leq f_{ij} \leq u_{ij}$.

- **the problem of the maximum flow for an arc** (i, j): Find a feasible flow in G for which the value of f_{ij} is maximum.

Without loss of generality (see p.45 and p.297 in (Ahuja et al., 1993)), and to overcome notation difficulties, we will consider that:

- if (i, j) is an arc of G then (j, i) is not an arc of G.
- all boundaries of capacities are nonnegative integers.

In fact, if all the upper bounds and all the lower bounds are integers and if there exists a feasible flow, then for any arc (i, j) there exists a maximum flow from j to i which is integral on every arc in G (See (Lawler, 1976) p113.)

The **value graph** (Laurière, 1978) of an non-binary constraint C is the bipartite graph $GV(C) = (X(C), D(X(C)), E)$ where $(x, a) \in E$ iff $a \in D(x)$.

4.2 The Alldifferent Constraint

4.2.1 Consistency and Arc Consistency. We have the relation (Régin, 1994):

Proposition 1 *Let C be an alldifferent constraint. A matching which covers X in the value graph of C is a tuple of $T(C)$.*

Therefore we have:

Proposition 2 *A constraint C=alldifferent(X) is consistent iff there exists a matching that covers $X(C)$ in $GV(C)$.*

From proposition 2 and by the definition of arc consistency, we have:

Proposition 3 *A value a of a variable x is consistent with C if and only if the edge $\{x, a\}$ belongs to a matching that covers $X(C)$ in $GV(C)$.*

Proposition 4 (Berge, 1970) *An edge belongs to some but not all maximum matchings, iff, for an arbitrary maximum matching, it belongs to either an even alternating path which begins at a free node, or an even alternating cycle.*

Proposition 5 *Given a bipartite graph $G = (X, Y, E)$ with a matching M which covers X and the graph $O(G, M) = (X \cup \{s\}, Y, Succ)$, obtained from G by orienting the edge in M from their y-endpoint to their x-endpoint, the edge not in M from their x-endpoint to their y-endpoint, and by adding an arc from every free node of Y to every matched node of Y. Then, we have the two properties*
1) Every directed cycle of $O(G, M)$ which does not contain an arc from a free node of Y to a matched node of Y corresponds to an even alternating cycle of G, and conversely.
2) Every directed cycle of $O(G, M)$ which contains an arc from a free node of Y to a matched node of Y corresponds to an even alternating path of G which begins at a free node, and conversely.

proof

1) G and $O(G, M)$ are bipartite and by definition of $O(G, M)$ the first property holds.
2) $O(G, M)$ is bipartite therefore all directed cycles of $O(G, M)$ are even. An even alternating

path which begins at a free node y_f in Y, necessarily ends at a matched node y_m in Y, because all nodes of X are matched and in $O(G, M)$ there is only one arc from a node x in X to a node in Y: the matching edge involving x. Hence, by definition of $O(G, M)$ there is an arc from s to y_f and an arc from y_m to s, so every even alternating path of G is a directed cycle in $O(G, M)$. Conversely, a directed cycle involving s can be decomposed into a path from a free node y_f in Y to a node y_m in Y and the path $[y_m, s, y_f]$. Since the cycle is even, the path is also even and it corresponds to an alternating path of G by definition of $O(G, M)$. Therefore the property holds.

From this proposition we immediately have:

Proposition 6 *Arc consistency of an alldifferent constraint C is established by computing M a matching which covers $X(C)$ in $GV(C)$ and by removing all the values (x, a) such that $(x, a) \notin M$ and a and x belong to two different strongly connected components of $O(GV(C), M)$.*

proof: By definition of the strongly connected components, there exists a cycle between two nodes belonging to the same strongly connected components. Therefore, from Proposition 5 the proposition holds.

4.2.2 Complexity. Let m be the number of edges of $GV(C)$, and $n = |X(C)|$ and $d = |D(X(C))|$. A matching covering $X(C)$ can be computed, or we can prove there is none, in $O(\sqrt{n}\, m)$ (Hopcroft and Karp, 1973). The search for strongly connected components can be performed in $O(m+n+d)$. Hence arc consistency for an alldifferent constraint can be established in $O(m+n+d)$.

Moreover, consider M a matching which covers X and suppose that k edges of the value graph are deleted (this means that k values have been removed from the domain of their variables). Then a new matching which covers X can be recomputed from M in $O(\sqrt{k}\, m)$ and arc consistency can be established in $O(m + n + d)$.

It is important to note that arc consistency may remove $O(n^2)$ values (Puget, 1998). Consider an alldifferent constraint defined on $X = \{x_1, ..., x_n\}$ with the domains: $\forall i \in [1, \frac{n}{2}]$, if i is odd then $D(x_i) = [2i - 1, 2i]$ else $D(x_i) = D(x_{i-1})$; and $\forall i \in [\frac{n}{2} + 1, n]$ $D(x_i) = [1, n]$. For instance, for $n = 12$ we will have: $D(x_1) = D(x_2) = [1, 2]$, $D(x_3) = D(x_4) = [5, 6]$, $D(x_5) = D(x_6) = [9, 10]$, $D(x_7) = D(x_8) = D(x_9) = D(x_10) = D(x_11) = D(x_12) = [1, 12]$. Then, if arc consistency is established, the intervals corresponding to the domains of the variables from x_1 ro $x_{\frac{n}{2}}$ will be removed from the domains of the variables from $x_{\frac{n}{2}+1}$ to x_n. That is, $2 \times \frac{n}{2}$ values will be effectively removed from the domains of $(n - (\frac{n}{2} + 1))$ variables. Therefore $O(n^2)$ values are deleted. Since m is bounded by n^2, the filtering algorithm establishing arc consistency for the alldifferent constraint can be considered as an optimal algorithm.

The complexities are reported here:

	Consistency checking	Establishing Arc consistency
From scratch	$O(\sqrt{n}\,m)$	$O(m+n+d)$
After k modifications	$O(\sqrt{k}\,m)$	$O(m+n+d)$

Two important works carried out for the alldifferent constraint must be mentioned. (Melhorn and Thiel, 2000) have proposed a very efficient filtering algorithm establishing bound consistency for the sort and alldifferent constraint. A linear complexity is reached in a lot of practical cases (for a permutation, for instance). (Stergiou and Walsh, 1999) made a comparison between different filtering algorithm associated with the alldifferent constraints and showed the advantages of this constraint in practice.

4.2.3 Some Results. A graph-coloring problem consists of choosing colors for the nodes of a graph so that adjacent nodes are not the same color. Since we want to highlight the advantages of the filtering algorithm establishing arc consistency for the alldifferent constraint we will consider only a very special kind of graph for this example.

The kind of graph that we will color is one with $n * (n-1)/2$ nodes, where n is odd and where every node belongs to exactly two maximal cliques of size n.

For example, for $n = 5$, there is a graph consisting of the following maximal cliques:
$c0 = \{0,1,2,3,4\}$, $c1 = \{0,5,6,7,8\}$, $c2 = \{1,5,9,10,11\}$
$c3 = \{2,6,9,12,13\}$, $c4 = \{3,7,10,12,14\}$, $c5 = \{4,8,11,13,14\}$

The minimum number of colors needed for this graph is n since there is a clique of size n. Consequently, our problem is to find out whether there is a way to color such a graph in n colors.

We compare the results obtained with the alldifferent constraint and without it (that is only binary constraints of difference are used). Times are expressed in seconds:

	clique size							
	27		31		51		61	
	#fails	time	#fails	time	#fails	time	#fails	time
binary $\neq$	1	0.17	65	0.37	24512	66.5	?	$> 6h$
alldiff	0	1.2	4	2.2	501	25.9	5	58.2

These results show that using global constraints establishing arc consistency is not systematically worthwhile when the size of the problem is small, even if the number of backtracks is reduced. However, when the size of problem is increased, efficient filtering algorithm are needed.

4.3 The Global Cardinality Constraint

4.3.1 Consistency and Arc Consistency. A GCC C is consistent iff there is a flow in an directed graph $N(C)$ called the value network of C (Régin, 1996):

Definition 19 *Given $C = gcc(X, l, u)$ a GCC; the* **value network** *of C is the directed graph $N(C)$ with lower bound capacity and upper bound capacity on each arc. $N(C)$ is obtained from the value graph $GV(C)$, by:*

- *orienting each edge of $GV(C)$ from values to variables. For such an arc (u, v): $l_{uv} = 0$ and $u_{uv} = 1$.*
- *adding a node s and an arc from s to each value. For such an arc (s, a_i): $l_{sa_i} = l_i$, $u_{sa_i} = u_i$.*
- *adding a node t and an arc from each variable to t. For such an arc (x, t): $l_{xt} = 1$, $u_{xt} = 1$.*
- *adding an arc (t, s) with $l_{ts} = u_{ts} = |X(C)|$.*

Proposition 7 *Let C be a GCC and $N(C)$ be the value network of C; the following two properties are equivalent:*

- *C is consistent;*
- *there is a feasible flow in $N(C)$.*

sketch of proof: We can easily check that each tuple of $T(C)$ corresponds to a flow in $N(C)$ and conversely. ⊙

Definition 20 *The* **residual graph** *for a given flow f, denoted by $R(f)$, is the digraph with the same node set as in G. The arc set of $R(f)$ is defined as follows:*

$\forall(i, j) \in U(G)$:

- *$f_{ij} < u_{ij} \Leftrightarrow (i, j) \in U(R(f))$ and has cost $rc_{ij} = c_{ij}$ and upper bound capacity $r_{ij} = u_{ij} - f_{ij}$.*
- *$f_{ij} > l_{ij} \Leftrightarrow (j, i) \in U(R(f))$ and has cost $rc_{ji} = -c_{ij}$ and upper bound capacity $r_{ji} = f_{ij} - l_{ij}$.*

All the lower bound capacities are equal to 0.

Proposition 8 *Let C be a consistent GCC and f be a feasible flow in $N(C)$. A value a of a variable x is not consistent with C if and only if $f_{ax} = 0$ and a and x do not belong to the same strongly connected component in $R(f)$.*

proof: It is well known in flow theory that the flow value for an arc (a, x) is constant if there is no path from a to x in $R(f) - \{(a, x)\}$ and no path from x to a in $R(f) - \{(x, a)\}$. Moreover, $u_{ax} = 1$ thus (a, x) and (x, a) cannot belong simultaneously to $R(f)$, hence f_{ax} is constant iff there is no cycle containing (x, a) or (a, x) in $R(f)$. That is, if x and a belong to different strongly connected components. ⊙

The advantage of this proposition is that all the values not consistent with the GCC can be determined by only one identification of the strongly connected components in $R(f)$.

4.3.2 Complexity. For our problem, a feasible flow can be computed in $O(nm)$ therefore we have the same complexity for the check of the constraint consistency. Moreover flow algorithms are incremental.

The search for strongly connected components can be done in $O(m+n+d)$ (Tarjan, 1983), thus a good complexity for computing arc consistency for a GCC is obtained.

Corollary 1 *Let C be a consistent GCC and f be a feasible flow in $N(C)$. Arc consistency for C can be established in $O(m + n + d)$.*

Here is a recapitulation of the complexities:

	Consistency	Arc consistency
From scratch	$O(n\,m)$	$O(m + n + d)$
After k modifications	$O(k\,m)$	$O(m + n + d)$

4.3.3 Some results. This section considers the sport-scheduling problem described in (McAloon et al., 1997) and in (Van Hentenryck et al., 1999). The problem consists of scheduling games between n teams over $n-1$ weeks. In addition, each week is divided into $n/2$ periods. The goal is to schedule a game for each period of every week so that the following constraints are satisfied:

1 Every team plays against every other team;

2 A team plays exactly once a week;

3 A team plays at most twice in the same period over the course of the season.

The meeting between two teams is be called a *matchup* and takes place in a *slot* i.e. in a particular period in a particular week.

The following table gives a solution to this problem for 8 teams:

	Week 1	Week 2	Week 3	Week 4	Week 5	Week 6	Week 7
Period 1	0 vs 1	0 vs 2	4 vs 7	3 vs 6	3 vs 7	1 vs 5	2 vs 4
Period 2	2 vs 3	1 vs 7	0 vs 3	5 vs 7	1 vs 4	0 vs 6	5 vs 6
Period 3	4 vs 5	3 vs 5	1 vs 6	0 vs 4	2 vs 6	2 vs 7	0 vs 7
Period 4	6 vs 7	4 vs 6	2 vs 5	1 vs 2	0 vs 5	3 vs 4	1 vs 3

In fact, the problem can be made more uniform by adding a "dummy" final week and requesting that all teams play exactly twice in each period. The rest of this section considers this equivalent problem for simplicity.

The sport-scheduling problem is an interesting application for constraint programming. On the one hand, it is a standard benchmark (submitted by Bob Daniel) to the well known MIP library and it is claimed in (McAloon et al., 1997) that state of the art MIP solvers cannot find a solution for 14 teams. The model presented in this section is computationally much more efficient. On the other hand, the sports scheduling application demonstrates fundamental features of constraint programming including global and symbolic constraints. In particular, the model makes heavy use of arc consistency for the GCCs.

The main modeling idea is to use two classes of variables: team variables that specify the team playing on a given week, period, and slot and the matchup variables specifying which game is played on a given week and period. The use of matchup variables makes it simple to state the constraint that every team must play against each other team. Games are uniquely identified by their two teams. More precisely, a game consisting of home team h and away team a is uniquely identified by the integer $(h1) * n + a$.

These two sets of variables must be linked together to make sure that the matchup and team variables for a given period and a given week are consistent. This link is ensured by a constraint whose set of tuples is explicitly given. For 8 teams, this set consists of tuples of the form $(1, 2, 1)$ (which means that the game 1 vs 2 is the game number 1), $(1, 3, 2)$, ..., $(7, 8, 55)$.

The games that are played in a given week can be determined by using a round robin schedule. As a consequence, once the round robin schedule is selected, it is only necessary to determine the period of each game, not its schedule week. In addition, it turns out that a simple round robin schedule makes it possible to find solutions for large numbers of teams.

The basic idea is to fix the set of games of each week, but without fixing the period of each game. The goal is then to assign a period to each game such that the constraints on periods are satisfied. If there is no solution then another round robin is selected.

The constraints on periods are taken into account with the global cardinality constraints. For every period a GCC is defined on the team variables involved in the period. Every value is associated with the two integers: 0 and 2 if the dummy week is not considered, otherwise if the team variables of the dummy week are involved the two integers are equal to 2.

The search procedure which is used consists of generating values for the matchups in the first period and in the first week, then in the second period and the second week, and so on. The results obtained by this method implemented with ILOG Solver are given in the following table. Times are expressed in seconds and the experiments have been run on a Pentium III 400Mhz machine. As far as we know, this method gives the best results for this problem.

#teams	8	10	12	14	16	18	20	24	30	40
#fails	10	24	58	21	182	263	226	2,702	11,895	2,834,754
time	0	0	0.2	0.2	0.6	0.9	1.2	10.5	138	6h

5. Global Constraints and Over-constrained Problems

Global constraints have been proved to be very useful in modelling and in improving the resolution of CSPs. This section aims to show that they can also be useful to model and to improve the resolution of over-constrained problems.

A problem is over-constrained when no assignment of values to variables satisfies all constraints. In this situation, the goal is to find a compromise. Violations are allowed in solutions, providing that such solutions retain a practical advantage. Therefore, it is mandatory to respect some rules and acceptance criteria defined by the user. Usually the set of initial constraints is divided into two sets: the hard constraints, that is the ones that must necessarily be satisfied, and the soft constraints, that is constraints whose violation is possible. A *violation cost* is generally associated with every soft constraint. Then, a global objective related to the whole set of violation costs is usually defined. For instance, the goal can be to minimize the total sum of costs. In some applications it is necessary to express more complex rules on violations, which involve several costs independent from the objective function. Such rules can be defined through meta-constraints (Petit et al., 2000). In order to model easily the part of the problem containing the soft constraints a global constraint involving the soft ones can be defined.

Moreover, in practice, among some other possibilities, two important types of violation costs can be identified:

• The violation cost depends only on the fact that the constraint is violated or not. In other words, either the constraint is satisfied and the violation cost is equal to 0, or the constraint is violated and the cost is equal to a given value. That is all the possible violations of a constraint have the same cost.

• The violation cost depends on the way the constraint is violated. The violation is quantified, thus we will call it quantified violation cost. For instance, consider a cost associated with the violation of a temporal constraint imposing that a person should stop working before a given date: the violation cost can be proportional to the additional amount of working time she performs.

In this section we show two different advantages of the global constraints for solving over-constraint problems. First, we consider the Maximal Constraint

Satisfaction Problem (Max-CSP), where the goal is to minimize the number of constraint violations, and we show that Max-CSP can be simply and efficiently modeled by a new global constraint. Then, we show how a quantified violation cost can be efficiently taken into account for a constraint and how new global constraints can be designed. These new constraints are called global soft constraints.

For more information about over-constrained problems and global constraints the reader can consult (Petit, 2002).

5.1 Satisfiability Sum Constraint

Let $\mathcal{N} = (X, \mathcal{D}, \mathcal{C})$ be a constraint network containing some soft constraints. Max-CSP can be represented by a single constraint, called the Satisfiability Sum Constraint (SSC):

Definition 21 *Let $\mathcal{C} = \{C_i, i \in \{1, \dots, m\}\}$ be a set of constraints, and $S[\mathcal{C}] = \{s_i, i \in \{1, \dots, m\}\}$ be a set of variables and unsat be a variable, such that a one-to-one mapping is defined between $\mathcal{C}$ and $S[\mathcal{C}]$. A* **Satisfiability Sum Constraint** *is the constraint $ssc(\mathcal{C}, S[\mathcal{C}], unsat)$ defined by:*

$$[unsat = \sum_{s_i=1}^{m} s_i] \wedge \bigwedge_{i=1}^{m} [(C_i \wedge (s_i = 0)) \vee (\neg C_i \wedge (s_i = 1))]$$

The variables $S[\mathcal{C}]$ are used in order to express which constraints of $\mathcal{C}$ must be violated or satisfied: value 0 assigned to $s \in S[\mathcal{C}]$ expresses that its corresponding constraint C is satisfied, whereas 1 expresses that C is violated. Variable $unsat$ represents the objective, that is the number of violations in $\mathcal{C}$, equal to the number of variables of $S[\mathcal{C}]$ whose value is 1.

Throughout this formulation, a solution of a Max-CSP is an assignment that satisfies the SSC with the minimal possible value of $unsat$. A lower bound of the objective of a Max-CSP corresponds to a necessary consistency condition of the SSC. The different domain reduction algorithms established for Max-CSP correspond to specific filtering algorithms associated with the SSC.

This point of view has some advantages in regard to the previous studies:

1. Any search algorithm can be used. This constraint can be associated with other ones, in order to separate soft constraints from hard ones.
2. No hypothesis is made on the arity of constraints $\mathcal{C}$.
3. If a value is assigned to $s_i \in S[\mathcal{C}]$, then a filtering algorithm associated with $C_i \in \mathcal{C}$ (resp. $\neg C_i$) can be used in a way similar to classical CSPs.

Moreover, the best algorithms to solve over-constrained problems like PFC-MRDAC (Larrosa et al., 1998) and the ones based on conflict-sets detection (Régin et al., 2000; Régin et al., 2001) can be implemented as specific filtering algorithms associated with this constraint. A filtering algorithm based on a PFC-MRDAC version dealing only with the boundaries of the domain of the variable has also been described in (Petit et al., 2002).

Furthermore, an extension of the model can be performed (Petit et al., 2000), in order to deal with Valued CSPs. Basically it consists of defining larger domains for variables in $S[\mathcal{C}]$.

5.2 Global Soft Constraints

In this section we consider that the constraints are associated with quantified violation costs. This section is based on (Petit et al., 2001).

Most of the algorithms dedicated to over-constrained problems are generic. However, the use of constraint-specific filtering algorithms is generally required to solve real-world applications, as their efficiency can be much higher.

Regarding over-constrained problems, existing constraint-specific filtering algorithms can be used only in the particular case where the constraint must be satisfied. Indeed, they remove values that are not consistent with the constraint: the deletion condition is linked to the fact that it is mandatory to satisfy the constraint. This condition is not applicable when the violation is allowed.

However, domains can be reduced from the objective and from the costs associated with violations of constraints. The main idea of this section is to perform this kind of filtering specifically, that is, to take advantage of the structure of a constraint and from the structure of its violation to efficiently reduce the domains of the variables it constrains.

The deletion condition will be linked to the necessity of having an acceptable cost, instead of being related to the satisfaction requirement.

For instance, let C be the constraint $x \leq y$. In order to quantify its violation, a $cost$ is associated with C. It is defined as follows:

– if C is satisfied then $cost = 0$.

– if C is violated then $cost > 0$ and its value is proportional to the gap between x and y, that is, $cost = x - y$.

Assume that $D(x) = [90001, 100000]$ and $D(y) = [0, 200000]$, and that the cost is constrained to be less than 5. Then, either C is satisfied: $x - y \leq 0$, or C is violated: $x - y = cost$ and $cost \leq 5$, which implies $x - y \leq 5$. Hence, we deduce that $x - y \leq 5$, and, by propagation, $D(y) = [89996, 200000]$.

Such a deduction is made directly by propagating *bounds* of the variables x, y and $cost$. Inequality constraints admit such propagation on bounds without consideration of the domain values that lie between. Such propagation, which depends on the structure of the inequality constraint, is fundamentally more efficient than the consideration for deletion of each domain value in turn. If we ignore the structure of the constraint in the example, the only way to filter a value is to study the cost of each tuple in which this value is involved. Performing the reduction of $D(y)$ in the example above is costly: at least $|D(x)| * 89996 = 899960000$ checks are necessary. This demonstrates the gain that can be made by directly integrating constraints on costs into the problem and employing constraint-specific filtering algorithms.

Following this idea, our goal is to allow the *same* modeling flexibility with respect to violation costs as with any other constrained variable. The most natural way to establish this is to include these violation costs as variables in a new constraint network.

For sake of clarity, we consider that the values of the cost associated with a constraint C are positive integers. 0 expresses the fact that C is satisfied, and strict positive values are proportional to the importance of a violation. This assumption is not a strong restriction; it just implies that values of cost belong to a totally ordered set.

A new optimization problem derived from the initial problem can be solved. It involves the same set of hard constraints $\mathcal{C}_h$, but a set of disjunctive constraints replaces $\mathcal{C}_s$. This set of disjunctive constraints is denoted by $\mathcal{C}_{disj}$ and a one-to-one correspondence is defined between $\mathcal{C}_s$ and $\mathcal{C}_{disj}$. Each disjunction involves a new variable $cost \in X_{costs}$, which is used to express the cost of $C \in \mathcal{C}_s$. A one-to-one correspondence is also defined between $\mathcal{C}_s$ and X_{costs}. Given $C \in \mathcal{C}_s$, the disjunction is the following:

$$[C \wedge [cost = 0]] \vee [\bar{C} \wedge [cost > 0]]$$

$\bar{C}$ is the constraint including the variable *cost* that expresses the violation. A specific filtering algorithm can be associated with it. Regarding the preliminary example, the constraints C and $\bar{C}$ are respectively $x \leq y$ and $cost = x - y$:

$$[[x \leq y] \wedge [cost = 0]] \vee [[cost = x - y] \wedge [cost > 0]]$$

The new defined problem is not over-constrained: it consists of satisfying the constraints $\mathcal{C}_h \cup \mathcal{C}_{disj}$, while optimizing an objective defined over all the variables X_{costs} (we deal with a classical optimization problem); constraints on a variable *cost* can be propagated.

Such a model can be used for encoding directly over-constrained problems with existing solvers (Régin et al., 2000). Moreover, additional constraints on cost variables can be defined in order to select solutions that are acceptable for the user (Petit et al., 2000).

5.2.1 General Definitions of Cost. When the violation of a constraint can be naturally defined, we use it (for instance, the constraint of the preliminary example $C : x \leq y$). However, this is not necessarily the case. When there is no natural definition associated with the violation of a constraint, different definitions of the cost can be considered, depending on the problem.

For instance, let C be an alldifferent constraint defined on variables $var(C) = \{x_1, x_2, x_3, x_4\}$, such that $\forall i \in [1, 4], D(x_i) = \{a, b, c, d\}$. If we ignore the symmetric cases by considering that no value has more importance than another, we have the following possible assignments: (a, b, c, d), (a, a, c, d), (a, a, c, c), (a, a, a, c), (a, a, a, a).

Intuitively, it is straightforward that the violation of case (a, a, a, a) is more serious than the one of case (a, a, c, d). This fact has to be expressed through the cost.

Two general definitions of the cost associated with the violation of a non-binary constraint exist:

Definition 22 : Variable Based Violation Cost *Let C be a constraint. The cost of its violation can be defined as the number of assigned values that should change in order to make C satisfied.*

The advantage of this definition is that it can be applied to any (non-binary) constraint. However, depending on the application, it can be inconvenient. In the alldifferent example above we will have $cost((a, b, c, d)) = 0, cost((a, a, c, d)) = 1$, $cost((a, a, c, c)) = 2$, $cost((a, a, a, c)) = 2$, and $cost((a, a, a, a)) = 3$. A possible problem is that assignments (a, a, c, c) and (a, a, a, c) have the same cost according to definition 22. For an alldifferent involving more than four variables, a lot of different assignments have the same cost.

Therefore, there is another definition of the cost, which is well suited to constraints that are representable through a primal graph (Dechter, 1992):

Definition 23 *The* **primal graph** $Primal(C) = (var(C), E_p)$ *of a constraint C is a graph such that each edge represents a binary constraint, and the set of solutions of the CSP defined by*
$\mathcal{N} = (var(C), D(var(C)), E_p)$ *is the set of allowed tuples of C.*

For an alldifferent C, $Primal(C)$ is a complete graph where each edge represents a binary inequality.

Definition 24 : Primal Graph Based Violation Cost *Let C be a constraint representable by a primal graph. The cost of its violation can be defined as the number of binary constraints violated in the CSP defined by $Primal(C)$.*

In the alldifferent case, the user may aim at controlling the number of binary inequalities implicitly violated. The advantage of this definition is that the granularity of the quantification is more accurate. In the example, the costs are $cost((a, b, c, d)) = 0$, $cost((a, a, c, d)) = 1$, $cost((a, a, c, c)) = 2$, $cost((a, a, a, c)) = 3$, and $cost((a, a, a, a)) = 6$.

5.2.2 Soft Alldifferent Constraint. The constraint obtained by combining a variable based violation cost and alldifferent constraint, is, in fact, a k-diff constraint where k is the minimum value of the cost variable. Therefore, if the modification of k is dynamically maintained, which is relatively easy because it can only be increased, then we obtain a filtering algorithm establishing algorithm for this global soft constraint.

The constraint formed by the combination of a primal graph based cost and an alldifferent constraint is much more complex. A specific filtering algorithm for this constraint has been designed by (Petit et al., 2002). Its complexity is in $O(|var(C)|^2\sqrt{|var(C)|}Kd)$, where $K = \sum |D(x)|, x \in var(C)$ and $d = max(|D(x)|), x \in var(C)$.

6. Quality of Filtering Algorithms

In this section, we try to characterize some properties of a good filtering algorithm.

Section 3.2 presents a general filtering algorithm establishing arc consistency. From a problem P for which a method giving a solution is known, this algorithm is able to establish and maintain arc consistency of $C(P)$ in $nd \times O(P)$. Therefore, there is no need to develop a specific algorithm with the same complexity. Every dedicated algorithm must improve that complexity otherwise it is not worthwhile.

From this remark we propose the following classification:

Definition 25 *Let C be a constraint for which the consistency can be computed in $O(C)$. A filtering algorithm establishing arc consistency associated with C is:*

- **poor** *if its complexity is $O(nd) \times O(C)$;*
- **medium** *if its complexity is $O(n) \times O(C)$;*
- **good** *if its complexity is $O(C)$;*

Some good filtering algorithms are known for some constraints. We can cite the alldifferent constraint or the global cardinality constraint.

Some medium filtering algorithms have also been developed for some constraints like global cardinality constraint with costs, and symmetric alldifferent. Thus, these algorithms can be improved.

Good filtering algorithms are not perfect and the definition of the quality we propose is based on worst-case complexity. This definition can be refined to be more accurate with the use of filtering algorithms in CP, because the incrementality is quite important:

Definition 26 *A filtering algorithm establishing arc consistency is* **perfect** *if it always has the same cost as the consistency checking.*

This definition means that the complexity must be the same in all the cases and not only for the worst one. For instance, such an algorithm is not known for the alldifferent constraint, because the consistency of this constraint can sometimes be checked in $O(1)$ and the arc consistency needs at least $O(nd)$.

The only one constraint for which a perfect filtering algorithm is known is the constraint $(x < y)$.

Two other points play an important part in the quality of a filtering algorithm: the incrementality and the amortized complexity. These points are linked together.

The incremental behavior of a filtering algorithm is quite important in CP, because the algorithms are systematically called when a modification of a variable involved in the constraint occurs. However, the algorithm should not be focus only on this aspect. Sometimes, the computation from scratch can be much more quicker. This point has been emphasized for general filtering algorithms based on the list of supported values of a value (Bessère and Régin, 2001). An adaptive algorithm has been proposed which outperforms both the non-incremental version and the purely incremental version. This is one in which the consequences of the deletion of a value are systematically studied from the information associated with the deleted value and never from scratch. There are two possible ways to improve the incremental behavior of the algorithm:

- the previous computations are taken into account when a new computation is made in order to avoid doing the same treatment twice. For instance, this is the idea behind the last support in some general filtering algorithm algorithms.
- the filtering algorithm is not systematically called after each modification. Some properties that cannot lead to any deletions are identified, and the filtering algorithm is called only when these properties are not satisfied. For instance, this is the case for the model we present to solve the n-queens problem.

When a filtering algorithm is incremental we can expect to compute its amortized complexity. This is the complexity in regard to the number of deletions, or for one branch of the tree-search. This is why the complexity can be analyzed after a certain number of modifications. The amortized complexity is often more accurate for filtering algorithm. Moreover, it can lead to new interesting algorithms that are not too systematic. For instance, there is a filtering algorithm for the symmetric alldifferent constraint that is based on this idea. The filtering algorithm establishing arc consistency calls another algorithm A n times, therefore its complexity is $n \times O(A)$. Another algorithm has been proposed in (Régin, 1999b), which can be described as follows: pick a variable then run A, and let k be the number of deletions made by A. Then you can run A for k other variables. By proceeding like that the complexity is $O(A)$ per deletions. Of course, the algorithm does not necessarily establish arc consistency but this is possibly a good compromise.

7. Discussion

7.1 Incomplete Algorithms and Fixed-Point Property

Some global constraints correspond to NP-Complete problems. Hence, it is not possible to check polynomially the consistency of the constraint to establish arc consistency. Nevertheless, some filtering algorithms can be still proposed. This is the case for a lot of constraints: the cumulative constraint, the diff-n constraint, the sequence constraint, the stretch constraint, the global minimum distance constraint, the number of distinct values constraints, and so on. When the problem is NP-Complete the filtering algorithm considers a relaxation, which is no longer difficult. Currently, the filtering algorithms associated with such constraints are independent from the definition of the problem. In other words, a propagation mechanism using them will reach a fixed-point. That is, the set of values that are deleted is independent from the ordering according to the constraints defined and from the ordering according to the filtering algorithms called. In order to guarantee such a property, the filtering algorithm is based either on a set of properties that can be exactly computed (not approximated), or on a relaxation of the domains of the variables (that is, the domains are considered as ranges instead of as a set of enumerated values). The loss of the fixed-point property leads to several consequences: the set of values deleted by propagation will depend on the ordering along with the stated constraints and on the ordering along with the variables involved in a constraint. This means that the debugging will be a much more difficult task because fewer constraints can lead to more deleted values, and more constraints can lead to fewer deleted values.

In the future, we will certainly need filtering algorithms with which the fixed-point property of the propagation mechanism will be lost, because more domain-reduction could be done with such algorithms. For instance, suppose that a filtering algorithm is based on the removal of nodes in a graph that belong to a clique of size greater than k. Removing all the values that do not satisfy this property is an NP-Complete problem; therefore the filtering algorithms will not be able to do it. However, some of these values can be removed, for instance by searching for one clique for every node. The drawback of this approach is that it will be difficult to guarantee that for a given node the graph will be traversed according to the same ordering of nodes. This problem is closed to the canonical representation of a graph; and currently this problem is unclassified: we do not know whether it is NP-Complete or not.

7.2 Closure

In general, a filtering algorithm removes some values that do not satisfy a property. The question is "Should a filtering algorithm be closed with regard to this property?"

Consider the values deleted by the filtering algorithm. Then, the consequences of these new deletions can be:

- taken into account by the same step of the filtering algorithm;

• or ignored by the same step of the filtering algorithm.

In the first case, there is no need to call the filtering algorithm again and in the second case the filtering algorithm should be called again. When the filtering algorithm is good, usually the first solution is the good one, but when the filtering algorithm consists of calling another algorithm for every variable or every value, it is possible that any deletion calls the previous computations into question. Then, the risk is to have to check again and again the consistency of some values. It is also possible that the filtering algorithm internally manages a mechanism that is closed to the propagation mechanism of the solver, which is redundant.

In this case, it can be better to stop the filtering algorithm when some modifications occur in order to use the other filtering algorithms to further reduce the domains of the variable and to limit the number of useless calls.

7.3 Power of a Filtering Algorithm

Arc consistency is a strong property, but establishing it costs sometimes in practice. Thus, some researchers have proposed to use weaker properties in practice. That is, to let the user choose which type of filtering algorithm should be associated with a constraint. In some commercial CP Solvers, like ILOG-Solver, the user is provided with such a possibility. Therefore it is certainly interesting to develop some filtering algorithms establishing properties weaker than arc consistency. However, arc consistency has some advantages that must not be ignored:

• The establishing of arc consistency is much more robust. Sometimes, it is time consuming, but it is often the only way to design a good model. During the modeling phase, it is very useful to use strong filtering algorithms, even if, sometimes, some weaker filtering algorithms can be used to improve the time performance of the final model. It is rare to be able to solve some problems in a reasonable amount of time with filtering algorithms establishing properties weaker than arc consistency and not be able to solve these problems with a filtering algorithm establishing arc consistency.

• There is room for the improvement of filtering algorithms. Most of the CP solvers were designed before the introduction of global constraints in CP. We could imagine that a solver especially designed to efficiently handle global constraints could lead to better performance. On the other hand, the behavior of filtering algorithms could also be improved in practice, notably by identifying more quickly the cases where no deletion is possible.

• For binary CSPs,for a long time it was considered that the Forward Checking algorithm (the filtering algorithms are triggered only when some variables are instantiated) was the most efficient one, but several studies showed that the systematic call of filtering algorithms after every modification is worthwhile (for instance see (Bessière and Régin, 1996)). All industrial solver vendors aim to solve real world applications and claim that the use of strong filtering algorithms is often essential.

Thus, we think that the studies about filtering algorithms establishing properties weaker than arc consistency should take into account the previous points and mainly the second point. On the other hand, we think that it is really worthwhile to work on techniques stronger than arc consistency, like singleton arc consistency which consists of studying the consequences of the assignments of every value to every variable.

8. Conclusion

Filtering algorithms are one of the main strengths of CP. In this chapter, we have presented several useful global constraints with references to the filtering algorithms associated with them. We have detailed the filtering algorithms establishing arc consistency for the alldifferent constraint and the global cardinality constraint. We have tried to give a characterization of filtering algorithms. We have showed how the global constraint can be useful for over-constrained problems and notably, we have presented the global soft constraints. At last, the the filtering algorithms we presented are mainly based on arc consistency, we think that some interesting work based on bound-consistency could be carried out.

References

Ahuja, R., Magnanti, T., and Orlin, J. (1993). *Network Flows*. Prentice Hall.

Baptiste, P., Le Pape, C., and Peridy, L. (1998). Global constraints for partial CSPs: A case-study of resource and due date constraints. In *Proceedings of the International Conference on Pronciples and Practice of Constraint Programming, CP98*, pages 87–101, Pisa, Italy.

Beldiceanu, N. (2001). Pruning for the minimum constraint family and for the number of distinct values constraint family. In *Proceedings of the International Conference on Pronciples and Practice of Constraint Programming, CP01*, pages 211–224, Pathos, Cyprus.

Beldiceanu, N. and Carlsson, M. (2002). A new multi-resource cumulatives constraint with negative heights. In *Proceedings of the International Conference on Pronciples and Practice of Constraint Programming, CP02*, pages 63–79, Ithaca, NY, USA.

Beldiceanu, N. and Contejean, E. (1994). Introducing global constraints in chip. *Journal of Mathematical and Computer Modelling*, 20(12):97–123.

Beldiceanu, N., Guo, Q., and Thiel, S. (2001). Non-overlapping constraints between convex polytopes. In *Proceedings of the International Conference on Pronciples and Practice of Constraint Programming, CP01*, pages 392–407, Pathos, Cyprus.

Berge, C. (1970). *Graphe et Hypergraphes*. Dunod, Paris.

Bessière, C. and Régin, J.-C. (1996). Mac and combined heuristics: Two reasons to forsake fc (and cbj?) on hard problems. In *Proceedings of the International Conference on Principles and Practice of Constraint Programming, CP96*, pages 61–75, Cambridge, MA, USA.

Bessière, C. and Régin, J.-C. (1997). Arc consistency for general constraint networks: preliminary results. In *Proceedings of the International Joint Conference of Artificial Intelligence, IJCAI97*, pages 398–404, Nagoya.

Bessière, C. and Régin, J.-C. (1999). Enforcing arc consistency on global constraints by solving subproblems on the fly. In *Proceedings of the International Conference on Principles and Practice of Constraint Programming, CP99*, pages 103–117, Alexandria, VA, USA.

Bessière, C. and Régin, J.-C. (2001). Refining the basic constraint propagation algorithm. In *Proceedings of International Joint Conference on Artificial Intelligence, IJCAI01*, pages 309–315, Seattle, WA, USA.

Bleuzen-Guernalec, N. and Colmerauer, A. (1997). Narrowing a 2n-block of sortings in o(nlog(n)). In *Proceedings of the International Conference on Principles and Practice of Constraint Programming, CP97*, pages 2–16, Linz, Austria.

Carlier, J. and Pinson, E. (1994). Adjustments of heads and tails for the jobshop problem. *European Journal of Operational Research*, 78:146–161.

Caseau, Y., Guillo, P.-Y., and Levenez, E. (1993). A deductive and object-oriented approach to a complex scheduling problem. In *Proceedings of the International Conference on Deductive and Object-Oriented Databases DOOD'93*.

Caseau, Y. and Laburthe, F. (1997). Solving various weighted matching problems with constraints. In *Proceedings of the International Conference on Principles and Practice of Constraint Programming, CP97*, pages 17–31, Austria.

Damaschke, P., Müller, H., and Kratsch, D. (1990). Domination in convex and chrodal bipartite graphs. *Information Processing Letters*, 36:231–236.

Dechter, R. (1992). From local to global consistency. *Artificial Intelligence*, 55:87–107.

Focacci, F., Lodi, A., and Milano, M. (1999a). Cost-based domain filtering. In *Proceedings of the International Conference on Principles and Practice of Constraint Programming, CP99*, pages 189–203, Alexandria, VA, USA.

Focacci, F., Lodi, A., and Milano, M. (1999b). Integration of CP and OR methods for matching problems. In *Proceedings of the International Workshop on the Integration of Artificial Intelligence and Operations Research techniques in Constraint Programming for Combiantorial Optimization, CP-AI-OR 99*, Ferrara, Italy.

Henz, M., Müller, T., and Thiel, S. (2003). Global constraints for round robin tournament scheduling. *European Journal of Operational Research*, To appear.

Hopcroft, J. and Karp, R. (1973). $n^{5/2}$ algorithm for maximum matchings in bipartite graphs. *SIAM Journal of Computing*, 2:225–231.

ILOG (1999). *ILOG Solver 4.4 User's manual*. ILOG S.A.

Larrosa, J., Meseguer, P., Schiex, T., and Verfaillie, G. (1998). Reversible DAC and other improvements for solving MAX-CSP. *Proceedings of the National Conference on Artificial Intelligence, AAAI*, pages 347–352.

Laurière, J.-L. (1978). A language and a program for stating and solving combinatorial problems. *Artificial Intelligence*, 10:29–127.

Lawler, E. (1976). *Combinatorial Optimization: Networks and Matroids*. Holt, Rinehart and Winston.

Leconte, M. (1996). A bounds-based reduction scheme for constraints of difference. In *Constraint-96, Second International Workshop on Constraint-based Reasoning*, Key West, FL, USA.

McAloon, K., Tretkoff, C., and Wetzel, G. (1997). Sports league scheduling. In *Proceedings of ILOG user's conference*, Paris.

Melhorn, K. and Thiel, S. (2000). Faster algorithms for bound-consistency of the sortedness and the alldifferent constraint. In *Proceedings of the International Conference on Principles and Practice of Constraint Programming, CP'00*, pages 306–319, Singapore.

Micali, S. and Vazirani, V. (1980). An $O(\sqrt{|V|}|E|)$ algorithm for finding maximum matching in general graphs. In *Proceedings 21st FOCS*, pages 17–27.

Montanari, U. (1974). Networks of constraints : Fundamental properties and applications to picture processing. *Information Science*, 7:95–132.

Pesant, G. (2001). A filtering algorithm for the stretch constraint. In *Proceedings of the International Conference on Principles and Practice of Constraint Programming, CP01*, pages 183–195, Pathos, Cyprus.

Petit, T. (2002). *Modelization and Algorithms for solving over-constrained Problems*. PhD thesis, Université de Montpellier II.

Petit, T., Régin, J.-C., and Bessière, C. (2000). Meta constraints on violations for over-constrained problems. In *Proceeding ICTAI-2000*, pages 358–365.

Petit, T., Régin, J.-C., and Bessière, C. (2001). Specific filtering algorithms for over-constrained problems. In *Proceedings Proceedings of the International Conference on Principles and Practice of Constraint Programming, CP01*, pages 451–465, Pathos, Cyprus.

Petit, T., Régin, J.-C., and Bessière, C. (2002). Range-based algorithms for MAX-CSP. In *Proceedings of the International Conference on Principles and Practice of Constraint Programming, CP02*, pages 280–294, Ithaca, NY, USA.

Puget, J.-F. (1998). A fast algorithm for the bound consistency of alldiff constraints. In *Proceedings of the National Conference on Artificial Intelligence, AAAI-98*, pages 359–366, Menlo Park, USA.

Régin, J.-C. (1994). A filtering algorithm for constraints of difference in CSPs. In *Proceedings of the National Conference on Artificial Intelligence, AAAI94*, pages 362–367, Seattle, Washington.

Régin, J.-C. (1995). *Développement d'outils algorithmiques pour l'Intelligence Artificielle. Application à la chimie organique.* PhD thesis, Université de Montpellier II.

Régin, J.-C. (1996). Generalized arc consistency for global cardinality constraint. In *Proceedings AAAI-96*, pages 209–215, Portland, Oregon.

Régin, J.-C. (1997). The global minimum distance constraint. Technical report, ILOG.

Régin, J.-C. (1999a). Arc consistency for global cardinality with costs. In *Proceedings of the International Conference on Principles and Practice of Constraint Programming, CP99*, pages 390–404, Alexandria, VA, USA.

Régin, J.-C. (1999b). The symmetric alldiff constraint. In *Proceedings of the International Joint Conference in Artificial Intelligence, IJCAI'99*, pages 425–429, Stockholm, Sweden.

Régin, J.-C. (2002). Cost based arc consistency for global cardinality constraints. *Constraints, an International Journal*, 7(3-4):387–405.

Régin, J.-C., Petit, T., Bessière, C., and Puget, J.-F. (2000). An original constraint based approach for solving over constrained problems. In *Proceedings of the International Conference on Principles and Practice of Constraint Programming, CP00*, pages 543–548, Singapore.

Régin, J.-C., Petit, T., Bessière, C., and Puget, J.-F. (2001). New lower bounds of constraint violations for over-constrained problems. In *Proceedings of the International Conference on Principles and Practice of Constraint Programming, CP01*, pages 332–345, Pathos, Cyprus.

Régin, J.-C. and Puget, J.-F. (1997). A filtering algorithm for global sequencing constraints. In *Proceedings of the International Conference on Principles and Practice of Constraint Programming, CP97*, pages 32–46.

Régin, J.-C. and Rueher, M. (2000). A global constraint combining a sum constraint and difference constraints. In *Proceedings of the International Conference on Principles and Practice of Constraint Programming, CP00*, pages 384–395, Singapore.

Rossi, F., Petrie, C., and Dhar, V. (1990). On the equivalence of constraint satisfaction problems. In *Proceedings of the European Conference on Artificial Intelligence, ECAI'90*, pages 550–556, Stockholm, Sweden.

Simonis, H. (1996). Problem classification scheme for finite domain constraint solving. In *CP96, Workshop on Constraint Programming Applications: An Inventory and Taxonomy*, pages 1–26, Cambridge, MA, USA.

Stergiou, K. and Walsh, T. (1999). The difference all-difference makes. In *Proceedings of the International joint Conference on Artificial Intelligence IJCAI99*, pages 414–419, Stockholm, Sweden.

Tarjan, R. (1983). *Data Structures and Network Algorithms*. CBMS-NSF Regional Conference Series in Applied Mathematics.

Van Hentenryck, P. and Deville, Y. (1991). The cardinality operator: A new logical connective for constraint logic programming. In *Proceedings of the International Conference on Logic Programming, ICLP-91*, pages 745–759, Paris, France.

Van Hentenryck, P., Deville, Y., and Teng, C. (1992). A generic arc-consistency algorithm and its specializations. *Artificial Intelligence*, 57:291–321.

Van Hentenryck, P., Michel, L., L.Perron, and Régin, J.-C. (1999). Constraint programming in opl. In *Proceedings of the International Conference on the*

Principles and Practice of Declarative Programming, PPDP 99, pages 98–116, Paris, France.

Zhou, J. (1996). A constraint program for solving the job-shop problem. In *Proceedings of the International Conference on Principles and Practice of Constraint Programming, CP96*, pages 510–524, Cambridge.

Zhou, J. (1997). *Computing Smallest Cartesian Products of Intervals: Application to the Jobshop Scheduling Problem*. PhD thesis, Université de la Méditerranée, Marseille.

Chapter 5

EXPLOITING RELAXATIONS IN CP

Filippo Focacci
ILOG S.A.
9, rue de Verdun, BP 85, 94253 Gentilly, France
ffocacci@ilog.fr

Andrea Lodi
DEIS - Università di Bologna,
Viale Risorgimento 2, 40136 Bologna, Italy
alodi@deis.unibo.it

Michela Milano
DEIS - Università di Bologna,
Viale Risorgimento 2, 40136 Bologna, Italy
mmilano@deis.unibo.it

Abstract In this chapter, we present the integration of various forms of problem relaxation in Constraint Programming (CP). The main motivation for the integration proposed concerns the introduction in CP languages of some form of *optimality reasoning*. Indeed, CP languages provide effective and powerful tools for reducing the search space of the problem by removing infeasible values. However, they barely consider the problem objective function and implement a naive form of branch-and-bound which poorly reduces the search space to be explored. Relaxations can be integrated in CP in different ways. Some approaches propose to automatically translate the whole CP program in linear form, while some other focus only on global constraints. In this perspective, a further differentiation can be noted. Some approaches translate all global constraints in a unique linear store, while others embed an optimization component within each global constraint. In other approaches, the part of the problem to be relaxed is decided by the user that explicitly states it in the program. First, we provide an introduction on different kinds of relaxation and we discuss many integration approaches, how they exploit results coming from the relaxation, providing references to related bibliography. In the second part of the chapter, we consider global constraints as suitable software components to integrate relaxations in CP. We discuss, as a case study, a particular global constraint, the path constraint, and show different relaxations and their tightness.

Keywords: Relaxations, Linear Programming, Assignment, Global constraints

1. Introduction and Motivation

Constraint Programming (CP) provides effective filtering algorithms aimed at reducing the infeasible parts of the search space. When optimization problems are tackled, the basic CP scheme implements a naive form of branch-and-bound, i.e., each time a solution is found an additional constraint, called *bounding constraint* (see Chapter 1), is posted forcing that further solutions should improve the last one found. Unfortunately, in many cases this added constraint poorly propagates since the link between the objective function and the problem decision variables is quite loose and not effectively handled by the CP solver.

For example, in scheduling problems, when the objective function to be minimized is the sum of setup times, it does not depend directly on the value of the activities starting times (problem variables), but it depends on their relative positions. In this case, imposing the bounding constraint affects the activities' domains only at the very low levels of the search tree when most decisions have been taken.

Thus, to overcome this drawback, an additional pruning of the search space can be performed aimed at removing *sub-optimal* regions of the search space. This is the main type of pruning performed in ILP solvers. For this purpose a problem relaxation is in general used.

The concept of relaxation has been introduced in Chapter 1. Roughly speaking, a relaxation of a problem P is a problem $R(P)$ where some original constraints of P have been 'manipulated', so as the resulting problem $R(P)$ is 'easier' to solve. Moreover, the solution space of $R(P)$ contains the one of P. Thus, the optimal solution value of $R(P)$ provides a bound for P, and if the optimal solution itself is feasible for P, it is also optimal. In general, however, the solution of $R(P)$ violates some constraints of P. An important aspect of the relaxation is its tightness, i.e., how well its solution space approximates the one of the original problem, and how good the bound is. Thus, many relaxation techniques have been proposed.

A widely used and general relaxation for optimization problems is the linear (or continuous) relaxation, but many others have been introduced either as general techniques or problem specific. We discuss many of them in Section 2.

The use of relaxations in CP is not standard and different integration approaches have been proposed in the last decade. For instance, one can translate the whole CP program in a system of linear inequalities whose continuous relaxation is solved by a linear solver. This method has been described by (Rodosek et al., 1999). Note that the authors refer to Constraint Logic Programs, i.e., they also take into account (and linearize), beside constraints, the program clauses. In this case the computed bound refers to the whole program thus providing a global view of the problem at hand.

Another possibility relies on the use of global constraints, a fundamental tool for modeling CP problems, see Chapter 4. In this case, each global constraint can be *linearized* in a unique linear store. Again, the bound computed provides a global problem perspective. This approach has been investigated by (Refalo, 2000).

An important aspect of global constraint is that they represent sub-problems and hide the filtering algorithm. Thus, a second way of exploiting global constraints for providing bounds is to embed (and hide) within each global constraint an optimization component that solves a relaxation associated to the constraint and the corresponding filtering. In this case, the bound produced relies on a local view of the problem through each global constraint. Thus, the bound is generally weaker, but the optimization component can be designed by exploiting the specific structure of the global constraint, i.e., by using a special-purpose algorithm. This has been proposed by (Caseau and Laburthe, 1997), (Focacci et al., 1999) and (Régin, 1999).

The above mentioned approaches hide the relaxation and its generation to the user. On the other hand, many approaches provide the user the possibility of writing *extended* CP programs where linear constraints can be written and explicitly sent to a linear solver. In this perspective, we can cite (Beringer and De Backer, 1995), (Hooker, 1997) and (Heipcke, 1999).

In any case, the added relaxation provides results that can be exploited both for pruning purposes and for guiding search.

The rest of the Chapter is organized as follows: in Section 2, we introduce various forms of relaxations. In Section 3 possible integrations of relaxations in CP are discussed. Section 4 and 5 are devoted, respectively, to the exploitation of the relaxation to better propagate and to guide search. A case study is presented in Section 6, where the architecture of global constraints embedding a relaxation is presented along with an example of the Path constraint.

2. Integer Linear Programming and Relaxations

As anticipated in Chapter 1, Combinatorial Optimization problems can be effectively modeled through Integer Linear Programming (ILP). ILPs are typically represented as

$$\min\{z = c^T x \mid Ax \leq b,\ x \in Z_+^n\} \tag{5.1}$$

i.e., problems defined on a set of linear inequalities and a set of decision variables assuming non-negative integer values, with the objective of minimizing a linear function[1].

A classical solution method for ILPs is the so-called *branch-and-bound*. Again the method has been described in Chapter 1, so we only give here a rough description so as to introduce some notations used in the following.

The idea is to partition the ILP at hand, P, into a set of sub-problems $P^1, P^2, \ldots, P^\ell$ such that:

$$\bigcup_{i=1}^{\ell} P^i = P \quad \text{and} \quad P^i \bigcap P^j = \emptyset \quad \forall i, j \in \{1, \ldots, \ell\}$$

i.e., each solution of P is a solution of one and only one sub-problem P^i. This partition step represents a so-called *branching* step, and it is not difficult to see that it can be iterated: each sub-problem P^i can be in turn partitioned in sub-problems using the same scheme.

Each branching creates sub-problems which are more constrained than the original problem, i.e., they have in general a solution space which is smaller than the one of the partitioned problem. However, each sub-problem itself has the same theoretical complexity of the original problem (each sub-problem is itself an ILP), thus the reduction in size would allow to solve the sub-problem only when its solution space contains just the optimal solution.

This means that branching alone is not enough, thus a *bounding* step is applied to each sub-problem: a relaxation of the current sub-problem P^i is solved to optimality obtaining an optimistic evaluation of the optimal solution of P^i, say LB_{P^i}. If:

$$LB_{P^i} \geq UB$$

where UB is the current best feasible solution of the overall problem P, this means that no solution better of UB can be found neither in P^i nor in all the sub-problems possibly originated by its partition. Indeed, all these sub-problems would be more constrained and with a smaller solution space. Thus, there is no need to optimally solve P^i, and obviously to create sub-problems from it.

It should be clear now how crucial is to define good relaxations of ILPs: as much the optimal solution of the relaxation tightly approximates the optimal value of the original problem as effective the branch-and-bound framework would be. Indeed, if the bound is tight many useless sub-problems can be abandoned. Moreover, as anticipated in the previous section, since many relaxed (sub-)problems must be solved, the relaxation must be 'easy' to solve.

In the following sections the main techniques to obtain good relaxations for ILPs are described.

2.1 Continuous Linear Relaxation

Associated with the ILP of problem P in (5.1) are two polyhedra:

- $P_L := \{x \in I\!R^n_+ : Ax \leq b\}$
  (the feasible region of the continuous linear relaxation),
- $P_I :=$ conv $\{x \in Z^n_+ : Ax \leq b\}$
  (the convex hull of feasible integer solutions).

It turns out that minimizing a linear objective function $c^T x$ over the polyhedron P_L is much easier than solving P_I. Specifically, ILPs are in general NP-hard to solve while Linear Programming (LP) is polynomially solvable, even if the algorithm used to solve LPs in practice, the simplex algorithm, has an exponential time complexity in the worst case.

Then, by dropping integrality constraints on the variables we obtain the simplest possible relaxation which often fulfills the two requirements we indicated for relaxations: (i) it is easy to solve, and (ii) it gives a good approximation, i.e., good bounds. We denote this relaxation as $L(P)$.

Sometimes, however, the bound $z(L(P))$ obtained by solving $L(P)$ can be poor, and in these cases a general technique for improving it is the so-called *cutting plane* generation.

A cutting plane, or simply cut, is a linear inequality which is satisfied by all points in P_I, but which is not satisfied by all points in P_L. This means that if this inequality is added to the set of inequalities $Ax \leq b$, we obtain a new set $A'x \leq b'$, and a new polyhedron P'_L such that $P'_L \subseteq P_L$. This is another (continuous) linear relaxation of P since once the integrality constraints are imposed the set of feasible solutions is perfectly the same, i.e., P_I remains unchanged. However, since $P'_L \subseteq P_L$ optimizing $c^T x$ over P'_L gives a lower bound $z(L'(P))$ such that $z(L'(P)) \geq z(L(P))$, thus improving the approximation.

Obviously, only some of the cutting planes are useful to improve the bound depending on $c^T x$, and the idea to find them is to iteratively solve the problem of finding an inequality $\alpha^T x \leq \alpha_0$ such that:

$$\alpha^T x \leq \alpha_0 \quad x \in P_I \quad \text{but} \quad \alpha^T x^* > \alpha_0$$

where x^* is the optimal (typically fractional) solution of the current relaxation. If such a cut is found and added to the current linear relaxation, x^* will not be feasible anymore and the bound will possibly improve (increase for minimization problems).

The above problem is called *separation problem*, and its general form is as follows:

Given a vector $x^* \in I\!R^n$, find a linear inequality which is valid for P_I and violated by x^*, or prove that none exists.

When the original problem P we want to solve is NP-hard then the separation problem is also NP-hard (see, (Grötschel et al., 1988)). However, specific classes of cuts can be separated in polynomial time, or heuristic algorithms can be used for separation, thus the cutting plane generation technique incorporated in the branch-and-bound framework (see, (Padberg and Rinaldi, 1991)) had a strong impact in solving ILPs.

2.2 Structured Relaxations

In the previous section, we considered the simplest way of obtaining a relaxation of a problem P, the one in which some of the constraints are deleted. However, we must point out that the simple deletion of constraints does not necessarily produce a relaxation which is 'easy' to solve. On the other hand, as mentioned in Section 1, a relaxation can be originated in a manipulation of the constraints (or a set of them) not necessarily a deletion.

Moreover, the two requirements for a relaxation, i.e., it must be 'easy' to solve and it must be tight, are in contrast. Obviously, indeed, removing all constraints produces a very easy problem to solve whose optimal solution is in general too far away from the one of the original problem to be useful.

The linear relaxation is very often used because it is the most general one, and because only a very specific set of constraints, the integrality requirements, are neglected, while the ones which characterize the problem itself, i.e., the set $Ax \leq b$ is fully taken into account. This gives a problem which does not have a specific structure, and general-purpose algorithms for LPs must be used, but in the last decade the improvement in solving LPs has been impressive and even very large problems can be efficiently solved in practice.

Sometimes, however, the set of constraints of an ILP can be manipulated so as to recognize a specific structure which is widely known, polynomial, and studied, i.e., for which there are very effective special-purpose algorithms. These special-purpose algorithms have in general nice characteristics which are discussed through examples in Section 6.2, and allow the efficient exploration of an even huge number of sub-problems (nodes) in the branch-and-bound framework[2].

Obviously, the easiest form of manipulation to obtain a structured problem is again the deletion. Suppose of rewriting the ILP in (5.1) in the following way:

$$\min z = c^T x \tag{5.2}$$

$$A'x \leq b' \tag{5.3}$$

$$A''x \leq b'' \tag{5.4}$$

$$x \geq 0 \quad \text{integer} \tag{5.5}$$

where we have separated the set of linear inequalities $Ax \leq b$ in two sets: a first set (5.3) containing the so-called *nice* constraints, and a second set (5.4) containing the so-called *bad* constraints. The word 'nice' means that the relaxation (5.2), (5.3), and (5.5), i.e., the one obtained by dropping the 'bad' constraints (5.4), can be solved to optimality with a special-purpose algorithm since it is itself a well-known and studied problem[3].

However, this is not the only way of coping with the so-called 'bad' constraints and in the next two sections we discuss two additional methods for obtaining structured relaxations.

2.2.1 Surrogate Relaxation. Another way of coping with 'bad' constraints is manipulate them by producing a new constraint which is in turn 'nice'. Let the set of 'bad' constraints contain k inequalities, and π be a k-vector with non-negative components, $\pi_i \geq 0, \ \forall i$. Then, a new constraint can be produced by combining the k 'bad' constraints using the vector of multipliers π. Formally:

$$\sum_{i=1}^{k} \pi_i a_i^T x \leq \sum_{i=1}^{k} \pi_i b_i. \tag{5.6}$$

Obviously, the problem obtained by substituting constraints (5.4) with the constraint (5.6) is a relaxation since a solution satisfying (5.4) also satisfies (5.6) while the reverse does not necessarily hold. It is not difficult to see that the quality of the relaxation, i.e., its tightness, strongly depends on π: specifically, we obtain a different relaxation for each vector of multipliers π. Then, we denote as $S(P,\pi)$ the surrogate relaxation of problem P using π.

However, the surrogate manipulation is particularly interesting if constraint (5.6) can be added to the set of 'nice' constraints (5.3) with the property that the set remains 'nice', i.e., the relaxation is again a structured problem.

Obviously, the best of these relaxations is the one producing the highest bound, i.e., the following problem must be solved:

$$\max_{\pi \geq 0} z(S(P,\pi)) \tag{5.7}$$

where $z(S(P,\pi))$ is the optimal solution value of the relaxation $S(P,\pi)$.

2.2.2 Lagrangean Relaxation. Sometimes it is not possible to add a constraint to the set of 'nice' constraints without transforming it in a 'bad' set. Thus, another way of taking into account the 'bad' constraints instead of simply dropping them is described in the following. Let λ be a k-vector with non-negative components $\lambda \geq 0, \ \forall i$. Then, we drop constraints (5.4) but the objective function (5.2) is modified as follows:

$$\min z(Lg(P,\lambda)) = c^T x + \sum_{i=1}^{k} \lambda_i (a_i^T x - b_i) \tag{5.8}$$

where we denote as $Lg(P,\lambda)$ the Lagrangean relaxation of problem P using λ, and as $z(Lg(P,\lambda)$ its optimal solution value.

The violation of any of these constraints, the so-called *dualized* constraints, i.e., $a_i^T x - b_i > 0$ for some i, produces a positive term in the new objective function to be minimized (5.8), thus penalizing the minimization process. In other words, the 'bad' constraints are transformed into 'soft' constraints penalizing those solutions which do not satisfy them.

As in the case of surrogate relaxation, any vector λ produces a different relaxation and the best one is obtained by solving the following problem:

$$\max_{\lambda \geq 0} z(Lg(P,\lambda)). \tag{5.9}$$

A final remark concerns the test for optimality. In both linear and surrogate relaxations the objective function is not changed, so if the optimal solution of the relaxation is feasible with respect to the constraints that have been manipulated it is in turn optimal. For the Lagrangean relaxation, instead, the objective is changed from (5.2) to (5.8), so to prove optimality we require both feasibility and the fact that the contribution of the dualized constraints in the objective function is zero.

3. Integrating Relaxations in CP

It is widely recognized that integrating inference in the form of constraint propagation and relaxation in the form of Linear Programming can yield substantial advantages in CP languages.

Computing the optimal solution of a relaxation in a CP solver provides important pieces of information for pruning the search tree and for guiding search.

Despite of the type of technique adopted, the relaxation should return three pieces of information: (i) the optimal solution of the relaxed problem x^*; (ii) the optimal solution value LB (a lower bound for the original problem); and (iii) a gradient function $grad(X_i, j)$ measuring the variable-value assignment cost.

These pieces of information are exploited both for filtering purposes and for guiding the search toward promising (in terms of costs) branches.

Clearly, the tighter the relaxation the more effective its impact in the CP solver. As for the filtering performed by CP constraints, there should be a tradeoff between the time spent in computing these pieces of information and the resulting pruning. In addition, the filtering obtained from the relaxation interacts with the pruning achieved by traditional constraints.

Now the questions are: which relaxation to use, which part of the problem to relax and how to exploit the results coming from the relaxation. In this section so as in the rest of the paper we try to answer these questions.

3.1 Which relaxation to use

The particular relaxation to use depends on the problem to be solved. Many types of relaxations have been introduced in Section 2. However, they can be roughly divided in two main kinds: structured and non structured relaxations.

Structured relaxations are those for which (effective) special-purpose algorithms are available. On the other hand, the linear relaxation can be solved through a general-purpose solver, i.e., it is in turn non-structured. As mentioned, despite the great practical improvement in solving LPs in the last decade, the most used solver, the simplex algorithm, is still a non-polynomial algorithm.

More importantly, however, the fact of exploiting the special structure of a relaxation, often provides important features which are extremely useful in the CP context. Special-purpose methods are often *incremental*, i.e., require guaranteed short time to each update of the optimal solution. This is very important in CP. Indeed, the solution method provided by CP tools naturally requires extensive enumeration and the triggering mechanism implies that a weaker and very fast relaxation repeatedly solved (up to the fix point) at each search node can lead to more effective algorithms than a stronger (and slower) relaxation solved once at each node.

This suggests a preference for structured relaxations, but the situation is, in fact, not so clear. First, the linear relaxation is a general and natural method and LP solvers are easily available. Second, the linear relaxation provides a *global view* of the problem at hand. Third, decades of research on methods for improving linear relaxations provided both general-purpose and special-purpose techniques (e.g., the addition of cutting planes) for strengthening this relaxation providing better bounds. Finally, it is not always possible to extract a significant structured relaxation from a problem formulation.

An intermediate way between a structured and a general relaxation is the Lagrangean relaxation which has been used with different flavors either together with cutting plane generation ((Focacci et al., 2000b), see Section 6.2.6) or to link different structured relaxations ((Fahle and Sellman, 2003), see Section 3.2.1).

3.2 Which part of the problem

An important aspect of the integration of relaxations in CP concerns which part of the problem to relax. We can translate the whole CP program, only one part of it, for example a sub-problem which can be efficiently solved, or only global constraints. In general, however, we start from a CP model, we devise a corresponding ILP (sub-)model and we have to link the two via a mapping that defines a correspondence between the two models and allows their interaction.

One of the first works where a linear relaxation has been used in Constraint Programming is the one by (Rodosek et al., 1999). The authors propose a translation of the whole Constraint Logic Program in a single store of linear inequalities. First, all clauses are translated in an LP model. Atoms are maintained, but an additional binary variable (representing their truth value) is introduced in their arguments. Disjunction of clauses is treated by inserting auxiliary binary variables. Linear constraints remain the same and predefined constraints (like domain constraints and global constraints) are translated in the corresponding set of linear inequalities. Finally, constraints are introduced linking the integer variables usually contained in a CP program and binary variables considered in the ILP model.

After this translation, we have two stores: the CP store handled by the CP solver and the LP store handled by the ILP or Mixed-Integer (linear) Programming (MIP) solver. Indeed, the authors combine two solvers by performing local constraint propagation on finite domain variables and what they call *global constraint propagation* solving through the simplex algorithm the continuous relaxation of the ILP problem. The two solvers exchange bounds on variables, and the solution of the relaxed problem provides a bound on the original one. Concerning the search strategy the optimal solution of the relaxation provides some guidelines for branching: indeed, the most constrained variable is chosen, its fractional value rounded and assigned to the variable.

Instead of relaxing the whole program, only some parts of it can be relaxed. For example, in (Beringer and De Backer, 1995) the user decides which parts of the model to relax by explicitly distinguish which constraints should be handled by which solver. Having a syntactic separation between the two constraints (linear and finite domain) the user can decide to relax the whole model or only a part of it. Communication of values and bounds is ensured between the two solvers.

In the same perspective, (Hooker and Osorio, 1999) propose the MLLP framework (Mixed Logical/Linear Programming) that uses hybrid solver specific models where the user can separate the problem into a discrete part (FD store) and a continuous part (LP store). As before, this approach achieves domain reduction on the FD store and the generation of bounds and cutting planes on the LP store. Models have the following form

$$\min \quad c^T x \tag{5.10}$$

$$\text{s.t.} \quad h_i(y) \rightarrow A^i x \geq b^i \qquad i \in I \tag{5.11}$$

$$x \in R^n, y \in D \tag{5.12}$$

This scheme is based on the logical entailment: as soon as a partial assignment satisfies the antecedent of the logical implication the corresponding linear inequality is added to the LP store and the resulting relaxation is solved. This framework has been extended to include also global constraints in (Ottosson et al., 1999).

3.2.1 Relaxation of global constraints. In many approaches, global constraints play an important role for the integration since they embed propagation algorithms and are declarative representation of sub-problems which can be linearized/relaxed. There are two different ways of integrating an LP model as the result of the linearization of CP global constraints.

The first one relies on the automatic linearization of each single global constraint (see, (Refalo, 2000)). The corresponding set of linear inequalities is stored in the LP store which represents a general LP model and can be solved by an LP solver like the simplex algorithm. The resulting LP model represents the linearization of the original problem where all constraints are considered and linearized. The LP model is conceptually separated from the CP model even if the two models tightly interact through variable fixing, bound tightening and cutting plane generation (see, (Refalo, 1999)). The conceptual architecture of this framework is depicted in Figure 5.1.

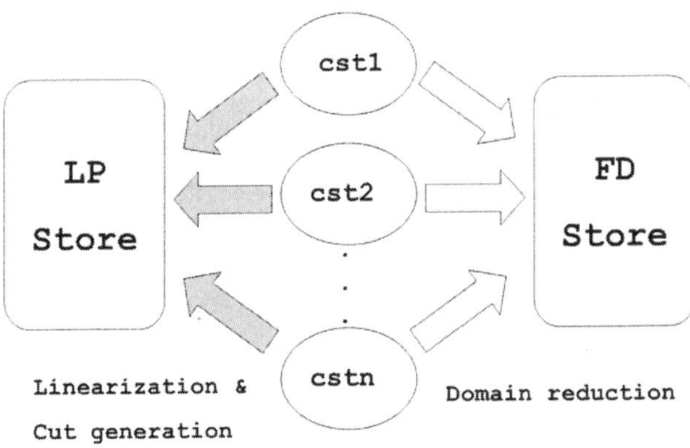

Figure 5.1. Linearization of all constraints in a single store

The second method, instead, relies on the use of relaxations within each global constraint, obtaining what we call *global optimization constraint*. A global optimization constraint, beside a filtering algorithm, embeds an optimization component representing a relaxation (possibly a linear relaxation) of the constraint itself. The optimization component provides pieces of information on the problem bound and on the variable-value assignment cost which can be used for filtering purposes. However, these pieces of information are *local* to the constraint. The conceptual architecture of this framework is depicted in Figure 5.2.

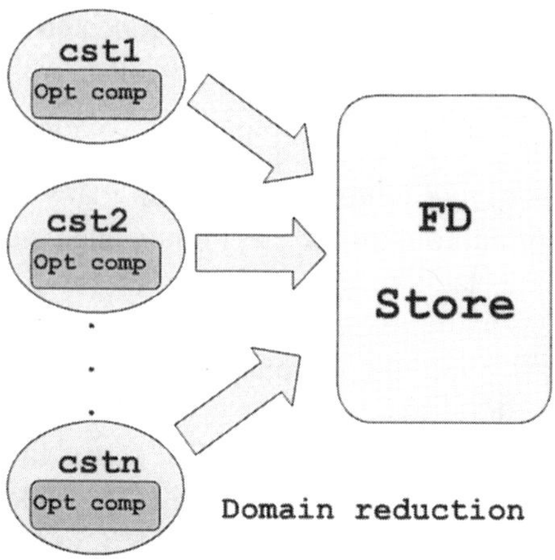

Figure 5.2. Optimization component in each constraints

Each approach has its advantages and drawbacks. Clearly, the first approach benefits of a *global view* of the problem we are solving. Indeed, since the set of linear inequalities are considered and solved all together, the bound on the problem is in general tighter than the one obtained by considering the linear component of each constraint separately and taking the best bound produced. Moreover, the generation of cutting planes in this approach can be done by exploiting the structure of the overall problem. In addition, changing the objective function of the problem is easy since the translation is independent on the objective function.

On the other hand, the resulting LP problem can be big and the LP computation can be too expensive with respect to the increased pruning achieved from the results provided. Finally, since the LP problem we consider has no structure, we can only use general-purpose techniques and algorithms for solving it.

The benefits of the second approach are mainly due to the fact that, by embedding a relaxation within a global constraint, we can exploit the structure of the problem declaratively represented by the constraint itself. This leads to several advantages. First, the dimension of the relaxation of a single constraint is in general smaller than the relaxation of the overall problem we are solving. Second, more important, we can exploit the structure and the properties of the problem represented by the constraint, thus obtaining a bound that could be computed more efficiently and, possibly, incrementally. This topic, already anticipated in Section 3.1, will be described in Section 6.2 by using a classical *Path* constraint as an example.

In summary, the advantage of embedding an optimization component within a global constraint concerns the fact that we are not forced to use a linear relaxation of the constraint, thus a linear solver for solving it.

The first approach is more *Operations Research-oriented*, where the linear relaxation of the overall problem is often generated. The second approach instead is more *Constraint Programming-oriented*, where each constraint represents a sub-problem which can be considered locally and used with other constraints in order to model and solve the problem. The filtering algorithm and the operational semantics of the constraint are hidden in the constraint in an software engineering fashion.

As depicted in Figure 5.3, an intermediate approach between considering a single relaxation of an overall problem or separate relaxations of its global constraints can be obtained by simultaneously taking into account the structure of more than one constraint so as to improve the quality of the information obtained. This idea is particularly suited in the context of cutting plane generation and Lagrangean relaxations and requires the design of an additional constraint merging the original ones. The topic is discussed in Section 6.2.5.

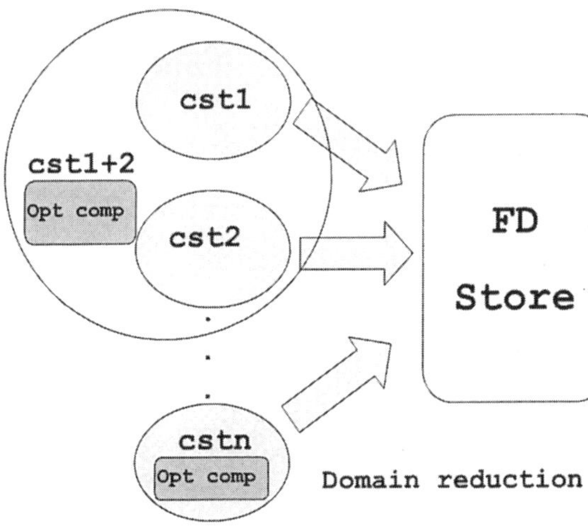

Figure 5.3. Merging of two constraints

Another interesting intermediate approach that in a sense merges the knowledge contained in two constraints is that proposed in (Fahle and Sellman, 2003). In that paper, the overall problem is modeled by using two structured components: from one side a knapsack constraint and on the other side a set of clique constraints. Relaxations of both components can be efficiently solved (see, (Fahle and Sellman, 2003) for details), but none of the two relaxations alone provides a good bound on the overall problem. Therefore, two Lagrangean relaxations are maintained merging the two components: in the first relaxation the knapsack constraint is dualized in the objective function, while in the second one the clique constraints are dualized. In this way, specialized (efficient) algorithms can be used and at the same time bounds with a much more global view of the problem are devised.

4. Relax to propagate

Despite the kind of technique we use for the relaxation, the unique requirement concerns its output. The relaxation should return three pieces of information:

- the optimal solution of the relaxed problem x^*;
- the optimal solution value LB. This value represents a lower bound on the objective function value;
- a gradient function $grad(X_i, j)$ measuring the variable-value assignment cost.

These pieces of information are exploited by the CP solver both for filtering purposes and for guiding the search toward promising (in terms of costs) branches.

In particular, in this section we consider the filtering aspect, while in Section 5 we discuss the impact of the relaxation on the search strategy.

Concerning filtering, beside the pruning on the objective function variable, a CP solver can implement the so-called cost-based filtering which is discussed in details in Section 4.2. This filtering removes from variable domains those values that are proven sub-optimal by exploiting the information provided by the relaxation. The output of the cost-based filtering algorithm is, as usual, a set of domain reductions (eventually, variable instantiations). In order to let the CP solver and the relaxation solver communicate, we need a mapping between the CP variables involved and the model exploited by the relaxation. This mapping depends on the chosen solver, and is discussed in the following.

4.1 Mapping

To enable the communications between a CP solver and the solver of the relaxation, we need to establish a correspondence (a mapping) between variables and constraints used in the relaxation and those used in the CP program. The mapping is aimed at allowing the communication and information exchange between the two components. Obviously, the mapping depends on the particular problem to be solved and to the specific relaxation used. Therefore, defining a general mapping between the CP model relaxation entities is clearly impossible.

Here we describe, as an example, the mapping between the CP formulation and a binary LP formulation previously suggested in (Rodosek et al., 1999). In CP, we have variables $X_1, ..., X_n$, ranging on domains $D_1, ..., D_n$, and cost c_{ij} of assigning value $j \in D_i$ to X_i. Obviously, the cost of each value not belonging to a variable domain is infinite. As an example, the problem we want to solve is to find an assignment of values to variables consistent with a global constraint, and whose total cost is minimal. Binary variables are $x_{ij} \in \{0, 1\}$: the mapping imposes that if the CP variable X_i is assigned to the value j, x_{ij} is equal to 1, $x_{ij} = 1 \iff X_i = j$. Constraint $\sum_{j \in D_i} x_{ij} = 1$ is part of the mapping and imposes that exactly one value should be assigned to each CP variable. This mapping is one-to-one, in the sense that to every feasible solution in the CP context corresponds a feasible solution in the (Integer) LP context. However, either due to the presence of the mentioned specific constraints or due to the intrinsic difficulty in solving ILPs, in most of the cases at this point, some of the constraints, e.g., the integrality requirement, are relaxed thus breaking this one-to-one correspondence. Thus, for every solution of the CP constraint there is exactly one correspondent solution of the LP model, while, if the integrality requirement on the x_{ij} variables is relaxed, there could be LP solutions x^* which do not correspond to any solution of the CP constraint, namely, all the LP solutions having at least one variable assuming a non-integer value.

Finally, if a relaxation is considered in the ILP context, the obvious link between the objective function variable of the CP model Z and the lower bound value LB computed by solving the relaxation is $LB \leq Z$.

Given the mapping between LP and CP variables, we know that the LP variable x_{ij} corresponds to the value j in the domain of the CP variable X_i. Thus, the reduced cost matrix $\bar{c}_{ij}$ provides information on CP variable domain values, $grad(X_i, j) = \bar{c}_{ij}$.

4.2 The Cost-Based Propagation

In this section we describe a general propagation derived from the information provided by the relaxation on the basis of the mapping presented above.

We have a first (trivial) propagation based on the optimal solution value LB of the relaxation. This value is a lower bound on the objective function Z of the overall problem we are solving. Since the objective function in CP is represented by a domain variable, the mapping contains the constraint $LB \leq Z$. Since LB is a constant, it updates the lower bound of the domain of Z. If LB is greater or equal than the upper bound of the domain of Z, a failure occurs. This kind of propagation generates a yes/no answer on the feasibility of the current node of the search tree; therefore, it does not allow any real interaction with the other constraints of the problem since, as mentioned, the constraint linking Z and problem decision variables is a non-primitive one.

More interestingly, we can implement the propagation from the gradient function $grad(X_i, j)$ towards the problem decision variables $X_1, ..., X_n$. The gradient function provides an optimistic evaluation on the cost of each variable-value assignment. Given this information, we can compute an optimistic evaluation on the optimal solution of a problem where a given variable is assigned to a given value, i.e., $LB_{X_i=j}$ (we can do this computation for each variable and all values belonging to its domain). Thus, if this evaluation is higher than the best solution found so far, the value can be deleted from the domain of the variable. More formally, for each domain value j of each variable X_i, we can compute a lower bound value of the sub-problem generated if value j is assigned to X_i as $LB_{X_i=j} = LB + grad(X_i, j)$. If $LB_{X_i=j}$ is greater or equal to the upper bound of the domain of Z, j can be deleted from the domain of X_i.

This filtering algorithm performs a real back-propagation from Z to X_i. Such a technique is known in Mathematical Programming as *variable fixing* (see Chapter 1). However, variable fixing in ILP context does not trigger in general an iterative process like constraint propagation, as happens in CP. Indeed, removing values from variable domains triggers constraints imposed on modified variables, and it appears therefore particularly suited for CP.

An important point which should be stressed is that this filtering algorithm is general and can be applied whenever the relaxation is able to provide these pieces of information. The filtering algorithm is independent on the structure of the relaxation.

4.3 Triggering Events

The CP solver propagation is triggered each time a modification in the domain of one variable involved is performed. We have to distinguish two cases:

- the modification happens in the domain of decision variable X_i;
- the modification happens in the domain of variable Z.

In the first case, traditional constraint propagation is triggered, possibly reducing other domain variables. The relaxation re-computation is triggered only if one of the values deleted from the domain belongs to its optimal solution x^*, i.e., to the optimal solution of the relaxation. Indeed, only in that case, the value of the optimal solution of the relaxation can change (increase for minimization problems). After the optimization component computation, the cost-based filtering algorithm is triggered since we have a new optimal solution of the relaxation and a new gradient function.

In the second case, instead, the cost-based domain filtering algorithm is triggered. Indeed, if the upper bound of the objective function is changed (decreased) we have the possibility that a variable domain is reduced through the gradient function-based propagation.

5. Relax to guide the search

In tree search approaches, the way the tree is explored greatly influences the algorithm performance. Therefore, it is important to define problem-dependent heuristics that efficiently explore the tree. In optimization problems, the optimal solution of a relaxed problem, the lower bound value, and the gradient function can be used for guiding heuristic decisions during the search. An optimization component can play the role of an *oracle* providing important information about the problem at hand. In the following we describe three ways of using the oracle.

5.1 Synchronous update of the oracle

We have seen in previous sections that a relaxation $R(P)$ for problem P can be computed during the exploration of the search tree. At any search node k, a sub-problem P_k is defined where P_k is the original problem plus a set of branching constraints BC to reach the node k, i.e., $P_k = P \cup BC$. In case of synchronous update of the oracle, a relaxation $R(P_k)$ of P_k is available at each node of the search tree. Therefore, $R(P_k)$ takes into account the branching constraints used to reach node k. In this case, the current relaxation $R(P_k)$ can be used to dynamically select the current variable and value selection of a branch-and-bound algorithm.

5.1.1 Select a branching object. Selecting a branching object means to define at each step of the tree search which branching constraint to use. A widely used branching object consists in selecting a variable and a value in its domain. However, other branching decisions can be taken, for example imposing an ordering constraint on two variables. In case of an available relaxation, we can exploit the results coming from its optimal solution to select the branching decision. We have identified two ways of exploiting the relaxation: one based on the violated constraints and one based on the gradient function.

Violated constraints. In CP the most commonly used heuristic exploits the *First-Fail* principle, selecting the variable with the smallest domain, or the variable that is involved in the tightest constraints. On the contrary, in a MIP-based branch-and-bound (using the continuous relaxation) often the variable selected for branching is the most fractional one (i.e., the one violating the most the integrality constraints which have been relaxed). If variables are binary, for example, it is therefore common to branch on the variable x_i whose value in the optimal solution of the continuous relaxation is closest to 0.5.

This technique is also used in the *probing* framework (see, (Ajili and El-Sakkout, 2003)). Here a relaxation based on LP or MIP is solved at every node and it is used as an oracle for a CP-based tree search. The branching decisions try to repair first the most violated constraints of the relaxed solution.

Variable selection is not the only branching decision. For example, ordering constraints can be imposed as branching constraints. As an example, consider a project scheduling problem with time windows where a set of tasks linked by temporal constraints must be scheduled within given time windows on resources with maximal capacity equal to one. In some approaches (e.g. (Cesta et al., 2002)) a branching decision consists in imposing a precedence relation among two selected tasks (i.e. task a before task b or task b before task a).

Regret Heuristic. Another way of exploiting the results coming from the relaxation for selecting the branching object is based on the gradient function. Refining the concept of regret, the gradient function can be used to select the branching variable following a *max-regret* heuristic. The regret of a variable X_i can be defined as the additional cost (*regret*) to be paid over the lower bound if X_i is *not* assigned to the value suggested by the optimal solution of the relaxation, say v_j. Clearly, this additional cost is not known, but the gradient function gives a good evaluation of the regret by considering the minimum value of the gradient (excluding $grad(X_i, v_j)$):

$$regret = \min_{v_h \in D(X_i), v_h \neq v_j} \{grad(X_i, v_h)\}.$$

The *max-regret* is a variable selection heuristic that suggests to assign first the variable with the highest regret so as to minimize the risk of paying a high cost if the best assignment becomes infeasible due to wrong heuristic decisions.

5.1.2 Select promising values. A branching decision often consists in assigning, to a given variable, one value in the domain of possible ones. The value should be chosen considering feasibility and optimality issues. Although the optimization component relaxes some of the original problem constraints, if a value belongs to its optimal solution it represents a good guess to be chosen as tentative value for the branching variable. This principle is the key idea of *probe backtrack* (see, (El-Sakkout and Wallace, 2000) and (Ajili and El-Sakkout, 2003)) where the solution of a relaxation is used to guide the search.

Alternatively, the gradient function can be used to evaluate the impact of value-variable assignments on the objective function so as to avoid making heuristic decisions that increases too much the lower bound. In this case, the gradient function can be used to rank domain values so as to try the most promising first (see, e.g., (Milano and van Hoeve, 2002)).

5.2 Asynchronous update of the oracle

As in the previous case, a relaxation $R(P)$ of a problem P can be computed during the exploration of the search tree. However, in some cases the relaxation $R(P)$ may be a complex combinatorial optimization problem itself, making its frequent re-computation computationally too expensive. Therefore, the relaxation can be computed either only at the root node or every time it *diverges* too much from the current assigned part of the problem. We recall that at each node k, we are exploring a problem P_k equivalent to the original problem P plus branching constraints BC used to reach the node. If the relaxation $R(P_k)$ is not computed, then it does not take into account BC. Therefore, it is likely to happen that the relaxation $R(P)$ we are considering for guiding the search performs assignments that violate BC. In this case, the information provided by $R(P)$ in the sub-tree rooted in k can be misleading.

5.2.1 Generation of static heuristics. In some cases (e.g., in scheduling and time-tabling applications) a relaxation computed at root node can provide a useful static heuristic that can be exploited in a CP branch and bound exploration. Consider for example a scheduling problem. A LP-based relaxation may provide time-buckets and resource allocation for all activities. These information can be used to create a static list of activities (i.e., problem variables) defining the order on which the CP engine takes its decisions.

In general, we can exploit the optimal solution of a relaxation to guide the search. However, if the relaxation is computed once for all at the root node, after some branching decisions, the current sub-problem we are exploring can diverge from the optimal solution of the relaxation. Therefore we can re-compute it after some branching decisions, so as to obtain a tradeoff between a good (semi)-static heuristic and the computational effort.

Moreover, all the methods mentioned in Section 5.1 can be adapted to the asynchronous update case, where the relaxation is re-computed once it gets too far from the instantiated part of the problem. Indeed, note that all these methods have a lot in common with *repair-based* methods, where an infeasible solution (given by a relaxation, or by other related problems) is iteratively repaired to satisfy all constraints.

5.2.2 Defining sub-problems and incomplete methods. The information provided by a relaxation can also be exploited in *incomplete* methods such as *Restricted Candidate List* and *Large Neighborhood Search* (see Chapter 9).

The former suggests to cut off uninteresting parts of the search tree a priori (leading to incomplete tree search). The gradient function can be used to find those assignments that would increase the most the objective function and that could, for this reason, be cut off. This method was used in scheduling context for a *min-sum tardiness* problem by (Focacci, 2001). In this problem, each activity is assigned to a position in the schedule. For each activity, some positions (the ones with the highest gradient) are removed, and the remaining sub-problem is (quickly) solved using branch-and-bound. The best solution of

this sub-problem is typically a very good solution of the original problem and is used as starting point for a complete tree search.
In as similar way, another approach where sub-problem definition has been applied in complete search is the one proposed in (Milano and van Hoeve, 2002), where the gradient function coming from a relaxation computed only at the root node is used to rank values and to explore promising sub-problems first.

In the Large Neighborhood Search context, instead, a good evaluation of the lower bound can be used to select promising areas for improvement, and the problem can be optimally solved on one or more of these areas (or neighborhoods). This method has been applied in (Caprara et al., 1998) for improving the solution of a *crew rostering* problem, and in (Focacci et al., 2000a) for scheduling problems with *sequence-dependent setup times*. In both applications the selected neighborhood is solved using CP branch and bound.

6. A case study: global optimization constraints for a *Path* constraint

In this section, we provide an example of the integration of a relaxation in a CP solver. In particular, we describe the approach where we embed an optimization component which is able to optimally, efficiently and incrementally solve a relaxation of the problem represented by the constraint itself. The optimal solution of the relaxation enables to perform domain filtering and to guide the search. The optimization component embeds effective OR techniques/algorithms and tightly interacts with the CP model.

First, we discuss the structure of global constraints obtained, called global optimization constraints. Then, we focus on the structure of a particular constraint, the $Path$ constraint and we show various relaxations which can be used for our purpose.

6.1 Global constraint architecture

In this section, we define the architecture of the global constraint we propose. Beside the problem decision variables $X_1, \ldots, X_n$ involved in the constraint, an additional parameter of the constraint is the variable representing the problem objective function, Z.

Conceptually, a traditional CP global constraint embeds a software component containing a filtering algorithm which prunes domains on the basis of *feasibility reasoning*. Indeed, a value is deleted from the domain of a variable if proven infeasible. Each constraint interacts with other constraints through modification of variable domains involved.

Intuitively, we want to perform pruning based on *optimality reasoning*. A value is deleted from the domain of a variable if proven sub-optimal. For this purpose, we extend the conceptual architecture of a CP global constraint with two additional components: an optimization component and a cost-based filtering algorithm.

The optimization component is a solver able to compute the optimal solution of a problem which is typically a relaxation of the problem represented by the global constraint. This relaxation depends on the declarative semantics of the global constraint and on the objective function of the problem. In general, the optimization component is based on OR techniques either special-purpose algorithms for structured problems, or general purpose LP algorithms. The conceptual architecture of the global constraint is depicted in Figure 5.4.

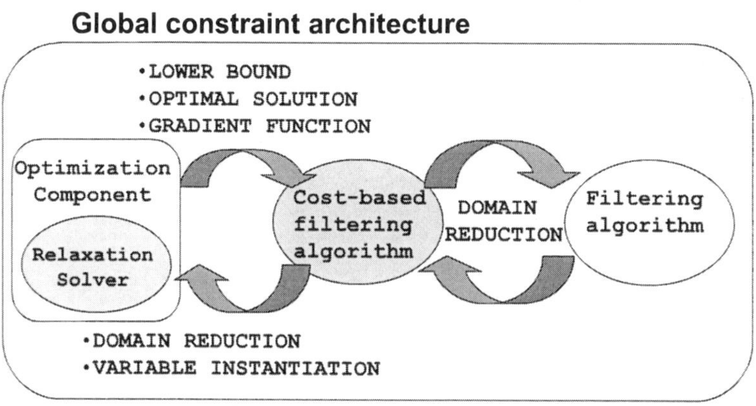

Figure 5.4. Global constraint architecture

In the following we discuss, as an example, several relaxations of the classical path constraint, a widely used constraint in CP solvers, and some corresponding optimization components.

6.2 The `Path` constraint case

Most commercial CP tools offer the global constraint *Path* which in its simplest form has the following syntax and semantics:

$Path(NextArray)$

where $NextArray$ is an array of $n+1$ variables of index $0, 1, \ldots, n$. Given a set of $n+1$ nodes $0, 1, \ldots, n$, the path constraint ensures that there exists a simple path involving the nodes: each node is visited only once, having only one direct *predecessor/successor* in the path. Each node i is modeled using a domain variable $Next_i$ ($Next_i = NextArray[i]$) identifying its successor node. If $Next_i = j$, node j directly follows node i in the path.

Clearly, the constraint can be used to model the *Traveling Salesman Problem* (TSP) of n cities and its variants. The TSP is the problem of finding an

Hamiltonian circuit visiting a set of cities and minimizing the travel distance. To be more general we consider the *asymmetric* version of the TSP. The TSP can be modeled with the $Path$ constraint by conventionally selecting a starting node and duplicating it to represent the ending node.

We describe the ILP model of the path constraint and some relaxations that can be embedded in the optimization component in order to provide information on costs. The resulting constraint is an global optimization constraint and its syntax is as follows:

$PathCost(NextArray, CostArray, Z)$

where $NextArray$ is again an array of domain variables representing the next node to be visited, $CostArray$ is the cost array containing the costs of connecting each pair of nodes, and Z is the objective function. The constraint is satisfied when $Next$ variables form one simple path involving all nodes.

6.2.1 An ILP model of the `Path` constraint. We consider a graph theory model based on a directed graph $G = (V, A)$, $|V| = n$ in which there is a variable x_{ij} and a cost c_{ij} associated to each $(i, j) \in A$. An ILP model for the `Path` is as follows:

$$z(Path) \quad = \min \sum_{i \in V} \sum_{j \in V} c_{ij}\, x_{ij} \tag{5.13}$$

$$\text{subject to} \quad \sum_{j \in V} x_{ij} = 1, \qquad i \in V \tag{5.14}$$

$$\sum_{i \in V} x_{ij} = 1, \qquad j \in V \tag{5.15}$$

$$\sum_{i \in S} \sum_{j \in V \setminus S} x_{ij} \geq 1, \qquad S \subset V : S \neq \emptyset \tag{5.16}$$

$$x_{ij} \geq 0, \qquad \forall (i, j) \in A \tag{5.17}$$

$$x_{ij} \text{ integer}, \qquad \forall (i, j) \in A \tag{5.18}$$

where $x_{ij} = 1$ if and only if arc (i, j) is in the optimal solution. Constraints (5.14) and (5.15) impose out-degree and in-degree of each vertex equal to one assuring each node (city) be visited exactly once, whereas constraints (5.16), imposing strong connectivity, are called *Subtour Elimination Constraints* (SECs).

Model (5.13)-(5.18) above is also the classical model of the asymmetric TSP (see, (Fischetti et al., 2002) for a recent survey, and (Gutin and Punnen, 2002) for a book on TSP). However, it appears as global constraint in many other optimization problems (e.g., in all TSP variants) of which it is a sub-problem.

As explained in Section 4.1, a mapping is needed between CP variables and ILP variables. Thus, the assignment $Next_i = j$ in the CP model corresponds to $x_{ij} = 1$. The costs c_{ij} is the cost of the arc (i, j) in the two models and the objective function is the same.

6.2.2 Continuous (linear) relaxation.

The continuous linear relaxation of model (5.13)-(5.18) is obtained, as usual, by removing the integrality requirement on the variables, i.e., constraints (5.18). However, SECs (5.16) are exponentially many and it is not possible, for interesting values of n, to include them all and solve a single LP.

In this case, in order to solve the continuous linear relaxation of the $PathCost$ constraint, SECs must be added *on the fly*, i.e., when they are required because violated from the optimal solution of the current relaxation. This method requires the solution of the so-called *separation problem* for SECs:

Given a vector $x^* \in I\!R^n$, find a SEC which is violated by x^*, or prove that none exists.

This suggests the following procedure to solve the linear relaxation of the `PathCost` constraint:

0. relax (remove) constraints (5.16) and (5.18);
1. solve the current LP through the simplex algorithm, x^* being the optimal solution;
2. run a procedure solving the separation problem for SECs;
3. if some violated SECs are found, add them to the current LP and return to step 1. Otherwise, stop.

It is well known that the separation problem for SECs can be solved in polynomial time by computing the *minimum capacity cut* in the graph induced by x^* (see, e.g., (Padberg and Rinaldi, 1990)), and the same holds for each following LP obtained by the iterative addition of SECs[4].

As mentioned in Section 2, the linear relaxation provides the pieces of information required for cost-based propagation. However, these pieces of information can be also provided by different (structured) relaxations, and this is shown in the following sections.

6.2.3 The linear Assignment Problem.

Another interesting relaxation of model (5.13)-(5.18) is the one obtained by removing constraints (5.18) but leaving the integrality requirements (5.18). The problem obtained is the so-called linear *Assignment Problem* (AP) (see, e.g., (Dell'Amico and Martello, 1997)) and can be seen as the graph theory problem of finding a set of *disjoint* sub-tours on a digraph $G = (V, A)$ such that all the vertices in V are visited once and the sum of the costs of selected arcs in A is a minimum.

Since the coefficient matrix of the AP is totally unimodular (see, (Papadimitiou and Steiglitz, 1982)), constraints (5.18) are redundant, in the sense that the optimal solutions of the AP and of its continuous relaxation coincide. This means, among other things, that AP can be solved in polynomial time by LP techniques, e.g., the ellipsoid method (see, again (Papadimitiou and Steiglitz, 1982)), but much more effective special-purpose algorithms can be used.

One of the most effective algorithms for AP is the so-called *Hungarian Algorithm* (see, e.g., (Carpaneto et al., 1988)), and it is based on the *primal-dual* method. The time complexity of the Hungarian Algorithm is $O(n^3)$, but interestingly enough, if a single assignment (i.e., an arc of a sub-tour) of the current optimal solution is removed (recall the discussion in Section 3.1) the re-computation of the optimal AP solution only requires $O(n^2)$ time.

Even more interestingly (but not surprisingly since the discussion on the total unimodularity), the Hungarian Algorithm provides exactly the pieces of information required by the cost-based filtering: the x^* solution (which is actually integral but possibly violating some SECs), the LB value, and a reduced cost $\bar{c}_{ij}$ for each pair of nodes (i, j).

6.2.4 The Minimum Spanning r-Arborescence. As for the AP, another structured relaxation of model (5.13)-(5.18) can be obtained by removing constraints (5.14), i.e., out-degree constraints and selecting a specific node, say $r \in V$. Indeed, the resulting problem is the so-called *Minimum Spanning r-Arborescence problem* (r-MSA), i.e., the graph theory problem of finding a partial graph $\bar{G} = (V, \bar{A})$ of a given digraph $G = (V, A)$ such that the in-degree of each vertex is exactly one and there is a path from vertex $r \in V$ to each other vertex.

It is not difficult to see that this problem can be polynomially solved in $O(n^2)$ time by finding the minimum spanning arborescence rooted at node r (see, (Fischetti and Toth, 1993)), and by finally adding the minimum cost arc entering vertex r. It is worth noting that the above relaxation can provide different values of the computed lower bound depending on the chosen node r. From our viewpoint, however, the most interesting thing is that, together with the optimal solution x^* (again integral but possibly violating some out-degree constraints), and the LB value, the reduced cost vector $\bar{c}$ can be also computed in $O(n^2)$ time, thus providing all the pieces of information in order to use r-MSA as optimization component[5].

6.2.5 Adding cutting planes. The procedure described in Section 6.2.1 was aimed at computing the linear relaxation of model (5.13)-(5.18) adding SECs only when required. From another viewpoint, however, it can be seen as a strengthening method for improving the bound given by the AP. Indeed, the first call to the simplex algorithm in step 1. of that procedure computes exactly the AP bound, while subsequent additions of SECs in the following steps improve it.

This is perfectly general, in the sense that, given any family $\mathcal{F}$ of inequalities which are valid for a specific problem and given a separation algorithm able to detect inequalities in $\mathcal{F}$, these inequalities, often called *cutting planes* or cuts, can be added to an initial relaxation in order to strengthen it. This general framework is illustrated in left-hand side of Figure 5.5. Thus, even the addition

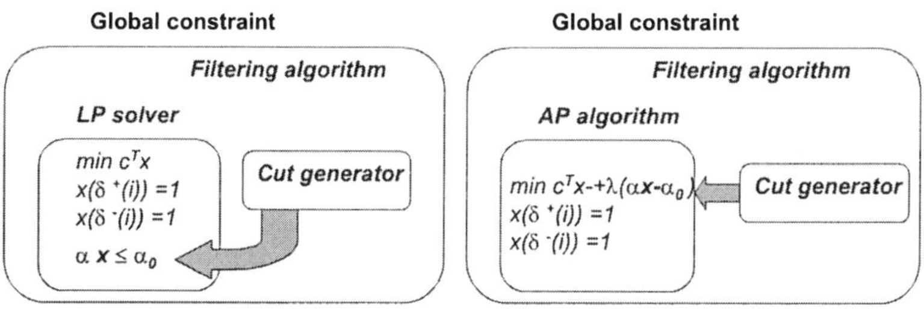

Figure 5.5. Addition of cuts in optimization constraints.

of only one cut to a structured relaxation results in a non-structured one, as in the AP case in each step 1. after the first. Then, the optimization component can only be an LP solver.

As anticipated in Section 3.2.1 and illustrated in Figure 5.3, an intermediate approach between considering a single relaxation of an overall problem or separate relaxations of its global constraints can be obtained by simultaneously taking into account the structure of more than one constraint. This is particularly suited in the context of cutting plane generation and is discussed in the following with the time constrained variant of the TSP, namely TSP with *Time Windows* TSPTW, as an example. TSPTW is the time constrained variant of TSP in which the visit of each city must be performed within a time window. Early arrivals are typically allowed, and the objective function is, in general, the minimization of the travel distance. In the TSPTW, the time windows may imply precedence constraints between nodes. These precedence constraints can be taken into account by another global constraint, that we call here $Prec$. When both the $Path$ and the $Prec$ constraints are present, instead of keeping them separate, we can exploit particular cutting planes devised for the overall problem called *Sequential Ordering Problem* (SOP) or *precedence constrained* TSP (see, e.g., (Balas et al., 1995)). In this setting, a new global constraint, say $PrecPathCost$, can be designed to take also into account information of precedences among nodes. A possible relaxation of this new global constraint is obviously the linear relaxation of the $PathCost$ constraint (by removing both integrality requirements and precedences) but this initial linear relaxation can be strengthened by separating and adding two families of inequalities which are specific for the SOP:

- *Predecessor inequalities* (π inequalities):

$$\sum_{i \in S \setminus \pi(S)} \sum_{j \in \bar{S} \setminus \pi(S)} x_{ij} \geq 1 \qquad S \subseteq V, \bar{S} = V \setminus S$$

  where $\pi(S)$ indicates the set of nodes which *precede* the nodes in S;

- *Successor inequalities* (σ inequalities):

$$\sum_{i \in \bar{S} \setminus \sigma(S)} \sum_{j \in S \setminus \sigma(S)} x_{ij} \geq 1 \qquad S \subseteq V, \bar{S} = V \setminus S$$

  where $\sigma(S)$ indicates the set of nodes which *follow* the nodes in S.

The separation problem for both π and σ families of cuts is solved heuristically[6] (see, (Balas et al., 1995)), and the cost-based filtering algorithm of $PrecPathCost$ benefits of both routing (TSP) and scheduling (SOP) viewpoints.

6.2.6 Lagrangean relaxation of cuts. A compromise allowing the exploitation of effective cuts so as the special structure of a given relaxation can be found in a technique based on the Lagrangean relaxation of the added cuts.

As usual the method is discussed by resorting to the $PathCost$ constraint as example. The structured relaxation we want to use is the AP, but at the same time we do not want to lose the chance of improving (increase) the AP bound with the addition of SECs (or, in the TSPTW case, of SOP inequalities). Thus, it is possible to apply the procedure described in Section 6.2.2, and at the end of the cut generation phase we have an LP, say AP$_S$, composed by the initial AP, and the set of added cuts S. The solution of AP$_S$, say $\bar{x}^*$, can be transformed into one, say $\widetilde{x}^*$, which is integer and has exactly the same objective function value, say LB_L, by relaxing in Lagrangean way the linear inequalities of set S. In particular, the optimal Lagrangean multipliers for the cuts are the dual values (associated to the cuts) computed by the linear solver. The problem obtained, say AP$_L$, is again an AP with a different cost vector, and, as mentioned, by solving AP$_L$ (through the Hungarian Algorithm) we have an integer solution whose value is LB_L.

More formally, consider the Lagrangean relaxation of the each cut in S in the form $\alpha^T x \leq \alpha_0$ in the objective function (5.13). We call λ ($\lambda \geq 0$) the optimal Lagrangean multiplier of the cut, i.e., the dual value associated to the cut in the LP solution[7]. The objective function of the Lagrangean relaxation becomes:

$$\min \sum_{i \in V} \sum_{j \in V} c_{ij} x_{ij} + \sum_{i \in S} \lambda_i (\alpha_i^T x - \alpha_{i0}) = -\sum_{i \in S} \lambda_i \alpha_{i0} + \min \sum_{i \in V} \sum_{j \in V} \widetilde{c}_{ij} x_{ij}$$

In the objective function, we have a constant factor and we can change the cost vector c into a new one $\widetilde{c}$ by adding the corresponding values coming from the relaxation of cuts as illustrated in the right-hand side of Figure 5.5.

The AP_L solution can be seen as an advanced starting point for the CP algorithm. This means that the cost-based filtering algorithm is from now on applied using AP_L as relaxation which modified through variable fixing derived at each step, but no more cuts are added nor Lagrangean multipliers are re-computed.

One of the main question now concerns the identification of a good point (if any) to stop the cut generation process. Indeed, in principle, there is no need of this Lagrangean relaxation of cuts, i.e., the cut generation can be done during the all search. However, our intuition beneath the Lagrangean framework is the following. We are confident, within a CP framework, to be able to explore efficiently even a huge number of branching nodes, thus we exploit cost-based propagation through a special-purpose, fast and incremental algorithm (the Hungarian method for AP). However, we want to reduce *preliminary* as much as possible the search tree by using cut generation, thus the best point to stop it is at the end of the root node.

This intuition in the TSPTW case is confirmed by the computational experiments in (Focacci et al., 2002). Other ways of integrating cutting planes within a CP framework are discussed, e.g., by (Refalo, 2000) and (Focacci et al., 2000b).

Before closing this section it is worth mentioning, however, that the Lagrangean relaxation of cuts described can have also some drawbacks during the search. The Lagrangean multipliers associated to the cuts are fixed to values that are optimal only at the root node, but could be 'far' from the optimal ones during the search, and no re-optimization (*subgradient optimization*) is performed. In particular, if during the search, a cut $\alpha^T x \leq \alpha_0$, which was tight at the root node (i.e., $\alpha^T x = \alpha_0$), become trivially satisfied by the current partial instantiation, i.e., $\alpha^T x < \alpha_0$, its contribution to the objective function is a *negative* value, hence producing a *penalty* with respect to the same solution where the cut is removed (recall that the best bound is the highest one). Thus, while at the root node, the bound produced by the Lagrangean relaxation is in general much better (higher) than the bound obtained by solving the AP relaxation (without cuts), during the search if cuts become no longer tight, the gain of the Lagrangean bound w.r.t to the pure AP one decreases. At some point, it may become worse than the pure AP bound since the Lagrangean multipliers are not re-optimized.

In any case, the bound obtained by solving AP_L is always a valid lower bound for the problem, and the above drawback can be reduced by performing some kind of *purging* in order to disable during the search those cuts that are no longer necessary, i.e., both not tight and trivially satisfied. Both general-purpose and special-purpose purging techniques are discussed in (Focacci et al., 2002).

Acknowledgements

Thanks are due to Farid Ajili and Mark Wallace for reading preliminary versions of this work and suggesting several improvements.

Notes

1. The equivalence among the ILP form (5.1) and many others (e.g., the maximization version) is trivial using linear algebra and rules for transformation can be found in Integer Programming textbooks, like (Nemhauser and Wolsey, 1988).
2. Enumerative algorithms based on the linear relaxation are, instead, typically designed for limiting the number of explored nodes by tightening the relaxation at each node through the addition of cuts.
3. Note that the linear relaxation can be considered as a special case of this framework in which the 'bad' set contains the integrality constraints.
4. The outlined procedure is a special case of classical *cutting plane methods*. The reader is referred to (Nemhauser and Wolsey, 1988) for complete definitions and algorithms.
5. Better LB values and more accurate gradient functions could be obtained combining within an *additive* framework AP and r-MSA as proposed by (Fischetti and Toth, 1992).
6. This means that in case no violated inequality is found by the separation procedure, there is no proof that such an inequality does not exist.
7. Since the cuts are in ≤ 0 form, the corresponding dual variables are non-positive, thus they must be inverted to be used as Lagrangean multipliers.

References

Ajili, F. and El-Sakkout, H. (2003). A probe-based algorithm for piecewise linear optimization in scheduling. *Annals of Operations Research*, 118:35–48.

Balas, E., Fischetti, M., and Pulleyblank, W. (1995). The precedence constrained asymmetric travelling salesman problem. *Mathematical Programming*, 68:241–265.

Beringer, H. and De Backer, B. (1995). Combinatorial problem solving in constraint logic programming with cooperating solvers. In Beierle, C. and Plümer, L., editors, *Logic Programming: formal Methods and Practical Applications*, pages 245–272. North Holland.

Caprara, A., Focacci, F., Lamma, E., Mello, P., Milano, M., Toth, P., and Vigo, D. (1998). Integrating constraint logic programming and operations research techniques for the crew rostering problem. *Software Practice & Experience*, 28:49–76.

Carpaneto, G., Martello, S., and Toth, P. (1988). Algorithms and codes for the assignment problem. *Annals of Operations Research*, 13:193–223.

Caseau, Y. and Laburthe, F. (1997). Solving small TSPs with constraints. In Naish, L., editor, *Proceedings of the Fourteenth International Conference on Logic Programming - ICLP'97*, pages 316–330. MIT Press.

Cesta, A., Oddi, A., and Smith, S. (2002). A constraint-based method for project scheduling with time windows. *Journal of Heuristics*, 8:109–136.

Dell'Amico, M. and Martello, S. (1997). Linear assignment. In Dell'Amico, M., Maffioli, F., and Martello, S., editors, *Annotated Bibliographies in Combinatorial Optimization*, pages 355–371. Wiley.

El-Sakkout, H. and Wallace, M. (2000). Probe backtrack search for minimal perturbation in dynamic scheduling. *Constraints - An International Journal*, 5:359–388.

Fahle, T. and Sellman, M. (2003). Constraint programming based lagrangian relaxation for the automatic recording problem. *Annals of Operations Research*, 18:17–33.

Fischetti, F., Lodi, A., and Toth, P. (2002). Exact methods for the asymmetric traveling salesman problem. In Gutin, G. and Punnen, A., editors, *The Travel-*

ing Salesman Problem and Its Variations, pages 169–205. Kluwer Academic Publishers.

Fischetti, M. and Toth, P. (1992). An additive bounding procedure for the asymmetric travelling salesman problem. *Mathematical Programming*, 53:173–197.

Fischetti, M. and Toth, P. (1993). An efficient algorithm for the min-sum arborescence problem on complete digraphs. *ORSA Journal on Computing*, 5:426–434.

Focacci, F. (2001). *Solving Combinatorial optimization problems in constraint programming*. PhD thesis, University of Ferrara, Italy.

Focacci, F., Laborie, P., and Nuijten, W. (2000a). Solving scheduling problems with setup times and alternative resources. In *Proceedings of the Fifth International Conference on Artificial Intelligence, Planning Systems - AIPS'00*. AAAI Press.

Focacci, F., Lodi, A., and Milano, M. (1999). Cost-based domain filtering. In Jaffar, J., editor, *Principle and Practice of Constraint Programming - CP'99*, LNCS 1713, pages 189–203. Springer-Verlag.

Focacci, F., Lodi, A., and Milano, M. (2000b). Cutting planes in constraint programming: an hybrid approach. In Dechter, R., editor, *Principle and Practice of Constraint Programming - CP'00*, LNCS 1894, pages 187–201. Springer-Verlag.

Focacci, F., Lodi, A., and Milano, M. (2002). A hybrid exact algorithm for the TSPTW. *INFORMS Journal of Computing*, 14:403–417.

Grötschel, M., Lovász, L., and Schrijver, A. J. (1988). *Geometric Algorithms and Combinatorial Optimization*. John Wiley & Sons, New York.

Gutin, G. and Punnen, A. (2002). *The Traveling Salesman Problem and Its Variations*. Kluwer Academic Publishers, Dordrecht.

Heipcke, S. (1999). An example of integrating constraint programming and mathematical programming. *Electronic Notes in Discrete Mathematics*, 1.

Hooker, J. N. (1997). Constraint satisfaction methods for generating valid cuts. In Woodruff, D. L., editor, *Advances in Computational and Stochastic Optimization, Logic Programming and Heuristic Search*, pages 1–30. Kluwer Academic Publishers.

Hooker, J. N. and Osorio, M. A. (1999). Mixed logical/linear programming. *Discrete Applied Mathematics*, 96-97:395–442.

Milano, M. and van Hoeve, W. J. (2002). Reduced cost-based ranking for generating promising subproblems. In Van Hentenryck, P., editor, *Principle and Practice of Constraint Programming - CP'02*, LNCS 2470, pages 1–16. Springer-Verlag.

Nemhauser, G. and Wolsey, L. (1988). *Integer and Combinatorial Optimization*. John Wiley and Sons.

Ottosson, G., Hooker, J. N., Kim, H. J., and Thorsteinsson, E. S. (1999). On integrating constraint propagation and linear programming for combinatorial optimization. In *Proceedings of the Workshop on integration of AI and OR Techniques in CP for Combinatorial Optimization Problems*, Ferrara, February 25-26.

Padberg, M. and Rinaldi, G. (1990). An efficient algorithm for the minimum capacity cut problem. *Mathematical Programming*, 47:19–36.

Padberg, M. and Rinaldi, G. (1991). A branch-and-cut algorithm for the resolution of a large-scale symmetric traveling salesman problems. *SIAM Review*, 33:60–100.

Papadimitiou, C. H. and Steiglitz, K. (1982). *Combinatorial Optimization*. Prentice-Hall, Inc., San Francisco.

Refalo, P. (1999). Tight cooperation and its application in piecewise linear optimization. In Jaffar, J., editor, *Principle and Practice of Constraint Programming - CP'99*, LNCS 1713, pages 375–389. Springer-Verlag.

Refalo, P. (2000). Linear formulation of constraint programming models and hybrid solvers. In Dechter, R., editor, *Principle and Practice of Constraint Programming - CP'00*, LNCS 1894, pages 369–383. Springer-Verlag.

Régin, J.-C. (1999). Arc consistency for global cardinality constraints with costs. In Jaffar, J., editor, *Principle and Practice of Constraint Programming - CP'99*, LNCS 1713, pages 390–404. Springer-Verlag.

Rodosek, R., Wallace, M., and Hajian, M. T. (1999). A new approach to integrating mixed integer programming and constraint logic programming. *Annals of Operations Research*, 86:63–87.

Chapter 6

HYBRID PROBLEM SOLVING IN ECLiPSe

Farid Ajili
IC-Parc, Imperial College London
London SW7 2AZ, United Kingdom.
f.ajili@imperial.ac.uk

Mark G. Wallace
IC-Parc, Imperial College London
London SW7 2AZ, United Kingdom.
mgw@imperial.ac.uk

Abstract Recent advances in merging Operations Research (OR) models and methods in Constraint Programming (CP) have stressed the need for programming language implementations which support and facilitate the development of hybrid solvers for combinatorial optimization problems. An important requirement on these implementations is that they distinguish the solver-independent conceptual model (one model for multiple solvers) from a design model which delegates subproblems to tailored solvers. ECLiPSe is a platform for building hybrid algorithms where different co-operative solvers are used in combination. The first part of this chapter presents some of the language ingredients which support CP-OR hybridization and how they can benefit the integration of heterogeneous solvers. The second part of the chapter illustrates ECLiPSe through an implementation of a generic hybrid algorithm applied on a general resource-constrained scheduling problem with a widely applicable objective function. We show how the hybrid search can be elegantly programmed in ECLiPSe.

Keywords: languages, constraints, mathematical programming, hybrid solvers

1. Introduction

1.1 Modelling Formalisms for Constraints and Mathematical Programming

Combinatorial satisfaction/optimization problems are commercially critical yet notoriously hard to solve. Constraint Programming (CP) and Operations Research (OR) are two alternative approaches for modelling and solving them. This chapter describes how constraint logic programming can be used to write hybrid CP-OR algorithms for solving combinatorial problems.

Historically the modelling interfaces to CP algorithms and to OR algorithms have been quite separate and very different.

Mathematical modelling formalisms have been used for some years as the end-user interface to mathematical programming solvers which employ variations on the Simplex and Barrier methods (Dantzig, 1963; Williams, 1999). These formalisms are highly structured, and very compact. Every variable is numeric, and every variable is global. There is no clustering of the variables or constraints. There is no hierarchical structuring.

CP languages are more complex and less compact (Marriott and Stuckey, 1998; Van Hentenryck, 1989). Besides the built-in constraints, these languages enable the user to construct new constraints using predicates. Models are constraint programs, and include predicates which may be defined in terms of other predicates. Variables may be local to a predicate. Constraint handling is, traditionally, limited to consistency techniques (Van Hentenryck, 1989; Tsang, 1993). Search is controlled by programmer-encoded strategies.

We shall show how a constraint programming language can be interfaced to both kinds of constraint solvers: mathematical programming solvers and consistency techniques. Moreover we shall show that having a common interface to the different kinds of solvers enables us to combine the solvers and build hybrid algorithms that can outperform the separate solvers.

1.2 Requirements on Hybrid Modelling Formalisms

Constraint programming has the expressive power to provide a front-end to mathematical programming packages. The trouble is that it has too much expressive power. It can model symbolic constraints, and logical combinations of constraints, for example, that cannot be handled within a mathematical programming framework.

A system has been implemented which maps constraint programming models onto mathematical programming packages by automatically converting symbolic constraints and logical combinations of constraints into integer/linear constraints (Rodosek et al., 1999). However there are many cases where such constraints are better handled by consistency techniques than by integer linear programming solvers.

The requirement is for a formalism sufficiently expressive to capture all constraints expressible as predicates, sufficiently flexible to map the different constraints to the solvers which can best handle them, and sufficiently integrated to enable the search routine(s) to communicate with all the different solvers, and exploit the solvers to drive the best possible search heuristics. This formalism should be simple enough that a user can "plug and play" with different combinations of solvers and search, so they can experiment with different hybrids until he or she has found the most efficient and scalable solution.

1.3 ECLiPSe and Piecewise Linear Probing

At IC-Parc we have used the ECLiPSe system (Group, 2002) to support modelling and solving of large scale combinatorial optimization problems using hybrid algorithms. ECLiPSe includes a domain solver *ic* that admits finite and infinite domains, and offers a common interface to the major commercial integer programming packages (through the *eplex* library). We shall give some examples showing how ECLiPSe meets the requirements outlined above.

We shall show how these facilities are exploited in the design and implementation of a complex hybrid algorithm which includes a linear solver handling linear relaxations of the problem, a solver called *repair* that performs label propagation on the values the variables take at the relaxed optimum solution, and consistency techniques handled by the *ic* solver. The search algorithm uses the linear relaxation both for heuristics and for the stopping criteria. It is hoped that this example will illustrate the advantages of the hybridizations made possible by combining CP and mathematical programming. Moreover we hope the example will point up the huge potential for further research into the discovery and exploitation of novel techniques for hybridization.

This chapter is organized as follows:

- Section 2 outlines the characteristics of CP and OR and briefly mentions some of the potential benefits of their merging.

- Section 3 presents some of the requirements on a "hybrid language" such as ECLiPSe. We basically focus on how these requirements can be supported by the language.

- Section 4 presents the constraint logic programming language ECLiPSe as a platform to accommodate hybrid CP-OR solvers. We show that the language support meets the requirements previously stated.

- Section 5 presents a probe-based hybrid algorithm for a generic problem: resource-constrained scheduling with piecewise linear optimization. The implementation of the algorithm, which can be adapted to a wide range of combinatorial optimization problems, illustrates the use of the ECLiPSe ingredients for programming a hybrid search.

- Section 6 concludes the chapter.

2. Integration of Constraints and Operations Research

CP and OR are two alternative approaches for modelling and solving combinatorial satisfaction/optimization problems. After briefly presenting their main characteristics and some potential benefits of their merging, we outline the forms of hybridization that will be illustrated and exploited in the rest of this chapter.

2.1 Constraint Programming

Because of the NP-hardness of many combinatorial optimization problems, it is very unlikely that a *general* polynomial-time solving technique will ever be found for them. In many areas of combinatorial optimization, specializing general-purpose algorithms for the problem at hand and exploiting the problem structure can lead to significant and sometimes spectacular efficiency improvements.

CP is a well-established paradigm that facilitates the design and specialization of such algorithms (Marriott and Stuckey, 1998). This paradigm is distinct from others, since it seeks to focus firmly on the use of consistency techniques (Tsang, 1993; Van Hentenryck, 1989) to tackle the satisfaction of symbolic and disjunctive constraints. Indeed, these constraints are at the heart of the problem complexity. A CP method usually relies on a tailored backtrack search strategy which is assisted by constraint propagation to infer new information about decision variables (e.g., pruning domains, detecting fixed variables, tightening bounds).

CP languages, such as ECLiPSe, offer an expressive and compact framework, where global constraints (e.g., *alldiff*, *cumulative*, *cycle*, *flow*) capturing interesting substructures in the model are modelled in a very natural way which is very close to the problem description. These high-level constraints support a more global satisfaction reasoning and, therefore, a sharper propagation (See Chapter 4 for more details). In a CP system, global constraints communicate only through variables' domains. This makes the algorithm prototyping and development easier since each newly added global structure needs only to implement its own constraint propagation rules regardless of existing constraints.

2.2 Operations Research

OR techniques, which are presented in Chapter 1, are often rooted in integer linear programming (Dantzig, 1963; Williams, 1999).

They target the use of mathematical models to analyse and solve decision problems. Problems are formulated as mathematical models and are solved to determine the best (optimum) decision among a combinatorial set of alternative decisions. The quality of each decision is measured by a cost function. The expressivity of the mathematical model is severely restricted. In addition to the convexity of the cost function, it requires the (possibly unbounded) variables' domains to be numeric and the constraints to be expressed as a finite conjunction of linear arithmetic equations and inequations. Integer linear programming approaches take advantage of desirable structural properties (e.g., unimodularity, polyhedral properties) of the solution set of a model. By contrast to CP which counts on local consistency, efficient linear programming techniques perform powerful global feasibility/optimality reasoning by using the strengths of bounding, relaxation and duality techniques. Often, they offer reliable search guidance (e.g., relaxed cost, reduced costs, shadow prices) and sharper bounds that speed up the convergence towards a provably good/optimal solutions.

2.3 Hybridization context

Several publications have compared OR and CP on modelling capabilities as well as performance considerations, among others (Ken Darby-Dowman et al., 1997; Rodosek et al., 1999; Smith et al., 1995; Van Hentenryck and Carillon, 1988). These comparisons and several attempts to combine them, have supported the (by now prevalent) view that their hybridization can lead to an especially effective solution to some combinatorial optimization problems which outperforms stand-alone CP and/or OR methods. Of course, this success relies on preserving their corresponding advantages.

The benefits of a tight hybridization of CP and OR methods are increasingly recognized and the integration of OR techniques has been placed at the top of CP research agenda. In recent years, researchers have successfully established evidence that their merging is beneficial in many problem domains both at the modelling and at the solving levels. This evidence is presented in many publications including (Beringer and De Backer, 1995; Bockmayr and Kasper, 1998; El Sakkout and Wallace, 2000; Focacci et al., 1999; Gervet and Wallace, 2001; Hooker et al., 1999; Milano et al., 2000; Ottosson et al., 1999; Ottosson et al., 2002; Refalo, 1999; Rodosek and Wallace, 1998).

In (Hooker et al., 1999; Ottosson et al., 1999; Ottosson et al., 2002) it is stated how the complementary weaknesses and strengths of CP and OR enable their integration to benefit the solving of combinatorial optimization. Several attempts to combine OR and CP have focused on how symbolic mixed constraints operating on shared variables can be exploited to bridge the gap between the two sides to tighten relaxation and to make it better than in integer linear programming (Bockmayr and Kasper, 1998; Milano et al., 2000; Ottosson et al., 1999; Refalo, 1999).

This paper starts from the perspective which considers that a hybrid algorithm comprises:

- A global (backtrack) search procedure which controls and co-ordinates the search.
- One or more constraint solvers that, after each search step, perform co-operative reasoning to detect infeasibility, or extract information to guide and/or prune the search. These solvers act in support to the search routine on specific and well-controlled subproblems. They do not implement a simple preprocessing and they interact with search during execution: these subproblems are dynamically re-shaped.

This hybridization form does not cover all other hybrid algorithms available in ECLiPSe, but for reasons of clarity we will focus on hybrid algorithms which fit this pattern. The solver detailed in Section 5 obeys this form.

3. Language Ingredients for Hybrid Solvers

3.1 Conceptual Models and Design Models

To express any combinatorial optimization problem the modelling formalisation must include:

- Decision variables
- Values (which can be assigned to the variables)
- Constraint definitions
- Constraint impositions.

This formalism enables us to simply state the problem, without necessarily making any commitments as to how it will be solved. We term such a simple problem statement a "conceptual" model of the problem.

In this section we discuss what is needed to map a problem, formalized as a conceptual model, down onto an algorithm, and how to make it possible to map the same conceptual model onto different alternative algorithms. Most importantly, since our focus is on hybrid algorithms, the language must be able to map a conceptual model onto different constraint solvers and search routines in a way that allows them to cooperate in finding solutions. We term result of mapping the conceptual model down to an underlying (hybrid) algorithm, the "design" model.

3.2 Context

Historically the formalisms for stating and solving combinatorial optimization problems have been closely tied, and there has not been a clear distinction between the conceptual model and the design model. In mathematical programming, for example, the constraints available for stating and solving a problem are deliberately limited. The constraint definitions (for linear numeric constraints and integer constraints) are built into the language, and therefore there is no need to define new constraints. Consequently the mathematical modelling languages have no concepts supporting constraint definition.

Correspondingly there is a choice of a few alternative solvers for handling linear constraints, and integrality constraints are handled by search. The design model simply maps all the linear constraints to the selected solver, and offers further parameters to control the search.

Constraint programming languages, historically, do offer the facility to define a new constraint as a finite set of value tuples. The handling of these constraints is, typically, by one of a standard set of propagation algorithms. Other constraints, for example numeric constraints, have a built-in constraint behaviour.

3.3 Creating a Design Model - Introduction and Example

In principle any conceptual model can be transformed into a design model by:

- Associating one or more constraint solvers with each constraint
- Defining a search algorithm which adds, or changes, variable assignments or constraints at each step
- Defining the conditions under which each solver is triggered.

To support this design model the language implementation must then:

- Communicate the necessary information between the different solvers
- Communicate the necessary information between the search engine and the solvers
- Control the search engine and solvers.

As a toy example consider the problem "Find X where $X + 2 * Y + Z = 10$ and $2 * Y + Z = 5$, and where all the variables are non-negative integers". Assuming numerical expressions, equations and inequations are built into our language, the conceptual model can be written as:

```
X >= 0, Y >= 0, Z >= 0,
```

```
integer(X), integer(Y), integer(Z),
X+2*Y+Z = 10, 2*Y+Z = 5
```

Clearly the solution is $X = 5$, but for illustration we can show two design models for this problem.

First suppose we have a solver called SIMPLEX that handles linear constraints, and a search routine called MIP, that takes a set of integer variables as a parameter. The MIP search alternately calls SIMPLEX and branches on non-integer variables.

The SIMPLEX/MIP design model is as follows:

```
send {X >=0, Y >= 0, Z >=0,
      X+2*Y+Z = 10, 2*Y+Z = 5 } to SIMPLEX
search MIP {X,Y,Z}
```

To run this design model, the language implementation must enable the solution to the linear problem found by SIMPLEX to be exported to the MIP search routine.

Now suppose we have another solver called INTERVAL that performs interval reasoning on numeric constraints, and for integer variables tightens their interval bounds to integer values. Suppose we have a search routine SPLIT that splits the domain of each variable in half until a given tolerance (we will use 0.5 in this example).

A design model using both solvers is as follows:

```
send {X >= 0, Y>=0, Z>=0,
      X+2*Y+Z = 10, 2*Y+Z = 5 } to SIMPLEX and INTERVAL
send {integer(X), integer(Y), integer(Z)} to INTERVAL
search SPLIT 0.5 {X,Y,Z}
```

To be able to execute a hybrid design model such as this, the underlying system must support communication between the solvers. Otherwise the SIMPLEX solver would repeatedly solve the linear constraints without tightening the bounds, and the INTERVAL solver would reduce the ranges of X, Y and Z to respectively 1..9, 0..2 and 1..5, but would not be able to make any further inference when the range of X was reduced by the SPLIT search routine.

For this hybrid algorithm the underlying system must:

- Propagate the upper and lower bounds from INTERVAL to SIMPLEX.
- When SIMPLEX detects infeasibility, to propagate failure to the SPLIT routine.

3.4 Information Passed between Solvers

In this section we outline what information must be communicated between what solvers in order to support different kinds of hybrids.

Firstly, variable upper and lower bounds can be handled, and tightened (i.e. propagated), by both linear solvers and domain solvers. Thus when one solver (linear or domain) tightens a bound, this information is passed to the other solvers.

Cooperation between linear and domain solvers can also be achieved through:

- Global constraints achieving stronger domain pruning and tighter linear relaxations (i.e., see for instance (Bockmayr and Kasper, 1998; Milano et al., 2000; Ottosson et al., 1999; Refalo, 1999)).
- Variable fixing (Williams, 1999).
- Sensitivity analysis (i.e., inference based on reduced-costs for instance (Focacci et al., 1999; Milano and van Hoeve, 2002)).
- Master/subproblem decompositions (i.e., Column generation (Junker et al., 1999) and Benders decomposition (Eremin and Wallace, 2001)).
- Lagrangian relaxation which exploits the duality to approximate the subproblem constraints within the cost function of the master problem (Benoist et al., 2001; Sellmann and Fahle, 2003).
- Cutting planes procedures which derive valid linear constraints to the domain solver (Focacci et al., 2000; Refalo, 1999).

The *global constraints* are domain constraints, whose active behaviour tightens the bounds on the variables. They are designed to use (some of) the same variables that are used in the integer/linear model. Thus variables which must be introduced for the purposes of integer/linear modelling can also be made available to the domain solver. An example of such a global constraint, for *piecewise linear* modelling, is discussed later. The language support needed for global constraints is:

- Shared variables between the domain solver and the linear solver, or the facility to dynamically propagate information between different variables playing the same modelling role in the different solvers.
- Domain constraints that involve both discrete and continuous variables (reflecting the two kinds of variables occurring in an integer/linear model).

Variable fixing requires the fixed value to be communicated from the linear solver to the domain solver.

Sensitivity analysis is another mechanism by which qualitative information (e.g., reduced costs, dual values) produced by the linear solver, is used to derive domain reductions in the domain solver.

Master/subproblem decompositions require the programmer to explicitly distinguish the different constraints and subproblems, and to specify what information is passed to the linear solver at each iteration.

Lagrangian relaxation requires the programmer to identify which constraints to relax, and how to systematically update the cost function coefficients (or multipliers) at each iteration.

Cutting planes require new linear constraints, derived by the linear solver, to be extracted by the interface routine, and posted to the domain solver.

3.5 Combining Solvers and Search Engine

3.5.1 Information Passed from Search Engine to Solvers. A constructive search step is a step at which a new constraint, or set of constraints, is posted to one or more solvers. For this the search engine must be able to post constraints to any set of solvers.

Greedy search uses only constructive steps. Tree search uses constructive steps down any single branch, but when moving up, or "across" the tree, the tree search algorithm instead returns to a previously encountered node. In this case it restores the constraints to the state they were in when the node was previously encountered. To support tree search, the search engine must be able to restore the constraint solvers to a previous state, either via some stored record of that state, or by re-posting all the same constraints to the same solvers again.

A local search step involves both the removal of a set of constraints from the solvers, and the posting of another set of constraints. Any further information propagated from the newly removed constraints must also be removed. In case different solvers are communicating propagated information, this must also be removed. Because this removal is expensive to implement, local search is typically only supported in conjunction with solvers which perform limited propagation. An example is to admit label propagation only, and where propagation "loops" are precluded. This kind of propagation can be used to maintain *invariants* (Michel and Van Hentenryck, 1997).

3.5.2 Information Passed from Solvers to Search Engine. The search engine can use a wide variety of information from the solvers, such as:

- Variable assignments (produced by domain propagation, or variable fixing)
- Value quality estimates (e.g., based on reduced costs as in (Milano and van Hoeve, 2002), number of constraints violated by that assignment, etc.)

- Estimates about the most difficult constraints to solve/variables to assign (e.g. based on duality information from the linear solver, domain sizes from the domain solver, etc.)

In this chapter we will encounter a particular form of information used by the search engine, which we call the "tentative value" of a variable. It may conflict with the constraints in one or more solvers. It may be returned by a solver, as a side effect of constraint solving. For example a linear solver returns a particular (tentative) optimal assignment of values to variables. This tentative assignment may violate the non-linear problem constraints that are not handled by the linear solver.

3.6 Requirements on the Language used for Design Models

3.6.1 Variable Attributes. All these forms of cooperation, or "hybridization forms", can be supported by adding procedures in the language which extract, process and forward the relevant information.

The procedures could forward this information by updating the data structures in the target solver or the search procedure. In this case, each new solver and each new search routine would require new procedures to forward information to it, *from each other solver*.

Instead some constraint programming languages allow information to be attached to each problem variable. This information serves as an external interface to the different solvers and the search engine. Information forwarded from the solvers is sent to the variable, where it is recorded as an *attribute* of the variable. Other solvers, and the search engine, then access the relevant attributes of the variable. In principle the forwarding of information to the variable attributes can be done eagerly, as soon as the information is available, or by need, whenever another solver or search procedure accesses the attribute.

3.6.2 Control Requirements. These procedures must be invoked at appropriate times during problem solving. The control issue - when to invoke which procedure - is complex, as reflected in the ongoing research on control of hybrid solvers as in (Monfroy, 1998) for instance.

If this control issue was explicit in the design model, then every time the user changed the search routine, or added another solver, new procedure calls would have to be added and inserted into the best positions in the search procedures and propagation iterations. Consequently the design model would have to undergo a quite radical rewrite.

To maintain a clear and natural relationship between the conceptual model and the design model, it is necessary to keep the control issue *implicit*. This leads to an important requirement on design model formalisms and their implementation. The user should simply name the hybridization forms, in the design model, and the implementation should manage the control issues.

This implicit control is managed by the constraint programming paradigm of constraint-driven *demons*. A demon is a procedure that wakes as soon as a certain condition holds, typically as a result of a constraint being added to a solver, and the subsequent handling of the constraint by the solver. On completing its execution, a demon re-suspends itself, if necessary.

These demons are perfectly adapted to communicating information between solvers, automatically as soon as it becomes available. Moreover they are also ideal for communication between the solvers and the search routine. At each step, the search routine posts constraints to solvers, and the demons then return the required information to the search routine.

Constraint programs also use demons to implement constraint handling behaviour. Just as constraint handling causes events, which wake communication demons, so also communication causes events (the arrival of new information) which wake constraint handling demons.

In programs involving a lot of demons, a single event may wake many of them simultaneously. Some of the woken demons may be expensive to execute and yield relatively little information, others may be computationally cheaper and more useful. For example the demon which copies all the relaxed optimum values from the linear solver to the search engine may be very expensive, if there are many variables. However the demon which tightens the domain of a single variable may be very cheap, and if it detects inconsistency it could save the expensive copying demon from needing to be executed at all. In this case the best performance would be achieved by scheduling the domain tightening demon to execute before the value copying demon, and not the other way around.

To ensure that the demons are executed in the right order, some form of priorities amongst the demons should be specified. This may be built into the demons themselves, or given by the user in case (s)he desires more fine-grained control. Typically, the priorities might be used to require that all the cheap subproblems are solved before the expensive ones (e.g., bound propagation or cut accumulation is performed before invoking the linear subproblem solver).

3.7 Infeasibility Detection

Correctness guarantees for the design model mean that any assignment returned as a result of running the design model must satisfy all the constraints stated in the conceptual model. To guarantee correctness, therefore, the constraint logic programming paradigm does not admit infeasible assignments.

If, as a result of a search step, any solver detects infeasibility, this search step is immediately undone. The design model should then link subproblems in a way which allows passing infeasibility detection as search progresses.

Local search, and hybrid algorithms, often make assignments which violate some of the constraints in some of the solvers, as in (Kamarainen and El Sakkout, 2002) for instance. This can be supported, even within a framework that guarantees correctness, by using the variable attributes to record the assignment, rather than instantiating the variable. During search some variables may be instantiated, while other may only have one or more "tentative" assignments recorded as attributes of the variable.

Tentative assignments may violate the constraints in the different solvers, without these solvers reporting infeasibility.

4. ECLiPSe as a Platform for Building Hybrid Algorithms

This section introduces the Constraint Logic Programming (CLP) platform ECLiPSe (Group, 2002). ECLiPSe is designed to be more than an implementation of CLP: it also supports mathematical programming and stochastic programming techniques. The crucial advantage of ECLiPSe is that it enables the programmer to use a combination of algorithms appropriate to the application at hand. The platform offers a conceptual modelling language for specifying the problem clearly and simply, in a way that is neutral as to the algorithm which will be used to solve it. Based on the conceptual model it is easy to construct alternative design models, also expressed in ECLiPSe. A design model is a runnable program, whose

execution employs a specific combination of algorithms. Thus the platform supports experimentation with different hybrid algorithms.

Various different constraint handling facilities are available as ECLiPSe libraries. We will introduce the following three libraries below:

- The domain solver *ic*.
- The linear solver interface *eplex*, which interfaces ECLiPSe to external linear constraint solvers, including Xpress-MP (Optimization, 2001) and CPLEX (ILOG, 2001).
- The *repair* solver which performs label propagation on tentative values.

The same constraint can be treated concurrently by several different handlers.

ECLiPSe supports both constructive and local search, and allows them to be combined into hybrid search techniques. It is designed to support the fast development of specific hybrid algorithms tuned to the problem at hand. It is not assumed that the first algorithm implemented by the application developer is guaranteed to be the best one: rather ECLiPSe provides a platform supporting experimentation with different hybrid algorithms until an appropriate one is found which suits the particular features of the application. The tutorial (Group, 2002) provides an overview of ECLiPSe, its modelling facilities, the different solvers available, and the program development environment.

In the next section, we shall explore ECLiPSe as a problem modelling language by illustrating the conceptual and design models of the previous example.

4.1 Modelling in ECLiPSe

Constraint logic programming (CLP) is an excellent formalism for conceptual modelling. For each feature in section 3.1 above, there is a CLP construct:

- **Decision variable:** Logic variable
- **Values:** Variable domain
- **Constraint definition:** Predicate definition
- **Constraint imposition:** Predicate invocation (i.e., goal).

To map a conceptual model to a design model, it is only necessary to send each constraint to one or more appropriate solvers, and to add a search routine. For the toy example above the conceptual model is:

```
toy_conceptual(X) :-
    [X,Y,Z]::0..inf,
    X+2*Y+Z = 10,
    2*Y+Z = 5.
```

In this conceptual model, the fact that the variables are non-negative integers is expressed in their domain specification.

To turn this into the second design model given above, we simply send the constraints to the chosen solvers and add the required search routine:

```
toy_design(X) :-
% Send constraints to solvers
    ic:([X,Y,Z]::0..inf),
    [ic,eplex]:(X+2*Y+Z =:= 10),
    [ic,eplex]:(2*Y+Z =:= 5),
% Set up external linear solver
    eplex_solver_setup(min(0), _, [], 5, [bounds]),
% Search for values for X, Y and Z
    Vs=[X,Y,Z],
    search(Vs,0,input_order,indomain_split,complete,[]).
```

The domain specification, including the integrality constraint on the variables, is sent to the *ic* solver. The linear constraints are posted to both *ic* and *eplex*. The variable bounds are automatically communicated between the different solvers. The external linear solver is then initialised by *eplex_solver_setup* with a constant optimization function (*min(0)*), and some other control information.

The search routine performs domain splitting on X, Y, Z until a value is found. Both the interval and linear solvers are woken by changes in the bounds of any of the variables imposed by the search routine. If at any point either solver detects an infeasibility, the search automatically backtracks to the last choice point and tries another alternative.

4.2 The Domain Solver: *ic*

The *Interval Constraints* library *ic* implements a unifying solver for two kinds of constraints:

- finite domain constraints
- interval constraints

Each constraint handled by *ic* is dealt with separately: the only communication between them is via the bounds of their shared variables. The class of mixed constraints that can be expressed and propagated in *ic* is wider than the one allowed by pure arc consistency or interval constraints. It therefore allows the application of tailored propagation algorithms to well-studied subproblems.

ic supports integer and real variables, which can be initialised by invoking `integers(Vars)` or `reals(Vars)`. A real or integer domain can be associated with one or more variables by giving its lower and upper bounds, or in the case of an integer variable, a list of integers. Either bound can be infinite, and a domain can be constructed from a sequence of subranges thus: `Var :: [0, 10..inf]`. Mathematical comparisons, such as `ic:(2*X-Y > X*X-4)` are supported.

ic constraints can be either:

- logically combined like:

  `ic:(X>Y => X>Y+3.35),`

- or *reified*, turning them into boolean-valued expressions, e.g.:

  `ic:(TruthValue =:= (X>Y+3)),`

- or "mixed", e.g.:

  `ic:(B=:=((Start =< Time) and (Time < Start + Duration))).`

ic supports a variety of global constraints, such as *alldiff* and *element*. Additional functionality (such as global constraints for scheduling, and branch and bound search) are available from other libraries.

For real numbers, *ic* provides safe rounding (i.e., guaranteed bounds),

```
?- integers(X), reals(Y), X::0..10,
   ic:(X>sqrt(Y+8)), ic:(Y>X-sqrt(2)).
```

```
Y = Y{2.59 .. 92.0}
X = X{4 .. 10}
```

4.3 The linear solver interface: *eplex*

The *eplex* library enables external mathematical programming solvers, such as Xpress-MP and CPLEX, to be integrated into an algorithm encoded in ECLiPSe.

A constraint is posted to *eplex*, in the same syntax as it is posted to *ic*, `eplex:(2*X-Y >= 3)`. This makes it easy and natural to post the same constraint to both solvers, thus: `[ic,eplex]:(2*X-Y >= 3)`.

An extra facility of *eplex* is support for multiple linear problem instances. This is done simply by giving the instance a name, such as *my_inst*, and declaring it as a solver instance: `eplex_instance(my_inst)`. Posting constraints to the instance is simply done by: `my_inst:(X-Y>=1)`.

At any time before, or during constraint setup, the linear solver, or solver instance, can be setup using `eplex_solver_setup`. It can then be explicitly called using `eplex_solve(Opt)`, or `my_inst:eplex_solve(Opt)` for my specific instance. Alternatively the solving can be performed by a demon, in response to events specified when the solver is set up, as in the example:

```
eplex_solver_setup(min(X+3*Y),Opt,[method(dual)],5,[bounds])
```

The parameters are as follows:

1 the optimization function to be minimized (`min(X+3*Y)`)

2 the optimal value (`Opt`)

3 a list of options, all of which are defaulted in this example, except the algorithm to be applied by the external solver ([`method(dual)`])

4 the priority of the `eplex_solve` demon (5)

5 the triggering events (here a change in variable bounds wakes the demon called `eplex_solve`).

The external solver solves a linear relaxation of the problem unless integrality constraints are explicitly posted to *eplex* or the solver instance, e.g. `my_inst:integers([X,Y,Z])`. If any variables are constrained to be integral, in this way, then the external solver will use its own MIP search mechanism to find an optimal solution which meets the specified integrality constraints.

Information can be extracted from the external linear solver, or instance, using a set of standardized access functions, e.g.,
`my_inst:eplex_var_get(Var, solution, Val)`
extracts the value Val of the variable Var at the current optimum.

4.4 The *repair* solver

With the *repair* library each variable can be given a *tentative* value. This tentative value can be set repeatedly to different values during forward search, not just on backtracking. The value is set using, for example, `Var tent_set 3`, and retrieved using `Var tent_get Val`.

The constraint $X > 2$ can be posted to the repair library:
`(X>2) r_conflict my_conflicts.`
In this example, `my_conflicts` names a key against which conflicts involving this constraint will be recorded, so allowing the conflict constraints to be separated into different groups. To elicit the conflict constraints associated with this key we write:

```
conflict_constraints(my_conflicts,List)
```

A functional constraint, such as $Y = X+1$, can either be monitored for conflicts using `(Y =:= X+1) r_conflict my_conflicts`, or it can be handled by propagating tentative values from the right-hand-side to the left-hand-side, by writing this: `Y tent_is X+1`. If the tentative value of X is set to 3, now, the tentative value of Y will automatically be set to 4. Under this treatment the constraint becomes an *invariant* (Michel and Van Hentenryck, 1997).

Using the repair library, tentative values can be used to focus a tree-search on conflict variables. Each search step uses *repair* to access the violated constraints (using `conflict_constraints(key,ConfList)` for each relevant key `key`). Given a violated constraint selected from the list, it chooses a variable from that constraint and labels it. If at any point there are no constraints in conflict, the remaining variables can all be labelled to their tentative values.

Constraints can be sent both to the repair library and to other constraint solvers to build hybrid algorithms. In this way the same constraint can be monitored for conflicts, with respect to the tentative values of its variables, and used for propagation on the domains of the same variables.

4.5 Attributed Variables and Demons in ECLiPSe

Every variable in an ECLiPSe model has a number of attributes associated with it that can be used to hold information about the variable and the constraints imposed on it.

Which attributes there are depends upon which solvers are being used by ECLiPSe. If only the interval solver *ic* is being used, for example, then the only attributes present are those associated with the *ic* solver. These attributes record the domain of the variable and its size, as well as lists of demons that should be woken when the domain is tightened in different ways. If *eplex* is also being used, then attributes associated with the *eplex* solver will also be present. Associated with the *repair* solver is an attribute called the "tentative value" of the variable.

In the algorithm presented in Section 5 of this chapter, the variable values at the relaxed linear optimum computed by the linear solver *eplex* are recorded as tentative values of the variables. This is specified in the ECLiPSe design model by associating a value forwarding procedure with the linear solver, within the *eplex_solver_setup* call.

We present the code of this forwarding procedure to illustrate how information is extracted from the linear solver and used to update the tentative values of the shared variables:

```
set_ans_to_tent :-
          eplex_get(vars,Vars),
          eplex_get(typed_solution,Solution),
          Vars tent_set Solution.
```

The procedure *eplex_get* extracts the specified information from the external linear solver: in this case we extract the complete set of variables known by the linear solver, and the complete relaxed solution. Now this solution is simply used to update the tentative values of all the variables in this set, by the procedure *tent_set*.

The algorithm presented later in this chapter is a hybrid involving three solvers: *ic*, *eplex* and *repair*. Each solver has associated demons, to propagate domain tightenings in *ic*, to wake the external linear solver in *eplex*, and to maintain the invariants in *repair*. The order in which these demons wake and execute, between search steps, does not affect the correctness of the algorithm, but it does influence its efficiency.

The best performance is achieved when all *ic* propagation is completed before the linear solver executes, and the invariants are only maintained after *ic* propagation and linear constraint solving is complete.

This behaviour is achieved in the ECLiPSe design model by assigning lower priorities to the *eplex* and *repair* demons than to the *ic* demons. For example the priority of the *eplex* demon is set in *eplex_solver_setup* as described above.

5. Programming a Hybrid Search in ECLiPSe

In this section, we present an hybrid ECLiPSe algorithm in the context of dynamic scheduling arising in a commercial transportation application. Its details can be found in (Ajili and El Sakkout, 2003; Ajili and El Sakkout, 2001). This work could be seen as a generalization of (El Sakkout and Wallace, 2000), which considered a convex minimal perturbation objective function, to handle any possibly non-convex cost function provided that it is piecewise linear. The piecewise linear objective criteria captures most of the common scheduling cost

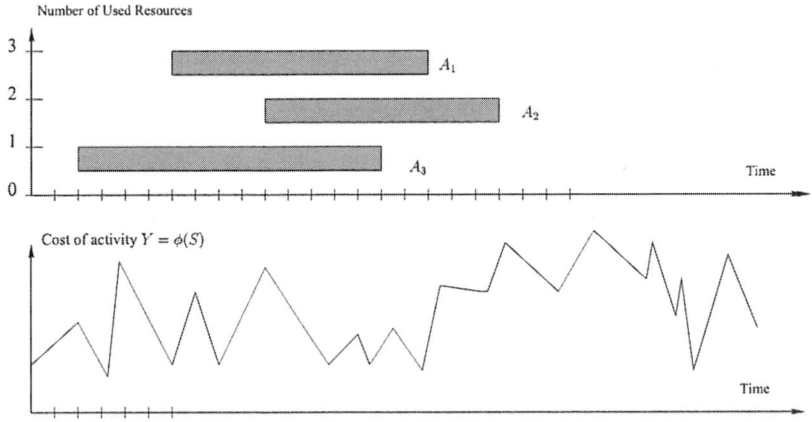

Figure 6.1. The input schedule and the PL cost function ϕ of an activity.

functions, such as minimal perturbation (El Sakkout and Wallace, 2000) and earliness/tardiness criteria studied in (Beck and Refalo, 2003).

We show how the design and development of the algorithm exploit the "glue" provided by the ECLiPSe language to tightly merge the *ic*, *eplex* and *repair* solvers presented at Section 4.

5.1 An Illustrative Example

This section briefly illustrates the problem of scheduling with piecewise linear (PL) optimization through a simplified example. We consider the operational schedule whose resource profile is depicted in Figure 6.1. It consists of 3 activities A_1,A_2 and A_3. They utilize three *unit-capacity* resources over a discrete time axis. The resources are uniform machines: there is one resource type. We assume that each activity requires a single resource. Let tu denote a time unit in this section. In addition, it is required that the placement of activities over time satisfies the following temporal constraints:

T_0) The respective durations of A_1, A_2 and A_3 are 11, 10 and 13.

T_1) The *distance* between the starts of A_1 and A_2 is smaller than 3 tu.

T_2) A_3 must end 6 tu after the start of A_2.

T_3) A_2 must end strictly after A_1.

In addition, it is assumed that the cost A_i depends only on its start time and this dependency is captured by a PL function ϕ_i like, for instance, the curve ϕ of Figure 6.1. The schedule cost is additive: it equals the sum of the 3 activity costs. Clearly, the schedule requires 3 machines. The schedule is *feasible*: it respects the resource capacity limit (3) and satisfies T_0, T_1, T_2 and T_3.

Due to changes in the schedule dynamic environment (e.g., new business regulations, unforeseen events like machine failures, etc), the following new scheduling requirements have to be met:

R_1) Only two resources are available (i.e., saving one machine).

R_2) Meet the relations T_0 and T_1 (T_2 and T_3 are retracted), plus the new relations:

- T_4) The start of an activity should differ from the "old" one by up-to 5 tu. This sets a "time window" for an activity.
- T_5) A_1 must start not more than 10 tu after the end of A_2
- T_6) A_3 should end less than 35 tu after the start of A_2.

Because of insufficient machines and/or new requirements, the original schedule is no longer feasible. An algorithm for this scheduling problem with PL optimization seeks to either prove the infeasibility of the conjunction of R_1 and R_2, or restore the schedule feasibility in an optimal way. By re-timing the activities, the algorithm will find, if there is any, the schedule which minimizes the cost.

This problem can be formulated as a constraint satisfaction and optimization problem. Let S_i (resp. E_i) denote the start (resp. end) time variable of activity A_i. Let s_i (resp. e_i) be its "old" start (resp. end) time in the original schedule. The goal is then to minimize $\Phi = \sum_{i=1}^{3} Y_i = \sum_{i=1}^{3} \phi_i(S_i)$ subject to three sets of constraints:

1 The *temporal constraints*:

T_0: $E_1 - S_1 = 11, E_2 - S_2 = 10, E_3 - S_3 = 13$
T_1: $S_2 - S_1 \leq 3, -3 \leq S_2 - S_1$
T_4: $\forall i \in \{1,2,3\}, s_i - 5 \leq S_i, S_i \leq s_i + 5$
T_5: $S_1 \leq E_2 + 10$
T_6: $S_2 + 35 \geq E_3$.

2 The *PL constraints* $Y_i = \phi_i(S_i)$, one per activity. They state that the point (S_i, Y_i) should lie on the PL curve ϕ_i.

3 The disjunctive *resource constraint*:

$$resource(Activities, NR)$$

where NR is the resource limit and $Activities$ is the list of activities. Each element of $Activities$ has the form *task(start_var,duration)*. This global constraint states that at each time point of the horizon, no more than $NR = 2$ elements in $Activities$ can be in progress simultaneously.

5.2 Problem Modelling

As illustrated previously, a scheduling problem with PL optimization, named Sched, is a conventional scheduling problem and a system of PL constraints that constrain the cost function. Sched is modelled as a constraint satisfaction whose constraints are:

1 the set of temporal constraints TC,

2 the set of piecewise linear constraints PL, and

3 the resource constraints RC.

The satisfaction of Sched consists of fixing the start/end times of the N activities and the cost variables Y_i such all the previous constraints hold. The optimality problem of Sched consists of finding a solution which minimizes $\Phi(Y_1, \ldots, Y_N)$. A start/end variable is called *temporal*. Since the time is discrete, the temporal variables should be assigned integers.

As the example suggests, TC does not have an arbitrary syntax. Each constraint in TC has one the following forms:

- $u\ R\ c$ (bounding constraints)
- $u\ R\ v\ \pm\ c$ (distance constraints)
  where $R \in \{ =, <, >, \leq, \geq \}, integer(c)$ and $u,\ v$ are temporal variables.

This defines a tractable class of relations which can capture many complex requirements that exist in many common scheduling problems. According to (Dechter et al., 1991), it corresponds to a *Simple Temporal Problem*.

The domain of the PL function ϕ is partitioned into segments on which ϕ is linear. The boundaries of the segments are its *breakpoints* (a_i, b_i). The time points a_i are assumed integers. The cost points b_i are real numbers. It is well known that ϕ is convex iff its slope sequence is sorted in increasing order. See (Fourer and Gay, 1995) for representations of PL function.

From now on, C denotes a constraint $Y = \phi(X)$ whose breakpoints are (a_i, b_i). In addition, $min(X)$ (resp. $max(X)$) designates the lower (resp. upper) bound of variable X.

5.3 Hybrid Probe-based Algorithm

In (El Sakkout and Wallace, 2000), the relaxed (or *probing*) problem consists of TC and the linear constraints stating that the cost equals the sum of the absolute values of the changes of the "old" start times. It showed that the probing problem matrix is *unimodular*. Thus, solving it by pure Linear Programming is sufficient to ensure the integrality of the temporal variables. The algorithm presented here uses the fact that TC plus (a relaxation of) PL has a similar desirable structure, which can be exploited by the search.

Below, the ECLiPSe encoding of the algorithm is summarized. It first builds the conceptual model. `design_model/4` maps the model constraints into different solvers. The *inference phase* is triggered when *ic* constraints are setup. `probe_tent_values/1` achieves the *probing phase* which solves by *eplex* to optimality a relaxation, named $\mathcal{M}$, of the model. Next section discusses the choice of $\mathcal{M}$.

The search routine `repair_label/2` subsequently repairs the "hard" resource violations by sending more "easy" temporal constraints to *eplex* and *ic*. The *feasibility phase* monitors the resource constraints against violation.

```
search(ProblemInput) :-
    conceptual_model(ProblemInput,Vars,TC,PL,RC),
    design_model(TC,PL,RC,Key),
    % Probing Phase
    probe_tent_values(Vars),
    repair_label(Vars,Key).

repair_label(Vars,Key) :-
    % Resource feasibility
    conflict_constraints(Key,ConfConstrs),
    (ConfConstraints == [] ->
        true % Found a solution
    ;
        select_violated_constraint(ConfConstrs,ConfConstr),
        repair_violated_constraint(ConfConstr,R),
        % A backtrackable choice point
        % It triggers the inference phase
        (
           [ic,eplex]:(R)
        ;
           negation(R,NotR),
           [ic,eplex]:(NotR)
        )
    ),
    % re-assign ''better values'' to tentative attributes
    probe_tent_values(Vars),
    repair_label(Vars,Key).
```

When a solution is found, the search is re-initialized using the ECLiPSe Branch and Bound library. This is not captured above. At the time the resource usage is the highest (bottleneck), the search strategy selects two activities and constrains them not to overlap by posting to *eplex* and *ic* solvers a *temporal constraint* R=($u \leq v$) between their start/end variables u and v. On backtrack, R is revoked and its negation $u \geq v + 1$ is posted to ensure completeness. Both R and its negation respect the syntax of TC. R ensures that the next assignment (or probe) will be "better" in the sense that the current bottleneck will be reduced.

5.3.1 The Algorithm Design Model. TC are shared by *ic* and *eplex*. Each constraint in PL is sent to *ic* and is used for propagation. `map_plc/1` maps each C of PL into *eplex* constraints.

This is discussed in Section 5.4. The resource constraints RC, which have a weak linear relaxation are respectively setup as *ic* and *repair* constraints by `cumulative` and `map_rc`. Thus, $\mathcal{M}$ comprises TC and the transformed PL constraints.

```
design_model(TC,PL,RC,resource_cts):-
        % Map the temporal constraints
        (foreach(C, TC)
         do
           [ic,eplex]:(C)
        ),
        % Map the piecewise linear constraints
        (foreach(C, PL)
         do
          ic: (C),
          map_plc(C)
        ),
        % Map the resource constraint(s)
        (foreach(C,RC)
         do
          C=resource(Activities,NR),
          ic:cumulative(Activities,NR),
          map_rc(C,resource_cts)
        ).
```

5.3.2 Inference phase. Each search node enforces local consistency of its *ic* constraint store to tighten the variables bounds. The store is updated during search. The temporal variables are represented as finite domain *ic* variables. The Y_is are continuous *ic* variables. In addition to *ic* interval propagation, the predicate `cumulative` which is supported by *ic* is using the resource constraints as a global structure to perform propagation on the temporal variables (Aggoun and Beldiceanu, 1993; Group, 2002). The PL constraints are sent to *ic* as well. Indeed, we implement a module which achieves interval propagation using each piecewise linearity constraint C individually. The module typically prunes the range of Y, whenever a tightening of the domain of X happens, and vice-versa.

5.3.3 Probing phase. The purpose of this phase is to "suggest" a probe to the resource feasibility phase. The probing consists of minimizing Φ subject to $\mathcal{M}$. The probe is partially consistent: it might violate the resource constraints. The probe generator (or prober) is implemented as an *eplex* solver demon which is woken repeatedly by `probe_tent_values`. When triggered by `eplex_solve`, the demon collects all the newly posted constraints and solves new *eplex* store. `set_ans_to_tent` forwards the extracted probe in the tentative attributes of variables (See Section 4.5).

```
probe_tent_values(Vars) :-
        % Collect constraints posted after the last run
        % and solve the probing problem
        eplex_solve(Cost),
        % ''forward'' probe found by eplex
        set_ans_to_tent.
```

5.3.4 Resource feasibility. The (resource) constraints that were encoded in *repair* and *ic* are automatically monitored for feasibility *w.r.t.* the current probe. All the violated constraints will be found in the conflict set annotated `resource_cts`. The resource profiles are automatically constructed assuming that each activity starts at the time suggested by the probe recorded as the tentative value of its start variable.

Now we show how these constraints are setup in *repair*. For each couple of activities, `overlap_bool` defines an overlap boolean `B12` which is set iff the second activity is in progress at the tentative start time of the first one. `tent_prop` eagerly calls `Goal` using tentative values of the variables.

Once initiated, `tent_prop` will stay active and keep updating the tentative value of `B12` as soon the tentative value of `S1` or `S2` is changed by the probing phase.

```
overlap_bool(Act1,Act2,B12) :-
        Act1 = task with start_var: S1,
        Act2 = task with [
                                   start_var: S2,
                                   duration: D2
                               ],
        Goal=(S2 =< S1, S1 < S2 + D2),
        % if Goal ''eagerly succeeds'' then B12 tent_set 1
        % else B12 tent_set 0
        tent_prop(Goal,B12).
```

Assume that there is a limited number NR of resources. Below is the *repair* encoding of as many resource constraints as activities. They link the resource reasoning with the temporal variables. They basically state that, for each activity, the number of activities overlapping with its start time should not exceed NR.

```
map_rc(C,Key):-
     C=resource(Activities,NR),
     % Get all B12's s.t. overlap_bool(_Act1,_Act2,B12)
     get_all_overlap_bools(Activities,OverlapBools),
     % Impose the resource limit
     (foreach(Act1, Activities),
      param(OverlapBools,NR,Key)
      do
        % Get all B12's s.t. overlap_bool(Act1,_Act2,B12)
        get_act_overlap_bools(Act1,Bools)
        % Setup a repair constraint capturing resource limit
        (sum(Bools) =< NR) r_conflict Key
     ).
```

5.4 Probing Strategies

The two conditions the prober must fulfill 1) it satisfies $\mathcal{M}$ and 2) it finds a "good" solution $w.r.t.$ the cost Φ. To solve Sched, (Kamarainen and El Sakkout, 2002) used a prober based on local search which gave no guarantee of the probe optimality. Since the probe drives the search, it should suggest sufficiently good values. A design model where the probe is "poor" can lead to inefficient search. ECLiPSe is flexible enough to allow different mappings of the design model to be implemented and tested quite easily. To illustrate this flexibility, we implement the following probing options:

MIP-based prober *eplex* handles the piecewise-linearity constraints. Because they are, in general non-convex, the prober should be a MIP solver. *A priori* we are not sure that the temporal variables will be assigned discrete values by the MIP-based prober, unless we post their integrality constraints through *eplex*. This is not affordable, because it would impact efficiency. Actually, the corresponding $\mathcal{M}$ has a good structure: time variables will be usually assigned integral values in the optimal MIP solution (no MIP branching on their integrality violation). In fact, this is ensured provided that cost function bounds are excluded from $\mathcal{M}$. Indeed, they the structure of $\mathcal{M}$. This option is studied in section 5.5.

Linear prober $\mathcal{M}$ contains a *linear relaxation* of the piecewise linearity constraints. The integrality of the temporal variables is still an important issue. Again, it turned out that $\mathcal{M}$ in this form still has a good structure in the sense that, when solved by a Simplex, its optimal solution would necessarily assign integers to the temporal variables. This result holds even when the cost bounds are in $\mathcal{M}$. The feasibility phase should monitor and repair any PL constraint violated by the probe. This is studied in Section 5.6.

5.5 Mixed Integer Programming based Probing

Four methods to linearize *convex* PL constraints by the introduction of new variables and linear constraints exist in the literature (Fourer, 1992; Ho, 1985). Here, we consider a method called λ-model.

Special Ordered Set (SOS) constraints were introduced in (Beale and Tomlin, 1970) and are commonly used in mathematical programming (I.R. de Farias et al., 2000; Williams, 1999). A particular instance named "SOS of type 2" (SOS2) offers a natural formulation of PL constraints. A constraint SOS2(L), where $L = [x_1, \ldots, x_n]$ is a list, is satisfied iff at most two variables in L are assigned non-zero values, and if two variables are assigned non-zero values then they are adjacent. *eplex* supports SOS2 constraints as instance options, when the prober demon is setup (see Section 4.3).

The idea of the λ-model (Williams, 1999; Dantzig, 1963) is that any point of the convex hull of C is a linear convex combination of the breakpoints. For each breakpoint vertex (a_i, b_i), a variable λ_i, interpreted as its "weight". Each C will then be encoded in the design model as below. The user-defined predicate `send_to_eplex/1` keeps `sos2(L)` as a delayed goal, which will be collected when the *eplex* demon is setup.

```
map_plc(C) :-
        get_constraint_info(C,X,Y,APoints,BPoints,N),
        length(Lambdas,N),
        eplex:(Lambdas :: 0..1),
        eplex:(X =:= APoints*Lambdas),
        eplex:(Y =:= BPoints*Lambdas),
        eplex:(sum(Lambdas) =:= 1),
        send_to_eplex(sos2(Lambdas)).
```

If *ic* performs some inference, then the *eplex* store needs to be updated dynamically so as to prune the MIP search tree. The idea is that, as a result of *ic* bound changes, some *valid* cuts on the weights are detected and sent to *eplex*. This pruning is done during the inference phase and thus, *before* the probing. The flexible suspension mechanism of ECLiPSe enables the implementation of this as an event-driven pruning handler which suspends itself on completion, if X is non-ground. If X is fixed then Y and the weights can be fixed as well.

The encoding below does not capture this. The handler is woken as soon as a bound of X changes. It is given the same priority as *ic* propagation.

```
prune_probing(C):-
        get_constraint_start_var(C,X),
        (nonground(X)->
           generate_cuts(C),
           suspend(prune_probing(C),4,[X->min,X->max])
        ;
           true
        ).
```

We summarize the handler behaviour. The first rule captures that, if a segment $[a_i, a_{i+1}]$ is strictly on the left of $min(X)$, then its weight λ_i can be set to 0. A similar rule sets λ_{i+1} to zero if $max(X)$ is excluded by the segment $[a_i, a_{i+1}]$. These rules can be strengthened by generating further cuts on the weights. The idea is to infer an upper bound on λ_i by an interpolation of the bounds of X w.r.t its current segment.

```
generate_cuts(C):-
         get_segs_weights_startvar(C,Segments,Lambdas,X),
         dvar_dom(X,MinX,MaxX), % Get bounds of X
         (foreach(Segment,Segments),
          foreach(Lambda,Lambdas),
          param(MinX,MaxX)
          do
             (seg_is_excluded_by_bound(Segment,MinX,MaxX)->
                eplex: (Lambda =:= 0)
             ;
                 (seg_contains_bound(Seg,MinX)->
                     interpolate(Seg,MinX,MaxLambda1)
                     eplex:(Lambda =< MaxLambda1) % cut
                 ;
                    (seg_contains_bound(Segment,MaxX)->
                        interpolate(Seg,MaxX,MaxLambda2),
                        eplex:(Lambda =< MaxLambda2) % cut
                    ;
                        true
                    )
                 )
             )
         ).
```

5.6 Linear Relaxation based Probing

As in (Refalo, 1999), we encode C by a *linear* relaxation RM, which does not require new variables. The relaxation RM of a PL constraint C is the formulation of its *exact convex hull*: the smallest *convex hull* generated by its breakpoints and intervals of X and Y. The convex hull procedure is incremental: it maintains up-to-date the convex hull of each C during the search. As soon as an interval bound of X and/or Y is refined by *ic*, it reacts by locally strengthening the convex hull *eplex* formulation by sending a "minimal" set of linear cuts to *eplex*.

```
map_plc(C) :-
       setup_convex_hull(C,ConvexHullCts),
       (foreach(Ct,ConvexHullCts)
        do
          eplex:(Ct)
       ),
       % Keep ''up-to-date'' the convex hull
       suspend(update_convex_hull(C,X,Y),4,
               [X->min, X->max,Y->min, Y->max]).
```

Since the non-convexity of C is dropped from RM, further search branching is required to satisfy C. If the probe suggests tentative values V_1 and V_2 to X and Y which do not solve C (*i.e.* $\phi(V_1) \neq V_2$) then a "relaxation conflict" *w.r.t.* C is pending for repair. The probe may now violate PL as well as constraints in RC. The design model should then post them both to *repair*:

```
design_model(TC,PL,RC,resource_cts, piecelin_cts):-
       % as before
       ........
       % Post PL as repair constraints
       (foreach(C,PL)
         do
          (C) r_conflict piecelin_cts
       ).
```

The conflict set `piecelin_cts` will annotate the violations of PL. As in the case of a resource violation, the search non-deterministically resolves a violation of C by splitting the domain of X and posting $R = (X \leq V_1)$, and its negation $X \geq V_1 + 1$ on backtrack. Clearly, both of them preserve the syntax of TC. As before, the posted constraint would ask the *eplex* for a "better" probe.

```
repair_relaxations(Vars,piecelin_cts):-
       conflict_constraints(piecelin_cts,ConfPLConstrs),
```

```
        (ConfPLConstrs==[]
            true
        ;
            select_violated_pl_ctr(ConfPLConstrs,PLConstr),
            get_constraint_startvar_label(PLConstr,X,V1),
            (
               [ic,eplex]:(X =< V1)
            ;
               [ic,eplex]:(X >= V1 + 1)
            ),
            probe_tent_values(Vars),
            repair_relaxations(Vars,piecelin_cts)
        ).
```

Using ECLiPSe, different search strategies can be easily prototyped and experimented. We adopt the strategy which resolves resource conflicts *before* relaxation conflicts: it favours feasibility against optimality. Implementing an other strategy, for instance the one which takes the conflicts in the opposite order, would minimally change code. The search predicate is rewritten as follows:

```
search(ProblemInput):-
       % as before
       ......
       repair(Vars).

repair(Vars):-
        % First: resolve resource contentions
        repair_label(Vars,resource_cts),
        % If no resource conflict, repair the relaxations
        repair_relaxations(Vars,piecelin_cts),
        (
         (
          conflict_constraints(resource_cts,[]),
          conflict_constraints(piecelin_cts,[])
         )->
            true % Found a solution
        ;
            repair(Vars)
        ).
```

5.7 Evaluation and Performance Analysis

We report a comparison of the solver called H with the MIP-prober (H(lambda)) and relaxation-based prober (H(relax)). The solver is implemented in ECLiPSe

Class	Nb of tests	PL curves	R(%)	T	θ
C_1	24	"nearly convex"	0	197	0.5
C_2	24	"nearly convex"	30	197	0.5
C_3	24	"nasty"	0	197	0.5
C_4	24	"nasty"	30	197	0.5

Table 6.1. Definition of the four classes of experiments.

5.1.3, where CPLEX 6.5 is the linear solver. The evaluation was carried out on PCs with Pentium II 450 MHz processors. The timeout is 30 CPU minutes.

5.7.1 Setting up the benchmark instances. A data generator outputs a randomly-generated *feasible* schedule containing:

- A set of N activities, with their *fixed* start/end times over a horizon T. There is a single resource pool.
- A set TC of *feasible* binary temporal constraints. Its generation is guided by a density parameter θ which captures the likelihood of a temporal constraint between any pair of start/end times.
- A set of N objective functions defined over T. The complexity of function measures "how hard" is finding its minimum, regardless of the other constraints.

From any feasible schedule, we can create many benchmarks by varying the scenario parameters:

1. For each start variable, the maximum/minimum allowed time shift around its "old" value. This sets a "time window".
2. The amount of *resource reduction* R: the percentage of machines to be pulled out (*i.e.* saved) from the initial set.

To get a large-scale search space, the time window is set to T. The tightness of a test is defined by those of its resource and optimization parts.

R captures the tightness of the resource part: requiring larger resource saving could tighten aggressively the problem, and cause terrific perturbations to the "old" schedule.

The complexity of the PL curves reflects the tightness of the optimization part. A curve is easy if it is "nearly convex": its distance to convexity is small. Intuitively, this distance should be somehow proportional to the variations of the signs of its slopes. The more often the slope sign changes, the harder is the curve. We consider a curve "nearly convex" if it has *at most* three local

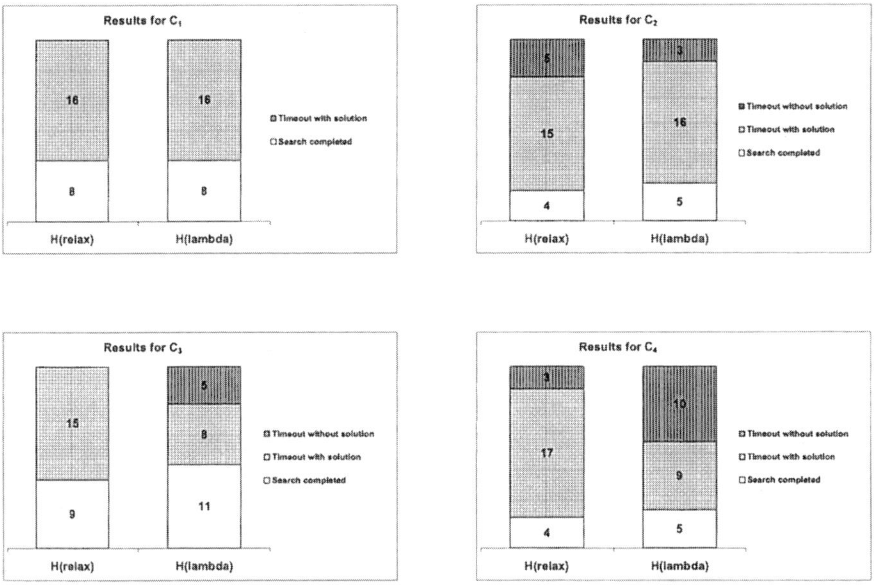

Figure 6.2. Results of running H(relax) and H(lambda) on the four test classes.

minima. This should restrict the sign variations in its slopes. This covers the convex case. A curve becomes "nasty", when unrestricted variations occur in the signs of its slopes (number of minima is greater than 3).

Depending on the complexity of their components , we generate 96 experiments, which are partitioned into four classes C_i, where $i \in [1..4]$. They are summarized in Table 6.1. Each C_i contains 8 instances with N activities, where N is in $\{30, 60, 90\}$.

5.7.2 Computational results. Figure 6.2, sketches the results. Initial evaluation has shown that H(lambda) is surprisingly ineffective without the pruning procedure. It significantly speed up the MIP search by providing a tighter linear relaxation. This shows the search sensitivity to constraint passing from *ic* to *eplex*.

Regarding the proof of optimality (Search completed), H(lambda) is usually, at least, as good as H(relax).

Run on C_1, H(lambda) and H(relax) behave equally well and they solved all the 24 tests. Since the resource subproblem has a sparse solution space, few choice points are required in both cases. Furthermore, the response times and the qualities of the probes are very close, because the PL curves are "easy". The same conclusion holds for C_2. The difference is that H(lambda) behaves slightly better. Indeed, H(relax) failed on 2 tests that H(lambda) solved. This could be seen as two counter examples to the "repair resources before relaxations" strategy. To avoid Simplex initialization overheads, it could be beneficial to solve all the easy relaxation conflicts by a single MIP, rather than fixing them at later stages by separate Simplex invocations.

Meanwhile, H(relax) solved all tests of $\mathcal{C}_3$. Since the resource constraints have a dense solution space and the curves are nasty, one would expect H(lambda) to be more effective. It is not the case, as H(lambda) failed on 5 tests. In the tree explored by H(lambda), the traversal of the branches that lead to failure is time-consuming because they require invoking an expensive MIP at each node. This is why on the 5 tests, H(lambda) times out before it explores the potentially successful branches which were pending for exploration.

The performance gap between the solvers gets clearly wider on $\mathcal{C}_4$, where the resources are more limited. The MIP search invests too much computational effort (the curves are nasty) in obtaining optimal probes that are not easily repaired to be feasible due to the limited resources. Thus, the expensive MIP probes are likely to be "revised" by the search. By contrast, the adopted search strategy which repairs resource before relaxation conflicts makes H(relax) more oriented towards resource feasibility: it delivers a first solution faster and, thus, enabling cost-pruning at early stages. In fact, the search strategy of H(lambda) should be interpreted from the opposite angle: the MIP solver is invoked to repair all the relaxation conflicts at once *before* focusing on relieving the resource contentions.

6. Conclusion

Combinatorial optimization models often embed well-structured and simpler sub-problems that are better suited for special-purpose OR and/or CP techniques. The automation of building CP-OR solvers is probably too ambitious. Following (Wallace and Schimpf, 2002), finding the "right" hybrid algorithm is by itself a "combinatorial meta-problem". ECLiPSe aims to aid the design, prototyping and evaluation of different hybrid solutions so as to quickly find the most appropriate for the problem at hand. This chapter considered how the constraint logic programming language ECLiPSe attempts to address some of the issues arising when it comes to programming a "hybrid search". We have seen how the language makes it possible for different solvers to work closely with each other during the search. The probe-based hybridization form used to solve the scheduling problem with piecewise linear optimization has illustrated how the language support is exploited.

Nevertheless, this form is not the only one which makes use of the language support for hybrid algorithms. At IC-Parc, ECLiPSe has been also used to accommodate other hybridization forms. These include using a finite domain solver to warm-start integer linear programming (Hajian et al., 1998) and embedding local search, Lagrangian relaxation, Benders decomposition and column generation into a constraint programming (Eremin and Wallace, 2001; Kamarainen and El Sakkout, 2002). Most of these forms are being applied at IC-Parc on a variety of innovative applications arising in the transportation and networking markets.

References

Aggoun, A. and Beldiceanu, N. (1993). Extending CHIP in order to solve complex scheduling and placement problems. *Journal of Mathematical Computer Modelling*, 17(7):57–73.

Ajili, F. and El Sakkout, H. (2001). LP probing for piecewise linear optimization in scheduling. Programme and papers presented at CPAIOR'01: `www.icparc.ic.ac.uk/cpAIOR01/`.

Ajili, F. and El Sakkout, H. (2003). A probe-based algorithm for piecewise linear optimization in scheduling. *Annals of Operations Research*, 118:35–48.

Beale, E. M. L. and Tomlin, J. A. (1970). Special facilities in a general mathematical system for non-convex problems using ordered sets of variables. In *Proceedings of the Fifth International Conference on Operations Research*, pages 447–454, London. Tavistock Publications.

Beck, J.C. and Refalo, P. (2003). A hybrid approach to scheduling with earliness and tardiness costs. *Annals of Operations Research*, 118:49–71.

Benoist, T., Laburthe, F., and Rottembourg, B. (2001). Lagrange relaxation and constraint programming collaborative schemes for travelling tournament problems. Programme and papers presented at CPAIOR'01: `www.icparc.ic.ac.uk/cpAIOR01/`.

Beringer, H. and De Backer, B. (1995). Combinatorial problem solving in constraint logic programming with cooperating solvers. In *Logic Programming: Formal Methods and Practical Applications*, pages 245–272. Elsevier.

Bockmayr, A. and Kasper, T. (1998). Branch and infer: A unifying framework for integer and finite domain constraint programming. *INFORMS Journal on Computing*, 10(3):287–300.

Dantzig, G. B. (1963). *Linear programming and extensions*. Princeton university Press.

Dechter, R., Meiri, I., and Pearl, J. (1991). Temporal constraint networks. *Artificial Intelligence*, 49:61–95.

El Sakkout, H. and Wallace, M. G. (2000). Probe backtrack search for minimal perturbation in dynamic scheduling. *Constraints*, 5(4):359–388. Special Issue on Industrial Constraint-Directed Scheduling.

Eremin, A. and Wallace, M.G. (2001). Hybrid benders decomposition algorithms in constraint logic programming. In *CP'2001*, volume 2239 of *LNCS*, pages 1–15. Springer.

Focacci, F., Lodi, A., and Milano, M. (1999). Cost-based domain filtering. In *CP'99*, volume 1713 of *LNCS*, pages 189–203. Springer.

Focacci, F., Lodi, A., and Milano, M. (2000). Cutting planes in constraint programming: an hybrid approach. In *CP'2000*, volume 1894 of *LNCS*, pages 187–201. Springer.

Fourer, R. (1992). A simplex algorithm for piecewise-linear programming iii: Computational analysis and applications. *Mathematical Programming*, 53:213–235.

Fourer, R. and Gay, D. M. (1995). Expressing special structures in an algebraic modelling language for mathematical programming. *ORSA Journal on Computing*, 7:166–190.

Gervet, C. and Wallace, M.G. (2001). Third international workshop on integration of AI and OR techniques in constraint programming for combinatorial optimization problems. Programme and papers presented at CPAIOR'01: `www.icparc.ic.ac.uk/cpAIOR01/`.

Group, The ECLiPSe (2002). *Introduction to ECLiPSe.*

Hajian, M.T., El-Sakkout, H., Wallace, M.G., Lever, J.M., and Richards, B. (1998). Towards a closer integration of finite domain propagation and simplex-based algorithms. *Annals of Operations Research*, pages 421–433.

Ho, J. K. (1985). Relationships among linear formulations of separable convex piecewise-linear programs. *Mathematical Programming Study*, 24:126–140.

Hooker, J. N., Ottosson, G., Thorsteinsson, E., and Kim, H. (1999). On integrating constraint propagation and linear programming for combinatorial optimization. In *In Proceedings of the Sixteenth National Conference on Artificial Intelligence*, pages 136–141. AAAI Press.

ILOG (2001). CPLEX. `www.ilog.com/products/cplex/`.

I.R. de Farias, Jr, Johnson, E. L., and Nemhauser, G. L. (2000). A generalized assignment problem with special ordered sets: a polyhedral approach. *Mathematical Programming*, 89:187–203. Issue:1.

Junker, U., Karisch, S.E., Kohl, N., Vaaben, B., Fahle, T., and Sellmann, M. (1999). A framework for constraint programming based column generation. In *CP'99*, volume 1713 of *LNCS*, pages 261–274. Springer.

Kamarainen, O. and El Sakkout, H. (2002). Local probing applied to scheduling. In *CP'2002*, volume 2470 of *LNCS*, pages 155–171. Springer.

Ken Darby-Dowman, K., Little, J., Mitra, G., and Zaffalon, M. (1997). Constraint logic programming and integer programming approaches and their collaboration in solving an assignment scheduling problem. *Constraints*, 1(3):245–264.

Marriott, K. and Stuckey, P. J. (1998). *Programming with Constraints*. MIT Press.

Michel, L. and Van Hentenryck, P. (1997). Localizer: A modelling language for local search. *Lecture Notes in Computer Science*, 1330.

Milano, M., Ottosson, G., Refalo, P., and Thorsteinsson, E.S. (2000). The benefits of global constraints for the integration of constraint programming and integer programming. In *Proceedings of the (AAAI-00)'sWorkshop on Integration of AI and OR Techniques for Combinatorial Optimization (AAAIOR)*. AAAI Press.

Milano, M. and van Hoeve, W.J. (2002). Reduced cost-based ranking for generating promising subproblems. In *CP'2002*, volume 2470 of *LNCS*, pages 1–16. Springer.

Monfroy, E. (1998). The constraint solver collaboration language of bali. In *In Proceedings of the International Workshop Frontiers of Combining Systems, FroCoS'98 (Amsterdam, The Netherlands, 1998)*.

Optimization, Dash (2001). XPRESS-MP. `www.dash.co.uk/`.

Ottosson, G., Thorsteinsson, E.S., and Hooker, J. N. (1999). Mixed global constraints and inference in hybrid CLP–IP solvers. In Heipcke, S. and G.Wallace, M., editors, *Proceedings of (CP-99)'s Workshop on Combinatorial Optimisation and Constraints*, volume 4 of *Elec. Notes in Discrete Mathematics*. Elsevier Science.

Ottosson, G., Thorsteinsson, E.S., and Hooker, J. N. (2002). Mixed global constraints and inference in clp-ip solvers. *Annals of Mathematics and Artificial Intelligence*, 34(4):271–290.

Refalo, P. (1999). Tight cooperation and its application in piecewise linear optimization. In *CP'99*, volume 1713 of *LNCS*, pages 375–389. Springer.

Rodosek, R., Wallace, M. G., and Hajian, M. T. (1999). A new approach to integrating mixed integer programming and constraint logic programming. *Annals of Operations Research*, 86:63–87. Special issue on Advances in Combinatorial Optimization.

Rodosek, R. and Wallace, M.G. (1998). A generic model and hybrid algorithm for hoist scheduling problems. In *Proceedings of the 4th International Conference on Principles and Practise of Constraint Programming*, pages 385–399, Pisa.

Sellmann, M. and Fahle, T. (2003). Constraint programming-based lagrangian relaxation for the automatic recording problem. *Annals of Operations Research*, pages 17–33.

Smith, B., Brailsford, S., Hubbard, P., and Williams, H.P. (1995). The progressive party problem: integer linear programming and constraint programming compared. In *CP'95*, pages 36–52, Marseille. Springer.

Tsang, E. (1993). *Foundations of Constraint Satisfaction*. Academic Press.

Van Hentenryck, P. (1989). *Constraint Satisfaction in Logic Programming.* Logic Programming Series. MIT Press, Cambridge, MA.

Van Hentenryck, P. and Carillon, J-P. (1988). Generality versus specificity: an experience with AI and OR techniques. In *AAAI*, Minnesota.

Wallace, M.G. and Schimpf, J. (2002). Finding the right hybrid algorithm - a combinatorial meta-problem. *Annals of Mathematics and Artificial Intelligence*, 34(4):259–270.

Williams, H.P. (1999). *Model building in mathematical programming.* Management Science. Wiley, 4th edition.

Chapter 7

CP BASED BRANCH-AND-PRICE

Kelly Easton
Department of Industrial and Manufacturing Engineering
Kansas State University
keaston@ksu.edu

George Nemhauser
Georgia Institute of Technology
gnemhaus@isye.gatech.edu

Michael Trick
Carnegie Mellon University
trick@cmu.edu

Abstract In this chapter, we present an overview of constraint programming based branch-and-price, a combined integer programming-constraint programming algorithm used to solve combinatorial problems formulated with a huge number of variables. We discuss a number of applications and implementation issues.

Keywords: branch-and-price, column generation, constraint programming

1. Introduction

Chapter 1 of this book presented the basic branch-and-bound algorithm for solving integer programming (IP) problems. In this chapter, we extend the branch-and-bound algorithm to accommodate problems that have a huge number of variables, typically a number that is exponential in the size of the input needed to describe the original problem. For reasons that will become clear later, the extended algorithm is called *branch-and-price*.

There are several reasons why an IP formulation with a huge number of variables may be preferred to a conventional compact formulation when the solution methodology is LP-based branch-and-bound. First, the effectiveness of this type of algorithm is highly dependent on the strength of the LP relaxation. Weak LP relaxations lead to excessive branching and long computation times. By allowing a huge number of variables, we can often create formulations that give stronger LP bounds. Second, huge formulations can encapsulate difficult modeling issues in the definition of the variables. For instance, formulating crew scheduling problems can be difficult since the definition of a feasible crew schedule involves a large number of constraints and requirements (Barnhart et al., 1999). By having a variable for every feasible crew schedule, the resulting formulations are generally much simpler, albeit with a huge number of variables. Third, for some problems, the compact formulation may have a symmetric structure that hampers the progression of a branch-and-bound algorithm because one can move to an equivalent solution after branching. An alternative formulation may add variables to avoid this symmetry.

The underlying idea of branch-and-price is that for an integer program with a huge number of variables, it is possible to start with a small subset of the complete set of variables (columns using LP terminology) and add new columns to the model throughout the branch-and-bound tree as necessary in order to find an optimal solution to the full integer program.

The branch-and-price algorithm begins with a *restricted master problem* at the root node of the search tree that includes only a small subset of the total set of possible columns. At each node, the optimality of the LP solution is checked by solving the *pricing problem*. The pricing problem is used to determine whether or not there exist columns with profitable reduced cost that are missing from the current formulation. Upon the addition of columns, the LP relaxation of the master problem is reoptimized. The algorithm iterates between solving the LP relaxation and solving the pricing problem until there are no more profitable columns to be added to the master problem. At this point, the algorithm proceeds as does the branch-and-bound algorithm. Specifically, if the LP relaxation is infeasible or its solution is worse than the current best integral solution, the node is cut off or *fathomed*. If the solution to the LP relaxation is integral and improves upon the current best solution, the current best is updated and the node is fathomed. Otherwise, the algorithm branches. When all nodes of the search tree are fathomed, the current best solution is optimal. If there is no current best solution, the problem is infeasible.

This approach raises several interesting issues. The first question is how to identify the necessary columns to add to the model throughout the tree. In an LP model, the only columns that can improve a basic solution are ones with profitable reduced cost. Thus, in order to prove LP optimality, it is enough to find at least one column with profitable reduced cost or show that none exist. Because there are a very large number of columns, this cannot be done by enumeration. However, for structured problems, it is frequently possible to devise a subproblem, *the pricing problem*, that can be solved efficiently to address this issue. The process of finding columns to add to the existing integer program is called *subproblem solving* or *column generation*.

There are many applications in which constraint programming (CP) is very effective for solving the pricing problem. This gives a natural way of integrating integer programming and constraint programming: IP is used to optimally combine subproblem solutions generated by CP. This approach, which to our knowledge was proposed by (Junker et al., 1999), has a wide range of application, most of which is barely explored at this point. CP has a number of attributes that make it a potent algorithm for subproblem solving. First, the pricing problem does not need to be solved to optimality; it is enough to find any column that potentially can be added to the basis. Thus the pricing problem can be viewed as more of a feasibility problem than an optimality problem. CP is well suited to solving highly constrained feasibility problems. It has a very broad modeling paradigm that may make the formulation more straightforward, potentially leading to greater efficiency in search and heuristic techniques. Although CP algorithms do not have the global perspective provided by LP relaxations, they can sometimes find a solution more quickly using logical implications and sparing the computational expense of LP. Secondly, it is common to generate several columns, rather than just one, each time the pricing problem is solved. Although this is possible to accomplish with IP, CP is particularly well suited to this task.

A second important issue raised by the branch-and-price approach is the branching strategy. Standard IP branching does not work well because the constraints it introduces conflict with subproblem solution methods. Branching rules need to be designed with the pricing problem in mind, leading to problem-specific branching strategies.

A third important issue concerns the LP termination criteria. Besides the usual use of LP in a branch-and-bound framework to provide bounds for fathoming nodes, in a branch-and-price framework the LP provides dual prices to be used in the pricing problem. There is a tradeoff between the quality of the dual values and the computational effort required to solve the LP. There are other related issues here as well. In order to obtain true LP bounds at every node, the pricing problem must prove that no more profitably-priced columns can be generated. However, it is not necessary to obtain true LP bounds at every node of the search tree. Furthermore, branch-and-price approaches are often used heuristically to obtain good solutions but not necessarily to prove optimality. This can have a substantial effect on the column generation process.

This chapter provides a general overview of CP based branch-and-price. By focusing on approaches that use CP methods to solve the pricing problem, it complements a survey paper by (Barnhart et al., 1998) that deals with branch-and-price approaches in general.

The remainder of the chapter is organized as follows. Section 2 presents three problems that illustrate the use of CP based branch-and-price. This section also includes a short discussion of other relevant literature. Section 3 discusses implementation issues including branching strategies, column management and parallel algorithms. Section 4 suggests some avenues for future research involving CP based branch-and-price.

2. Three Illustrative Examples

To motivate the issues involved in using a branch-and-price approach, particularly for cases in which CP is effective as a subproblem solver, we explore three examples.

2.1 The Generalized Assignment Problem

Our first problem, the Generalized Assignment Problem (GAP), has a very clear branch-and-price approach developed by (Savelsberg, 1997). In the GAP, there are n jobs that must be assigned to m machines. A job is assigned to exactly one machine. If job j is assigned to machine i, then it takes a_{ij} time units to process the job, and a cost c_{ij} is charged. Machine i has b_i time units available to process its jobs. While the GAP is a useful model in its own right, its primary use is as a subproblem in areas such as vehicle routing, location and resources planning.

Formulating the GAP as a compact integer program is straightforward. Define a binary variable x_{ij} to be 1 if and only if job j is assigned to machine i. Then we get the formulation

$$\min \sum_i \sum_j c_{ij} x_{ij} \tag{7.1}$$

$$s.t \quad \sum_i x_{ij} = 1 \quad \forall j \tag{7.2}$$

$$\sum_j a_{ij} x_{ij} \leq b_i \quad \forall i \tag{7.3}$$

$$x_{ij} \text{ binary } \forall i, j \tag{7.4}$$

The most complicated constraints in this formulation are the *knapsack constraints* of the form $\sum_j a_{ij} x_{ij} \leq b_i$ which ensure that the set of jobs assigned to any given machine does not violate the machine capacity. Since there are many approaches to solving knapsack problems, including integer programming, dynamic programming and constraint programming (see (Martello and Toth, 1990), (Trick, 2002) and (Fahle and Sellmann, 2000)), it is natural to try to reformulate the problem to exploit these algorithms.

Each solution of the GAP consists of a set of feasible assignments, one for each machine. We can solve the problem by virtually enumerating all the feasible assignments for all the machines then choosing exactly one assignment for each machine. We describe the k-th assignment for machine i by the vector $(f_1^{ki}, \ldots, f_n^{ki})$ where f_j^{ki} is 1 if and only if job j is scheduled on machine i in this assignment. Furthermore, we introduce a variable y_k^i indicating whether the k-th assignment for machine i is selected for the solution. Of course, there are a large number of y variables. Each y_k^i has a cost $d_k^i = \sum_j c_{ij} f_j^{ki}$, corresponding to the cost of that solution to the knapsack. Our expanded formulation now becomes

$$\min \quad \sum_i \sum_k d_k^i y_k^i \tag{7.5}$$

$$s.t \quad \sum_i \sum_k f_j^{ki} y_k^i = 1 \quad \forall j \tag{7.6}$$

$$\sum_k y_k^i = 1 \quad \forall i \tag{7.7}$$

$$y_k^i \text{ binary } \forall i, k \tag{7.8}$$

To solve this problem with the expanded formulation, we begin with a restricted set of solutions to the knapsack problems and solve the LP relaxation of the above problem restricted to those solutions, i.e. we solve the LP relaxation of the restricted master problem. One byproduct of solving this is a dual value π_j for each job j corresponding to its constraint in the first set in the formulation above and a dual value γ_i for each machine i corresponding to its constraint in the second set. To determine if there is an attractive knapsack solution for machine i to add to the restricted master problem, we solve the knapsack problem

$$\min \quad \sum_j (c_{ij} - \pi_j) z_j \tag{7.9}$$

$$s.t \quad \sum_j a_{ij} z_j \leq b_i \tag{7.10}$$

$$z_j \text{ binary } \forall j \tag{7.11}$$

If the optimal objective value for this problem is less than γ_i, we say that the *reduced cost* $\gamma_i - \sum_j (c_{ij} - \pi_j) z_j$ is negative. In this case, the z_j form a knapsack solution that could improve the current solution. We add its corresponding y_k^i to the restricted master problem and continue.

Note that we can solve the knapsack problems with whatever approach is most effective for these problems. Particularly, in GAP instances that include additional constraints on the job assignments, standard methods for solving the knapsack problems may be insufficient and constraint programming may be the most effective approach. This provides a natural way to integrate integer programming methods with constraint programming methods.

So far, we have worked only with the linear relaxation of the restricted master problem. To solve the integer program associated with the master problem, we have to provide mechanisms for branching in the case of fractional y_k^i variables.

The standard branching rule for branching from a linear relaxation solution is to choose a fractional y_k^i and to create two subproblems: one with $y_k^i = 0$ and one with $y_k^i = 1$. This approach does not work well for two reasons. First, and most importantly, it may be very difficult to add such constraints without complicating the pricing problem. Adding a constraint $y_k^i = 0$ is equivalent to prohibiting a particular knapsack in the pricing problem. This prohibition is difficult to model and generally is not supported by many specialized knapsack algorithms. Such a constraint is possible to add to a constraint programming formulation, though the propagation properties of such a constraint are not particularly strong. The second problem with simple branching is that the solution space is not divided equally between the two subproblems since the branch with $y = 1$ is very restrictive while the branch with $y = 0$ excludes almost nothing. Equal dividing is a characteristic that is generally associated with good branching rules.

For these reasons, we need to find a branching rule that is compatible with the pricing problem and better divides the solution space. For this problem, and many others, the branching rule associated with the compact formulation has a natural interpretation in the branch-and-price formulation. Take any solution $\{y_k^i\}$ to the restricted master problem. If this solution is not integral, it is easy to see that some job j must be assigned fractionally to more than one machine. For one such machine i, we can create two subproblems: in the first, job j is assigned to machine i; in the second job j is not assigned to machine i. These constraints are compatible with the pricing problem: in the first case, job j is forced into the knapsack for machine i and is forced out for any other machine; in the second, job j is not included in the machine i pricing problem.

With a combination of fast pricing problem solution methods and effective branching rules, this branch-and-price approach can be a very effective solution method for some classes of GAP instances.

2.2 The Traveling Tournament Problem

Our second problem, the Traveling Tournament Problem (TTP), was introduced by (Easton et al., 2001). Their solution approach, a CP based branch-and-price algorithm, is given in (Easton et al., 2002). The TTP requires finding a double round robin tournament, a set of games in which each of n teams plays each other team once at home and once away such the length of every home stand and road trip is between l and u inclusive and the total distance traveled by the teams is minimized. Common parameter assignments are $l = 1$ and $u = 3$. These values are assumed for the purposes of this example.

There are several compact formulations for the TTP. In the simplest model, the binary variable x_{tijs} is set to 1 if team t plays at team i in slot s then at team j in slot $s+1$ and 0 otherwise. The set of teams is represented by $T = \{1, \ldots, n\}$ and the set of slots is represented by $S = \{1, \ldots, 2n-2\}$. Each variable has an objective coefficient d_{ij} equal to the distance between teams i and j. This gives the following IP

$$\min \quad \sum_t \sum_i \sum_j \sum_s d_{ij} x_{tijs} \tag{7.12}$$

$$s.t \quad \sum_j \sum_s x_{tijs} = 1 \quad \forall t, i \neq t \in T \tag{7.13}$$

$$\sum_i \sum_j x_{tijs} = 1 \quad \forall t \in T, s \in S \tag{7.14}$$

$$\sum_i x_{tijs} = \sum_k x_{tjks+1} \quad \forall t, j \in T, \forall s \in 1 \ldots |S| - 1 \tag{7.15}$$

$$\sum_j x_{tijs} = \sum_k x_{iiks} \quad \forall t, i \neq t \in T, \forall s \in S \tag{7.16}$$

$$sum_{i:i \neq t} \sum_{j:j \neq i, j \neq t} \sum_{k:s \leq k \leq s+3} x_{tijk} \leq 3 \quad \forall t \in T, \forall s \in 1 \ldots |S| - 3 \tag{7.17}$$

$$\sum_{i:i \neq t} \sum_{j:j \neq i, j \neq t} \sum_{k:s \leq k \leq s+3} x_{tijk} \geq 1 \quad \forall t \in T, \forall s \in 1 \ldots |S| - 3 \tag{7.18}$$

$$x_{tijs} \text{ binary } \forall t, i, j \in T, s \in S \tag{7.19}$$

The first constraint requires each team to play each opponent exactly once. The second constraint requires each team to play exactly once in each slot. The third constraint provides continuity from one slot to the next. The fourth constraint requires an opponent being visited to be at home. The final two constraints enforce the upper bound on the size of road trips and home stands.

The LP relaxation for this model is fairly weak. There is a more complex compact formulation of the TTP that has a stronger initial LP relaxation; however, it does not continue to be strong after several variables have been fixed in the search tree. Reformulating the problem so that the variables are tours, i.e. complete single team schedules, produces a model with much stronger LP relaxations, albeit with an exponential number of variables.

Specifically, each tour j has a corresponding team k_j and a venue vector $v(j, s)$ giving the venue for team k_j in slot s. This implies that the vector $v(j, .)$ has exactly $n-1$ entries equal to k_j corresponding to home games with the remaining entries being equal to the other teams, one for each, for the away games. Each tour j has a cost c_j equal to the distance traveled in the tour. The binary variable y_j is 1 if and only if tour j is selected. This gives the following IP:

$$\min \sum_j c_j y_j \tag{7.20}$$

$$s.t \quad \sum_{j:k_j=i} y_j = 1 \quad \forall i \in T \tag{7.21}$$

$$\sum_{j:k_j=i,v(j,s)=i} y_j = \sum_{j:k_j\neq i,v(j,s)=i} y_j \quad \forall s \in S, \forall i \in T \tag{7.22}$$

$$y_j \text{ binary } \forall j \in \{1,\ldots,m\} \tag{7.23}$$

The first constraint says that exactly one tour must be chosen for each team; the second constraint requires that a team be at home if another team is visiting it.

To solve this problem, we form the restricted master problem with a small initial set of tours; the set of all tours that are optimal for each single team scheduling problem is an obvious choice. We then solve the LP relaxation of the restricted master problem generating dual values that correspond to the constraints above. Most importantly, the second set of constraints gives one dual value for each (team,slot) pair.

These dual values, or reduced costs, can be used to generate new tours. Since we are minimizing, any column with a net reduced cost that is less than zero may be added to the restricted master problem. The pricing problem can thus be viewed as a feasibility problem since there is no need to find the most negative reduced cost column. For this reason, CP is a good choice for the subproblem solver. Additionally, as mentioned above, CP can be easily used to generate multiple solutions. Adding sets of new columns to the restricted master problem is often more effective than adding just one at a time since it can reduce the number of iterations between the master problem and the pricing problem (Barnhart et al., 1998).

We solve the following constraint program, expressed as an OPL model, to determine whether or not there are beneficial columns available for addition to the master problem.

```
range Teams 1..nbTeams;
range Slots 1..nbSlots;
int Distance[Teams,Teams] = ...;
float rc[Teams,Slots] = ...;
var int travel[0..nbSlots] in 0..MAXT;
```

```
var int venue[0..nbSlots+1] in 1..nbTeams;
var int opp[Slots] in -nbTeams..nbTeams;
var float totcost;

solve {

/* Opponent variables */
   alldifferent(opp);
   opp[s]<>home;
   opp[s]<>-home;
   opp[s]<>0;

/* Limit length of road trips and home stands */
   forall (s in 1..nbSlots-3) {
      sum (j in 0..3) (opp[s+j]<0) <= 3;
      sum (j in 0..3) (opp[s+j]>0) <= 3;
      sum (j in 0..3) (opp[s+j]<0) >= 1;
      sum (j in 0..3) (opp[s+j]>0) >= 1;

   };

/* Distance calculations */
   forall (s in Slots) {
      opp[s]>0 => venue[s]=home;
      opp[s]<0 => venue[s]=-opp[s];
   };
   venue[0]=home;
   venue[nbSlots+1]=home;
   forall (s in 0..nbSlots) travel[s]=
      Distance[venue[s],venue[s+1]];

/* Generate only negative reduced cost columns */
   totcost = sum (s in 0..nbSlots) travel[s] +
      sum (s in Slots) rc[venue[s],s];
   totcost <= -EPS;
};
```

Note that solving the pricing problem involves running the above CP once for each team in the tournament.

In order to produce an optimal solution to the TTP, our algorithm must iterate between the master problem and the pricing problem until the above CP proves that no more negative reduced cost columns are available.

Since the TTP is an integer program, we need to discuss how to branch in the case of fractional variables in the solution to the LP relaxation of the master problem. We've discussed the reasons for why standard IP branching cannot be used in a branch-and-price algorithm. As in the GAP example above, the variables in the compact formulation of the TTP have a natural interpretation in the expanded formulation and can be used to imply a higher order branching strategy. (Easton et al., 2002) suggest three possible strategies: In the first, the higher order variables are the patterns indexed by team and slot. In other words, the solution space is divided into schedules in which team t is home in slot s and schedules in which team t is away in slot s. The second set of higher order branching variables is inspired by the *follow-on variables* often used in airlines applications. The follow-on branching variables for the TTP are indexed by triplets (t, i, j). The solution space is divided into schedules in which team t plays teams i and j consecutively and schedules in which t does not play i and j consecutively. The third set of higher order branching variables is games scheduled in particular slots. Branching on the triplet (t, i, s) divides the solution space into schedules in which team t plays at team i in slot s and schedules in which t does not play i in slot s.

In our experiments, we've found that branching on pattern set variables is, in general, the most effective. That is to say, for the majority of test instances we've used, pattern set branching provides the shortest run times. There are, however, some test instances for which one of the other two strategies provide better results.

The size of the TTP formulation grows very quickly with the number of teams. For an instance with 8 teams, there are over 80 million possible tours. In order to solve larger instances of the TTP, a parallel algorithm must be considered. Parallel implementation issues are addressed in Section 3 of this chapter.

2.3 The Social Golfers Problem

Consider a group of $4k$ golfers who play golf every week in k foursomes (each player playing exactly once per week). What is the greatest number of weeks they can play before a pair of golfers plays with each other twice? This problem appears in CSPLIB for the case of $k = 8$. It is known that 9 weeks is possible and 11 weeks is not, but it is unknown whether they can play for 10 weeks before a pair plays together twice.

There are a number of integer programs possible for this problem. Here we will define x_{ijt} to be a binary variable that is 1 if players i and j play together in week t. Let P be the players and T be the time periods for a particular number of weeks. The Social Golfers Problem (SGP) becomes a feasibility problem: are there binary values x_{ijt} such that

$$\sum_j x_{ijt} = 3 \quad \forall i \in P, t \in T \tag{7.24}$$
$$\sum_t x_{ijt} \leq 1 \quad \forall i, j \in P \tag{7.25}$$
$$x_{ijt} = x_{jit} \forall i, j \in P, t \in T \tag{7.26}$$
$$x_{ijt} + x_{jlt} - x_{ilt} \leq 1 \forall i, j, l \in P, t \in T \tag{7.27}$$
$$x_{ijt} \text{ binary } \forall, i, j \in P, t \in T \tag{7.28}$$

The first set of constraints says that every player plays with 3 others every week. The second set requires every pair of players to play together no more than once. The third set says that if i plays with j, then j plays with i. The final set are linear constraints saying "if i plays with j and j plays with l, then i plays with l."

Unfortunately, this formulation does not adequately address symmetries in the problem. In particular, for any solution, it is possible to permute weeks to get another equivalent solution. Symmetries often slow down solution methods since branch-and-bound will regenerate essentially identical solutions that differ only by symmetry.

A branch-and-price formulation can avoid this particular symmetry problem (there are others that need to be handled differently). Define a *pairing* to be a division of players into k foursomes. So, associated with pairing p are values a_{ij}^p which is 1 if i and j play together in pairing p and 0 otherwise. Let y_p be a binary variable equal to 1 if and only if the pairing y is selected. Now the SGP is

$$\max \quad \sum_p y_p \tag{7.29}$$
$$s.t \quad \sum_p a_{ij}^p y_p \leq 1 \quad \forall i, j \in P \tag{7.30}$$
$$\sum_{j:k_j=i, v(j,s)=i} y_j = \sum_{j:k_j \neq i, v(j,s)=i} y_j \quad \forall s \in S, \forall i \in T \tag{7.31}$$
$$y_p \text{ binary } \forall p \tag{7.32}$$

Note that the only constraint is that no pair of players play together more than once.

In the branch-and-price algorithm, the restricted master problem gives a dual value π_{ij} for each pair of players. The subproblem then is to find a division of players into foursomes that maximizes the sum of the π_{ij} values within the foursomes. This clustering problem can best be solved with methods that can be thought of as specialized constraint programming methods (Mehrotra and Trick, 1998).

Note that this formulation no longer depends on the exact assignment of pairings to weeks. In doing so, this formulation avoids the symmetry issues of the compact formulation.

2.4 Other Applications

The previous examples illustrate different reasons for taking a branch-and-price approach. In the Generalized Assignment Problem, the motivation is to exploit fast subproblem algorithms; for the Traveling Tournament Problem, the branch-and-price formulation is simpler, since much of the complexity is hidden in the definition of the variables, and provides more powerful LP relaxations; for the Social Golfers Problem, the branch-and-price formulation removes complicating symmetries.

There are many other problems for which a hybrid branch-and-price approach has been proposed. Here we provide a brief description of several applications for which a CP based branch-and-price approach has proved effective.

The airline industry is a fertile field for complex combinatorial problems. The Crew Pairing Problem (CPP) requires grouping flight legs into pairings so as to produce work assignments without associating them with specific crews. In the U.S., pairings are assigned to crew members through a bidding process; however, in other countries, the Crew Assignment Problem (CAP) is solved to find a schedule that dictates which crew members will man which flight pairings plus ground duties, reserve duties and off-duty blocks over a given period of time. Both the CPP and the CAP have traditionally been modeled as set partitioning problems. Current solution approaches fall into two categories: branch-and-price algorithms and approaches that break the problem down into many subproblems to optimize (see (Fahle et al., 2000) and (Klabian et al., 2001)).

(Fahle et al., 2000) solve the CAP for a European airline using a branch-and-price algorithm that employs CP as the subproblem solver. IP based branch-and-price algorithms for the CAP typically use dynamic programming to solve the pricing problem which is formulated as a constrained shortest path problem. In some cases, however, the complex rules associated with crew rostering cannot be fully expressed in a traditional shortest path context. (Fahle et al., 2000) solve their pricing problem with a specialized CP that includes two key global constraints. The first uses path computation to obtain a legal roster while ensuring that the overall reduced cost is negative. The second encapsulates a single source shortest path algorithm in order to lower bound the reduced cost of a roster (or partial roster) within the search tree. The goal in this case is to produce "good" rather than optimal solutions, so the CP used to solve the pricing problem does not need to be run to completion and the search can be limited in order to produce better columns more quickly.

(Chabrier, 1999) uses a CP based branch-and-price algorithm to solve the CPP, also finding good, rather than optimal, solutions. He uses a set covering model and obtains an integral solution at every node. CP is used to generate new pairings and modify existing ones.

Vehicle routing problems (VRPs) are easily decomposed into a master problem that combines individual routes into a complete schedule and a subproblem that generates single vehicle routings. (Rousseau et al., 2002) consider a VRP called the Profitable Tour Problem with two conflicting objectives: to minimize distance traveled and maximize the total amount of "prize" collected. They use CP to solve a resource-constrained shortest path problem on a general graph as the subproblem in their branch-and-price algorithm.

(Yunes et al., 2000) use a hybrid branch-and-price algorithm to solve both the crew scheduling and crew rostering problem for a large bus company in Brazil. While their crew rostering problem is similar is some ways to the CAP described above, its objective is very different. The goal in this case is to provide a schedule that requires a minimum number of crews. The pricing problem is solved using CP.

3. Implementation Issues

The above examples have raised numerous issues regarding the implementation of CP based branch-and-price algorithms. In this section, we discuss some of these issues in more detail.

3.1 Set Partitioning Versus Set Covering

Most branch-and-price algorithms are developed for *set partitioning* problems. The general set partitioning problem has a ground set of elements and rules that dictate how feasible subsets are constructed and how their costs are computed. The goal is to find a minimum cost partition of the ground set into feasible subsets. Let $x_{ij} = 1$ if element i is included in subset j and 0 otherwise. Let x_j denote the characteristic vector of subset j, i.e. the vector with entries x_{ij} for each element i. Let c_j denote the cost of subset j. Then the general partitioning problem can be expressed as

$$\min \quad \sum_j c_j x_j \tag{7.33}$$

$$s.t \quad \sum_j x_{ij} = 1 \quad \forall i \tag{7.34}$$

$$x_j \in S \tag{7.35}$$

$$x_j \text{ binary } \forall j \tag{7.36}$$

In some applications, however, the equality constraints in the above model can be replaced by inequality constraints requiring that each element be included in the solution at least once rather than exactly once. The resulting model is called a *set covering* formulation. The general covering problem is expressed as

$$\min \sum_j c(x_j) \tag{7.37}$$

$$s.t \sum_j x_{ij} \geq 1 \quad \forall i \tag{7.38}$$

$$x_j \in S \tag{7.39}$$

$$x_j \text{ binary } \forall j \tag{7.40}$$

A set covering formulation is preferred over a set partitioning formulation primarily because its LP relaxation is more stable and easier to solve. Secondly, it is trivial to construct a feasible integer solution to a set covering problem from a continuous solution.

Consider a vehicle routing problem formulated so that every customer is visited exactly once and a solution is a partition. Alternatively, it may be formulated so that every customer is visited at least once, and a solution is a cover. Deleting customers from a route, i.e. simply not visiting them, can only decrease the cost of the route. Thus, an optimal solution to the set covering problem will be an optimal solution to the set partitioning problem.

In some cases, the translation from a cover to a partition is not as straightforward, but the set covering model can still be used heuristically. In the CAP, extra crew members can't simply be deleted from flights; there is a substantial cost to "deadhead" airline's personnel. However, (Fahle et al., 2000) are able to make use of a set covering model to produce more meaningful dual values for the pricing problem. Additionally, (Chabrier, 1999) works with a set covering model to solve an airlines flight pairing problem with the concession that it does not produce provably optimal solutions.

3.2 Initial Solution

At the root node, the restricted master problem must provide a feasible LP solution so that dual information can be passed to the column generation subproblem. For some problems, the construction of an initial restricted master problem is not obvious. In this case, a two-phase method similar to that used to find an initial basic feasible solution for the simplex method may be used. Generally, artificial variables with big M costs are added forming an identity matrix, although it suffices to add a single artificial variable with big M cost and a column of all ones to guarantee feasibility. However, this simplification decreases the information feedback on the rows that cannot be covered. Since it is necessary to provide a feasible LP relaxation at every node, the artificial variables are usually maintained throughout the search tree.

In defining an initial restricted master problem, the primary goal is to provide a feasible LP relaxation. However, a secondary goal may be to construct the restricted master problem in such a way that "good" dual information is passed to the column generation process. For example, in a second paper on solving the CAP, (Sellmann et al., 2000) use a CP model to provide complete solution sets of columns for the initial restricted master problem with the goal of providing more accurate dual prices.

3.3 Column Management

In a minimization problem, any column with negative reduced cost can enter the basis, so it is not necessary to require the column with the most negative reduced cost. In a general CP based branch-and-price algorithm, the constraint program used to solve the pricing problem generates one or more negative reduced cost columns at a time. The algorithm iterates between solving the LP relaxation and the pricing problem until there are no more negative reduced cost columns to be added. In order to provide a true LP bound and prove optimality, the subproblem CP must prove that there are no more negative reduced cost columns available.

While some algorithms "price out" or generate columns at every node until the optimality criterion is met, other algorithms only price out when a node is about to be cut off, either when the value of the LP relaxation is worse than the best know integer solution or the LP relaxation is infeasible. There are two important advantages to this strategy. It may reduce the total number of times the pricing problem is solved and reduce the total number of columns added to the master problem. Keeping the master problem as small as possible may limit the time required to solve it. However, unless all negative reduced cost columns are generated at every node, the LP bounds are not "true" which interferes with a number of node selection strategies, including best bound. In order to mitigate this disadvantage, in some cases, it is possible to calculate independent lower bounds.

Another way to reduce the number of times the column generation CP is run is to allow it to generate a substantial number of columns and store these columns in a column pool. Then, when columns need to be added to the master problem, the column pool is checked first. If there are any negative reduced cost columns, they are added to the master problem, and the pricing problem doesn't need to be run at all. This highlights one of the advantages of using CP to solve the pricing problem. While IP models can be set up to generate multiple solutions, CP is particularly well suited to this task.

The goal of some algorithms is not to prove optimality but to simply generate the best possible solution within a reasonable amount of time. In this case, it is not necessary to obtain true LP bounds at any node in the search tree.

Obviously, there is a trade-off involving the number of columns added to the master problem. At one extreme, adding only one column at a time keeps the total number of columns in the master problem relatively low but could require numerous iterations to significantly affect the LP relaxation's bound. At the other extreme, adding all existing negative reduced cost columns at each iteration keeps the number of iterations relatively low but could overload the master problem with columns and increase the solution time for the LP relaxation.

One way to keep the master problem from becoming too large is to delete nonbasic columns with very positive reduced costs. This is usually done after the first iteration of solving the LP relaxation at a given node.

Another column management issue concerns duplicate columns. When the pricing problem is solved, it often generates columns that already appear in the master problem. There are a number of ways to deal with these columns. Some algorithms check each column as it is generated against the columns in the master problem and discard duplicates. Others check each column as it is selected from the column pool and discard duplicates. Still others do not check for duplicates when columns are added to the master problem. In the latter case, the algorithm might perform a sweep at intervals to delete a portion of the duplicate columns existing in the current master problem.

The problem of duplicate columns may be mediated by using a search strategy designed to produce significantly different columns. For example, a *limited discrepancy search* (Harvey and Ginsberg, 1995) follows the majority of choices given by a particular heuristic but allows a set number of "discrepancies." The number of discrepancies allowed is iteratively increased until a solution is found. Another possibility is to introduce an element of randomness into the search strategy so that many different sections of the search tree are explored.

3.4 LP Relaxation

The LP relaxation of the restricted master problem is used for two purposes: to generate the dual values used to "price out" new columns and to inform the branching scheme. Just as there are alternatives to LP for directing the branching strategy, there are other ways to generate prices for the pricing problem. For example, subgradient optimization in the context of lagrangian dual based relaxations could be used in place of the LP relaxations. See (Barnhart et al., 1998) for more discussion on the use of subgradient optimization in this context.

3.5 Branching

Typically, as discussed above, branch-and-price algorithms branch on sets of variables in a way that preserves the simplicity of the pricing problem and divides the solution space equally.

In order to affect this type of branching, it is common to define higher-order branching variables that represent sets of the original variables. In the TTP example above, higher-order pattern set variables are described. Specifically, the variable p_{ts} is equal to 1 if team t is home in slot s and equal to 0 if team t plays away in slot s. Branching on p_{ts}, creating one branch in which $p_{ts} = 1$ and one branch in which $p_{ts} = 0$, is equivalent to setting all columns in which team t has an away game in slot s equal to 0 in the first branch and setting all columns in which team t plays at home in slot s to 0 in the second branch. In the pricing problem, it is easy to fix team t to play at its home venue in slot s in one case and eliminate home from the domain of the venue variables for team t in slot s in the second case.

The higher-order branching strategy used in solving the TTP is a specific example of a general technique that may be described as "branching on rows." Originally proposed by (Ryan and Foster, 1981) for fully-enumerated set partitioning problems, the idea of branching on rows has nonetheless proven very useful in a column generation context.

The key idea in this method is to identify a pair of rows and a set of variables S such that both rows are simultaneously satisfied by setting any one of the variables in S equal to 1. Then in one branch, all the variables with positive coefficients in both rows are set to 0, and in the other branch, all the variables with exactly one positive coefficient in the two rows are set to 0. A branch-and-price algorithm, or more generally, any branch-and-bound algorithm, using this strategy will terminate after a finite number of branches because there are a finite number of pairs of rows.

Formulations with a huge number of variables tend to have a very limited number of rows; thus, branching on rows is a good strategy in that it may produce a relatively small search tree. The question remains, however, of how to select the pair of rows on which to branch at any given node.

There is a general strategy called *strong branching* that has been used successfully in a number of integer programming algorithms including branch-and-price (see (Easton et al., 2002)) to select good branching variables, or for our purposes, row-pairs. Strong branching uses the LP relaxation to test a set of branching candidates for their potential effect on the objective function. A set S of row-pairs containing fractional variables is chosen. Each row-pair is tested by performing a small number of dual simplex iterations of the current master problem for each of its two possible branches. The resulting objective values are used to compare the candidates in S and select a row-pair on which to branch.

Clearly, a critical consideration in selecting a branching strategy is how it will be incorporated into the pricing problem. This is one of the advantages of using CP as a subproblem solver. CP's broad, flexible modeling paradigm permits an exact specification of the column generation problem that may be easier to link to a branching strategy than a shortest path or knapsack model.

3.6 CP as a Subproblem Solver

While we have discussed some general characteristics of CP that may make it an effective algorithm for solving the pricing problem, it is important to point out that specific adaptation of the CP model could be critical to its success. For example, since only columns with negative reduced cost are feasible, the constraint propagation algorithm should exploit the cost information as much as possible in order to eliminate choices that lead to positive reduced costs.

Secondly, branch-and-bound algorithms have traditionally used powerful mathematical programming algorithms to solve subproblems formulated in a particular way, e.g. as knapsack or shortest path problems. These algorithms are often not able to handle additional complex constraints. As discussed above, CP's broad modeling paradigm allows it to succinctly express complex models. This is not to say, however, that CP should completely replace a traditional algorithm which is well suited to a fundamental part of the subproblem. Instead, the CP algorithm might incorporate another algorithms as an *encapsulated constraint*. In this way, traditional algorithms can be used within a CP subproblem solver to both find solutions and eliminate values that do not belong to any feasible solution.

3.7 Column Generation Heuristics

The most obvious way to generate columns in a hybrid branch-and-price algorithm is to run the CP for the pricing problem, producing columns as they arise as feasible integer solutions in the search tree. However, there are numerous heuristics used to either speed the column generation process or provide particularly "good" columns for addition to the master problem.

One idea is to add complete solutions to the restricted master problem as opposed to unrelated negative reduced cost columns. The goal is to help the restricted master problem find good integer solutions and to produce more robust dual information to pass back to the pricing problem. In some cases, a column with low but positive reduced cost might be added in order to provide complete solutions. This technique is used by (Sellmann et al., 2000) in building the initial restricted master problem for the CAP and by (Clements et al., 1997) as a column generation heuristic.

Another idea is to simply use the CP for the pricing problem in a heuristic way by fixing certain elements. In solving the TTP, it has proven useful to populate the column pool not only with columns that have negative reduced cost but also with columns that simply represent low distance tours. The CP used to solve the pricing problem can easily be adjusted to produce columns of this nature as well.

A third idea is to use a set covering model to supply columns for the set partitioning master problem. As discussed above, a set covering model is usually much easier to solve than its set partitioning counterpart. This technique is used by (Sellmann et al., 2000) in solving the CAP. Within the column generation subroutine, a set covering version of the master problem is solved. A CP algorithm is used to convert the solution from a cover into a partition. The repaired columns are then added to the master.

3.8 Combining Column and Row Generation

Some hybrid column generation algorithms not only generate columns but rows as well. In other words, they combine branch-and-price with another branch-and-bound based algorithm called *branch-and-cut*. In a branch-and-cut algorithm, classes of valid inequalities are left out of the master problem at the root node because there are too many of them to consider and most of them wouldn't be binding anyway. When the LP relaxation of the master problem obtains an optimal solution and it is infeasible for the full model, a *separation problem* is solved in order to find violated inequalities. If the separation problem yields violated inequalities, they are added to the master problem as *integrality cuts* and the LP relaxation is reoptimized. If no violated inequalities are found and the LP solution is fractional, the algorithm branches. *Feasibility cuts* may also be added at any node in the search tree to better define the polyhedral representation of the problem. A detailed description of branch-and-cut can be found in (Johnson et al., 2000).

The most critical question that arises in combining branch-and-price and branch-and-cut is how cut generation will affect the pricing problem. The addition of a cut leads to an extra dual value, and there may not exist an obvious way to incorporate this new dual information into the objective coefficients in the pricing problem.

Assuming this difficulty can be overcome, a question may arise as to whether or not to incorporate newly added columns into previously added cuts. For example, a version of the TTP includes a constraint preventing two games between the same two opponents to be played in two consecutive slots. This constraint is called the "no repeaters" constraint. It can be included in the model explicitly; however, it substantially increases the size of the model, the time required to solve the LP relaxation, and consequently, the total solution time for the problem.

Perhaps a more effective way to handle this constraint is to add a cut generation routine to the hybrid branch-and-price algorithm used to solve the TTP. In the resulting algorithm, when a solution is produced, it is checked for feasibility relative to the "no repeaters" constraint. If there exists a set of repeating games in slots s and $s + 1$, then a cut is added to the model specifying that at most one of the set of all columns with either of these games in either of these slots may appear in a solution. Now, a new column is generated. One option is to check all cuts to see if this new column should have a coefficient of 0 or 1 in each. The other option is to set the coefficient for this column to 0 in all cuts without further evaluation. Clearly, there is a trade-off between the computation time required to check all existing cuts to see if the new column might be included and the extra computation time spent to cut off a solution with a repeater involving the new column if one happens to arise.

The literature on branch-and-cut-and-price algorithms is rather sparse. For examples see (Nemhauser and Park, 1991) work on combining column and row generation to solve the edge coloring problem and (Clarke and Gong, 1996) application of a combined column and row generation algorithm to the capacitated network design problem. Also see the survey paper by (Barnhart et al., 1998) for further discussion on adding cut generation to a branch-and-price algorithm.

3.9 Parallel Implementation Issues

Branch-and-price algorithms are usually used to solve large-scale combinatorial problems. In some cases, these problems are so large that a parallel algorithm is required. In such a case, the key issue is the granularity of parallelism. Many constraint programming methods are inherently serial. Fortunately, the branch-and-price algorithm will generate a number of independent subproblems to be solved, so effective speedup is possible. For more on parallel implementations of branch-and-bound algorithms, see (Bixby et al., 1995), (Easton et al., 2002) and (Klabian et al., 2001).

4. Future Directions for CP Based Branch-and-Price

We have seen that constraint programming is an appealing approach for solving the subproblems in a branch-and-price algorithm for a number of reasons:

- Constraint programming may be the fastest way of solving complicated subproblems.

- It is only necessary to find a solution with negative reduced cost, turning the subproblem into a feasibility problem well suited to a constraint programming approach.
- It is useful to generate many solutions with negative reduced cost, again exploiting constraint programming's flexibility.
- Branching rules are less restricted due to the power of constraint programming models.

This leads to a number of interesting research directions:

- It would be useful to exploit the special structure of the "cost" constraint to better propagate restrictions of the form "This formula must be negative."
- It may be possible to find new branching rules that generate naturally strong constraint programs. The branching rules presented here were developed in the integer programming context. Are there branching rules that lead to stronger CP formulations?
- In combining branch-and-price with branch-and-cut, can constraint programming be used to find violated inequalities?

It is clear that the branch-and-price framework provides a strong and natural way to combine integer programming and constraint programming, exploiting the power of each. Future research can improve both the integer programming and constraint programming sides of the algorithm as well as their interaction. Furthermore, future research can serve to identify additional applications that can be effectively solved with this combined methodology.

References

Barnhart, C., E.L. Johnson, G.L. Nemhauser, M.W.P. Savelsbergh, and P.H. Vance, (1998). Branch-and-Price: Column Generation for Huge Integer Programs, *Operations Research*, 46, 316-329.

Barnhart, C., E.L. Johnson, G.L. Nemhauser and P. Vance, (1999). Crew Scheduling, *ITAL Handbook of Transportation Science*, R.W. Hall (ed.), Kluwer, 493-522.

Bixby, R., W. Cook, A. Cox and E. Lee, 1995. Parallel Mixed Integer Programming. Technical Report CRPC-TR95554, Center for Research on Parallel Computation, Rice University.

Caprara, A., F. Focacci, E. Lamma, P. Mello, M. Milano, P. Toth and D. Vigo, 1998. Integrating Constraint Logic Programming and Operations Research Techniques for the Crew Rostering Problem," *Software - Practice and Experience*, 28, 49-76.

Chabrier, A., 1999. A Cooperative CP and LP Optimizer Approach for the Pairing Generation Problem, *Proceedings of the International Workshop on Integration of Artificial Intelligence and Operations Research Techniques in Constraint Programming for Combinatorial Optimization problems, CP-AI-OR '99*, Ferrara, Italy.

Clarke, L.W. and P. Gong, 1996. Capacitated Network Design with Column Generation, Technical Report, Georgia Institute of Technology.

Clements, D.P., J.M. Crawford, D.E. Joslin, G.L. Nemhauser, M.E. Puttlitz and M.W.P. Savelsbergh, 1997. Heuristic Optimization: A Hybrid AI/OR approach, *Proceedings of the CP '97 Workshop on Constraint-directed Scheduling*.

Easton, K., G.L. Nemhauser and M.A. Trick, 2001. The Traveling Tournament Problem: Description and Benchmarks, *Proceedings of the International Conference on Principles and Practice of Constraint Programming - CP '01*, Springer Lecture Notes in Computer Science, 2239, 580-585.

Easton, K., G.L. Nemhauser and M.A. Trick, 2002. Solving the Traveling Tournament Problem: A Combined Integer Programming and Constraint Programming Approach, *Proceedings of the international conference on the Practice And Theory of Automated Timetabling, PATAT '02*.

Fahle, T. and M. Sellmann, 2000. Constraint Programming Based Column Generation with Knapsack Subproblems, *Proceedings of the International Workshop on Integration of Artificial Intelligence and Operations Research Techniques in Constraint Programming for Combinatorial Optimization problems, CP-AI-OR '00*, Paderborn, Germany, 33-44.

Fahle, T., U. Junker, S.E. Karisch, N. Kohl, M. Sellmann and B. Vaaben, 2000. Constraint Programming Based Column Generation for Crew Assignment, Technical Report tr-ri-99-212, University of Paderborn.

Harvey, W.D. and M.L. Ginsberg, 1995. Limited Discrepancy Search, *Proceedings of the Fourteenth International Joint Conference on Artificial Intelligence, IJCAI'95*, 607-615.

Hoffman, K.L. and M. Padberg, 1993. Solving Airline Crew Scheduling Problems by Branch-and-Cut, *Management Science*, 39, 657-682.

Johnson, E.L., G.L. Nemhauser and M.W.P. Savelsbergh, 2000. Progress in Linear Programming Based Branch-and-Bound Algorithms: An Exposition, *INFORMS Journal on Computing*, 12, 1-48.

Junker, U., S.E. Karisch, N. Kohl, B. Vaaben, T. Fahle and M. Sellmann, 1999. A Framework for Constraint Programming Based Column Generation, *Proceedings of the International Conference on Principles and Practice of Constraint Programming*, Lectire Notes in Computer Science, 1713, Springer, 261-274.

Klabjan, D., E. L. Johnson, and G. L. Nemhauser, 2001. Solving Large Airline Crew Scheduling Problems: Random Pairing Generation and Strong Branching *Computational Optimization and Applications*, 20, 73-91.

Martello, S. and P. Toth, 1990. *Knapsack Problems: Algorithms and Computer Implementations*, John Wiley.

Mehrotra, A. and M.A. Trick, 1998. Cliques and Clustering: A Combinatorial Approach, *Operations Research Letters*, 22, 1-12.

Nemhauser, G.L. and S. Park, 1991. A Polyhedral Approach to Edge Coloring, *Operations Research Letters*, 10, 315-322.

Rousseau, L.M., M. Gendreau, and G. Pesant, 2002. Solving Small VRPTWs with Constraint Programming Based Column Generation, *Proceedings of the International Workshop on Integration of Artificial Intelligence and Operations Research Techniques in Constraint Programming for Combinatorial Optimization problems, CP-AI-OR '02*, Le Croisic, France.

Ryan, D.M. and B.A. Foster, 1981. An Integer Programming Approach to Scheduling, in A. Wren (ed.) *Computer Scheduling of Public Transport Urban Passenger Vehicle and Crew Scheduling*, North-Holland, Amsterdam, 269-280.

Sellmann, M., K. Zervoudakis, P. Stamatopoulos and T. Fahle, 2000. Integrating Direct CP Search and CP Based Column Generation for the Airline Crew Assignment Problem, *Proceedings of the International Workshop on*

Integration of Artificial Intelligence and Operations Research Techniques in Constraint Programming for Combinatorial Optimization problems, CP-AI-OR '00, Paderborn, Germany.

Savelsbergh, M.W.P., 1997. A Branch-and-Price Algorithm for the Generalized Assignment Problem, *Operations Research* 45, 831-841.

Trick, M.A., 2002. A Dynamic Programming Approach for Consistency and Propagation for Knapsack Constraints, *Annals of Operations Research*, to appear.

Yunes, T.H., A.V. Moura and C.C. de Souza, 2000. Hybrid Column Generation Approaches for Solving Real World Crew Management Problems, *Lecture Notes in Computer Science*, 1753, 293-332.

Chapter 8

RANDOMIZED BACKTRACK SEARCH

Extending the reach of complete search methods

Carla P. Gomes
Dept. of Computer Science
Cornell University
Ithaca, NY 14853
gomes@cs.cornell.edu

Abstract

Randomized search methods have greatly extended our ability to solve hard computational problems. In general, however, we think of randomization in the context of local search. While local search methods have proved to be very powerful, in some situations they cannot supplant complete or exact methods due to their inherent limitation: Local search methods cannot prove inconsistency or optimality.

In recent years, we have seen the emergence of an active research area focusing on the study and design of complete randomized search procedures. In this chapter we introduce the ideas and concepts underlying such randomized procedures. In particular, we describe a new generation of backtrack-style methods that exploit randomization while guaranteeing completeness. We analyze the run time distributions of such methods and show that quite often they exhibit so-called heavy tails. *Heavy tailed* distributions are non-standard distributions that have infinite moments (*e.g.*, an infinite mean or variance). Such non-standard distributions have been observed in areas as diverse as economics, statistical physics, and geophysics. The understanding of the heavy-tailedness of complete backtrack search methods has led to the design of more efficient procedures. We describe such new techniques. In particular, we present different ways of randomizing complete backtrack style methods and show how restart and portfolio strategies can increase performance and robustness of such methods. We also provide formal models of heavy-tailed behavior in combinatorial search and results on performance guarantees for restart strategies.

Keywords: Complete backtrack search, randomization, heavy tails, restarts, portfolio strategies

1. Introduction

One of the most surprising discoveries in the area of algorithm design is that the incorporation of an element of randomness into the computational process can lead to a significant speedup over purely deterministic methods. For example, randomness lies at the core of highly effective methods for determining whether a given number is prime— a key problem in modern cryptography (Agrawal et al., 2002; Goldwasser and Kilian, 1986; Rabin, 1980; Solovay and Strassen, 1977)— and it is a key component in the popular simulated annealing method for solving optimization problems (Kirkpatrick et al., 1983). More generally, randomization and stochastic strategies have been very successful in local search. Local search methods are often used to solve large-scale combinatorial and optimization problems that are beyond the reach of complete search methods. Such methods start with an initial solution, not necessarily feasible, and improve upon it by performing small "local" changes. There are several ways of implementing local search methods, depending on the choice of initial solution, the types of "local" changes allowed, and feasibility and cost of (intermediate) solutions. Chapter 9 provides a detailed overview of local search.

Two of the key features of local search methods are their flexibility and their ability to sample large portions of the search space. However, local search methods are inherently incomplete, that is, they cannot establish inconsistency or prove optimality, which is critical in many real-world applications.

Given the inherent intractability of NP-complete problems, one often has to resort to exhaustive search procedures that explore the large space of potential solutions in order to find an actual solution (or prove that no solution exists); in the case of optimization problems, the goal is to prove the optimality of a solution. These methods are generally implemented as "backtrack-style" search procedures. The idea is to iteratively construct a solution. If it becomes apparent that the current solution path is neither feasible, nor optimal, the procedure backtracks to explore another solution path. This general strategy underlies complete search methods for solving large classes of combinatorial problems, such as Boolean satisfiability (SAT), constraint programming (CP), and integer programming (IP).

The performance of backtrack-style search methods can vary dramatically, depending on the way one selects the next variable to branch on (the "variable-selection heuristic") and in what order the possible values are assigned to a variable (the "value selection heuristic"). Branching heuristics play a central role in guiding backtrack search procedures toward regions of the search space that contain solutions. In some applications, in fact, backtrack search methods perform better than local search methods, largely as a result of a combination of highly tuned search heuristics and strong propagation mechanisms.

Unfortunately, quite often the branching heuristics provide incorrect search guidance, forcing the procedure to explore large subtrees of the search space that do not contain any solution. As a consequence, backtrack search methods exhibit a large variation in performance. For example, we see significant differences on runs of different heuristics, runs on different problem instances, and, for randomized backtrack search methods, runs with different random seeds. The inherently exponential nature of the search process appears to magnify the unpredictability of search procedures. In fact, it is not uncommon to observe a backtrack search procedure "hang" on a given instance, whereas a different heuristic — or even just another randomized run — may solve that instance quickly.

Various researchers studying the computational nature of search methods on combinatorial problems have informally observed this high variance in the performance of backtrack search methods. A related phenomenon is that of so-called exceptionally hard instances. An instance is considered to be *exceptionally hard*, for a particular search algorithm when it lies in the region where almost all problem instances are satisfiable (*i.e.*, the under-constrained area) but, for a given algorithm, is considerably harder to solve than other similar instances, and even harder than most of the instances in the critically constrained area (Gent and Walsh, 1993; Smith and Grant, 1995). (Selman and Kirkpatrick, 1996), observed that, by simply renaming the variables, one could solve exceptionally hard instances easily. Thus, the "hardness" does not necessarily reside in the instances themselves, but rather in the instances in combination the details of the deterministic algorithm.

The extreme variance in performance of search algorithms has led researchers studying the nature of computationally hard problems to use the median, rather than the mean, to characterize search difficulty, since the behavior of the mean tends to be quite *erratic*, while the median is generally much more stable (Cook and Mitchell, 1998; Gent and Walsh, 1993; Hogg et al., 1996; Kirkpatrick and Selman, 1994; Mitchell and Levesque, 1996; Vandengriend and Culberson, 1999). More recently, it has been shown that the full runtime distributions of search methods provide a better characterization of search methods than just the moments and the median and that they yield much useful information for the design of algorithms (Frost et al., 1997; Gomes et al., 1997; Gomes and Selman, 1997a; Hoos, 1998; Kwan, 1995; Rish and Frost, 1997). In particular, (Gomes et al., 1997), showed that the runtime distributions of complete backtrack search methods quite often exhibit *heavy-tails*. Heavy-tailed distributions are characterized by infinite moments (*e.g.*, they can have infinite mean, or infinite variance). Heavy tails have been observed in aggregated runtime distributions of backtrack search methods when considering a collection of instances of the same class (*e.g.*, random binary CSP instances generated with the same parameter space). They have also been encountered when running a randomized backtrack search procedure on the *same* instance several times, but where *randomization was only used to break ties in the variable and/or value selection.* Gomes *et al.* further showed how randomized *restarts* of search procedures can dramatically reduce the variance in the search behavior. In fact, they demonstrated that a search strategy with restarts provably eliminates heavy tails (Gomes et al., 2000). Consequently, the introduction of an element of randomness into a backtrack search procedure can lead to a family of runs with different runtimes, which can be exploited in conjunction with restart strategies in order to obtain a procedure with considerably better performance, *on average.*

The stringent completeness requirements of several applications has led to a tremendous growth in research aimed at boosting performance of complete backtrack-style methods. A good example is the area of software and hardware verification. Verification problems can be encoded as Boolean satisfiability (SAT) problems. The SAT community has embraced and extended randomization and restart techniques to boost performance of complete search methods for SAT. In fact, randomization and restart strategies are now an integral part of many state-of-the-art SAT solvers (Bayardo and Schrag, 1997; Goldberg and Novikov, 2002; Li, 1999; Lynce and Marques-Silva, 2002a; Moskewicz et al., 2001; Zhang, 2002). The world's fastest SAT solver at the present time, called Chaff, incorporates such techniques (Moskewicz et al., 2001). In Chaff, restarting is combined with clause learning, another technique central to Chaff's overall effectiveness. In clause learning, whenever a conflict (*dead end*) is reached, a *nogood* is recorded and carried over between restarts (Bayardo and Schrag, 1997; Marques-Silva and Sakallah, 1999; Lynce and Marques-Silva, 2002c)). Informally, a nogood provides information about inconsistent (partial) assignments inferred from the search performed thus far. Chaff can handle problem instances with over one million variables and four million constraints.

In summary, the study and design of complete randomized search procedures is an emerging and active research area. The understanding of the heavy-tailed nature of the distributions underlying such search methods has led to the design of more efficient backtrack search techniques. In this chapter we describe such new techniques. In section 2 we discuss different ways of randomizing complete backtrack-style methods. In section 3 we present formal models that provably exhibit heavy-tailed behavior. In section 4 we characterize in detail *heavy-* and *fat*-tailedness. In section 5 we provide empirical distributions of randomized backtrack search methods that illustrate heavy-tailed behavior. In sections 6 and 7 we show how restart and portfolio strategies can increase performance and robustness of randomized backtrack search methods. Our conclusions, together with suggested directions for future research, are presented in section 8.

2. Randomization of Backtrack Search Methods

An emerging area of research is the study of Las Vegas algorithms. These are *exact* randomized algorithms; they always return a model satisfying the constraints of the search problem, proving optimality in the case of optimization problems— or proving that no such model exists in the case of infeasible problems (Motwani and Raghavan, 1995). Contrast Las Vegas algorithms with, for example, Monte Carlo algorithms, randomized methods that can provide incorrect answers. The class of randomized local search methods is an example of a class of Monte Carlo algorithms, given their inherent incompleteness.

A randomized algorithm can be viewed as a probability distribution on a set of deterministic algorithms. The behavior of a randomized algorithm can vary even on a single input, depending on the random choices made by the algorithm. The classical adversary argument for establishing lower bounds on the runtime of a deterministic algorithm is based on the construction of a input on which the algorithm performs poorly. Therefore, different inputs may have to be constructed for each deterministic algorithm. When we consider a randomized algorithm, we are implicitly considering a choice of one algorithm at random from a whole family of algorithms, and while an adversary may be able to construct an input that foils one (or a small fraction) of the deterministic algorithms in the set, it is more difficult to devise inputs that are likely to defeat a randomly chosen algorithm. For many problems, randomized algorithms have been shown to be more efficient than the best-known deterministic algorithms, and, in general, they are simpler to describe than deterministic algorithms of comparable performance (Motwani and Raghavan, 1995).

We consider general techniques for randomizing backtrack search algorithms while maintaining completeness. Backtrack search provides a general framework for solving hard computational problems.

Backtrack search methods construct a solution incrementally. At each step, a heuristic is used to select an operation to be applied to a partial solution, such as assigning a value to an unassigned variable. Eventually, either a complete solution is found, or the algorithm determines that the current partial solution is inconsistent (or suboptimal), in which case the algorithm backtracks to an earlier point in its search tree.

Figure 8.1 illustrates the application of a simple backtrack search algorithm to the Boolean satisfiability problem, looking for an assignment to the variables A, B, and C that satisfies the formula $(A \vee \neg B \vee \neg C) \wedge (B \vee \neg C) \wedge (A \vee C)$. The figure shows two different executions of the backtrack search procedure. In the search tree shown on the left, the variable-choice heuristic picks the variables in the order A, B, then C, and always tries the value 0 before the value 1. Execution is traced along the left-most branch of the tree until a constraint is determined to be false, creating a dead-end (marked X). The algorithm then

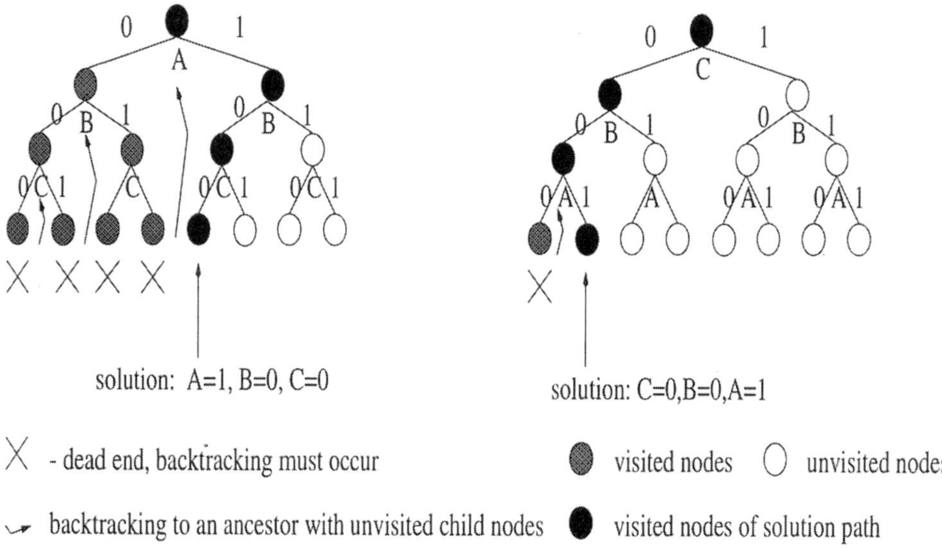

Figure 8.1. Two different executions of a backtrack search methods looking for an assignment to the variables A, B, C that satisfies the formula $(A \vee \neg B \vee \neg C) \wedge (B \vee \neg C) \wedge (A \vee C)$

backtracks up a level in the tree and tries the right branch (assigning 1 to the variable). To backtrack from a right branch, the algorithm goes up two levels, and so on. In the search tree shown on the right, the variable-choice heuristic picks the variables in the order C, B, then A, and, again, always tries the value 0 before the value 1. The execution depicted on the right is shorter than the one on the left.

There are several techniques to enhance the performance of backtrack-style search methods, in particular *look-ahead* and *look-back* techniques. Look-ahead techniques exploit information about the remaining search space. Pruning, cutting planes, and constraint-propagation mechanisms that enforce consistency on the basis of previous assignments are examples of look-ahead techniques. Look-back techniques, on the other hand, include *learning* and forms of *intelligent backtracking*, *i.e,* techniques to reason about the search space already explored. For example, in intelligent backtracking, the causes of inconsistency are analyzed at the time at which backtracking is to occur, so that branches that have been proved not to contain solutions can be skipped.

There are, therefore, several opportunities to introduce randomization into a backtrack search method, basically at the different decision points where the next operation to be performed by the backtrack solver is to be selected. In particular, randomization can be incorporated into the following operations:

- variable selection

- value selection
- look-ahead procedures
- look-back procedures
- backtrack-points
- target-point selection (at the time of backtracking)

In general, backtrack search procedures use *deterministic* heuristics to select the next operation. Highly tuned heuristics are used for variable and value selection. If several choices are heuristically determined to be equally good, then a deterministic algorithm applies some fixed rule to pick one of the operations, for example, using lexicographical order. Therefore, the most straightforward way to apply randomization is in this tie-breaking step: If several choices are ranked equally, choose among them at random (Gomes et al., 1997; Gomes et al., 2000). Even this simple modification can dramatically change the behavior of a search algorithm, as we will see below. If the heuristic function is particularly powerful, however, it may rarely assign the highest score to more than one choice. To handle this, we can introduce a "heuristic-equivalence" parameter to the algorithm. Setting the parameter to a value H greater than zero means all choices that receive scores within H percent of the highest score are considered equally good. This expands the choice set for random tie-breaking (Gomes et al., 1998a).

A more general procedure, which includes the tie-breaking and the H-percent equivalence-window procedures as special cases, imposes a probability distribution on the set of possible values for a given operation. For example, in the tie-breaking case, we are implicitly assigning a uniform probability to the values that are considered by the heuristic to be equally good, while in the case of using the H-percent window, we assign a uniform probability to all choices whose scores lie within H percent of the highest score. An alternative would be to impose a different distribution on the values within the H-percent window (for example, an exponential distribution), biasing the selection toward the highest score.

Random reordering is another way of "randomizing" an algorithm. This technique involves randomly reordering the input data, followed by the application of a deterministic algorithm (Motwani and Raghavan, 1995). (Walsh, 1999), applied this technique to study the runtime distributions of graph-coloring problems, using a deterministic, exact coloring algorithm based on DSATUR (Trick, 1996); see also section 5.4.

A key issue when introducing randomization into a backtrack search algorithm is that of guaranteeing completeness. We note that the introduction of randomness into the branching-variable/value selection does not affect the completeness of the backtrack search. Some basic bookkeeping ensures that the procedures do not revisit any previously explored part of the search space, which means that, unlike the use of local search methods, we can still determine inconsistencies. The bookkeeping mechanism involves keeping track of some additional information for each variable in the stack, namely which assignments have been tried thus far. This can be done relatively inexpensively.

Randomization of the look-ahead and look-back procedures can also be done in a relatively straightforward manner, for example, by randomizing the decision of whether to apply such procedures after each variable/value assignment. This technique can be useful, especially when look-ahead techniques are expensive. On the other hand, randomization of the backtrack points requires the use of data structures that are more complicated in general, so there may be a substantial increase in the time/space requirements of the algorithm. Lynce *et al.*, (Lynce et al., 2001; Lynce and Marques-Silva, 2002b)) discuss learning strategies and randomization of the backtrack points (specifically, random backtracking and unrestricted backtracking), as well as how to maintain completeness (see also section 6.6).

We should note that when implementing a randomized algorithm, one uses a pseudo random-number generator, which is in fact a deterministic algorithm for generating "random" numbers. By choosing different initial random seeds, we obtain different runs of the algorithm. For experimental purposes, it is important to save the "seed" given to the random-number generator, so that the same experiment can be replayed.

One can also speak of "deterministic randomization" (Wolfram, 2002), which expresses the fact that the behavior of some very complex deterministic systems is so unpredictable that it actually appears to be random. This notion of deterministic randomization is implicitly used, for example, in calculating "random" backtrack points by applying a deterministic formula to the clauses learned during search (Lynce et al., 2001; Lynce and Marques-Silva, 2002b). Another example is the restarting of (deterministic) backtrack search solvers that feature clause learning: Each time the solver is restarted, it behaves quite differently from the previous run (because of the additional learned clauses), and thus appears to behave "randomly" (Moskewicz et al., 2001; Lynce and Marques-Silva, 2002b).

As mentioned earlier, the performance of a randomized backtrack search algorithm can vary dramatically from run to run, even on the same instance. As a consequence, runtime distributions of backtrack search algorithms are often characterized by heavy tails. In the next section we provide formal models characterized by such non-standard distributions.

3. Formal Models of Heavy-Tailed Behavior

In order to provide an intuitive feel for the notion of heavy-tailedness, in this section we present several applications for which formal results demonstrate heavy-tailed behavior of the underlying distributions. We start by introducing the random-walk model, also known as the aimless wanderings of a "drunkard," which is characterized by an infinite mean. We then describe a tree-search model and indicate which values of the relevant parameters yield both an infinite mean and an infinite variance, and which yield just an infinite variance. We also discuss the notion of *bounded* heavy tails within the context of the tree-search model.

3.1 Random Walk

Consider a one-dimensional, symmetric random walk, where at each time step one takes a unit step up or down with equal probability (Feller, 1968; Feller, 1971); see figure 8.2 (top panel), where the quantities plotted on the horizontal and vertical axes are the number of time steps and the distance from the origin, respectively, and both are in units of 1000). One can show that after starting at the origin, with probability 1, the walk will eventually return to the origin. The *expected* time before return is infinite, however, and on average a walk of this sort will reach all possible points before its return. Another intriguing phenomenon involves the expected number of returns to the origin ("zero-crossings") in a given number of steps. Intuition would dictate that if in k steps one has l crossings on average, then in a walk that is m times as long, one would expect on average $m \times l$ crossings. It can be shown, however, that in $m \times k$ steps, one will observe only $\sqrt{m} \times l$ crossings, on average. This means that there can be surprisingly long periods between crossings (see figure 8.2 (top panel); note, for example, the long period between step 2544 and step 7632, during which time there is no crossing). In fact, when doing a series of r random walks, each terminating at the first crossing, some of the walks will be of the same order as the length of all the other walks combined (on average), *no matter what the value of r is.* Such events would normally be dismissed as outliers, but when dealing with heavy-tailed distributions, they are far from rare and are an inherent aspect of the distribution. These distributions are, therefore, good models for dealing with phenomena that exhibit extreme fluctuations.

A visual illustration of the heavy-tail behavior of the runtime distribution of a random walk is depicted in the bottom panel in figure 8.2. To generate this distribution we used simulation data for 10,000 runs of a symmetric random walk. For each run we recorded the number of steps X it took to return to the origin for the first time. If x is the number of time steps , let $f(x)$ denote the empirical probability that the walk returned to the origin in *exactly* x steps, and let $F(x)$ be the (cumulative) probability of returning in *at most* x steps (i.e., $f(x) = \Pr\{X = x\}$ and $F(x) = \Pr\{X \leq x\}$). In the figure, we plot the complement-to-one of the cumulative distribution, namely, $1 - F(x)$, which is just the probability that the walk did *not* return to the origin within x steps. (Note that $1 - F(x) = 1 - \Pr\{X \leq x\} = \Pr\{X > x\}$.) Of course, none of the runs could have returned to the origin after just one step, so $f(1) = 0$. Moreover, $f(2) = 0.5$, so we have $F(2) = 0.5$, which means that in our experiment the walk had a 50% chance of returning to the origin in at most two steps.

In the figure, we give a log–log plot of $1 - F(x)$. (In the figure, the walk data is given by the diagonal straight line.) As the figure shows, we obtain a nearly straight line for the tail of the distribution. This suggests that the function $1-F(x)$ has power-law decay, *i.e.*, we have $1-F(x) = \Pr\{X > x\} \sim Cx^{-\alpha}$, and thus the distribution is heavy tailed (see section 4.1.2). The slope of the line gives us an estimate of the index of stability, α, which in this case is equal to 0.5 (also known by rigorous analysis). The relatively high frequency of large outliers is clear from the figure. For example, although 50% of the walks return in just 2 steps, 1% of the walks take more than 5,000 steps to return to the origin, and about 0.1% take over 200,000 steps. In fact, several of the walks in our sample take almost $1,000,000$ steps.

To demonstrate how different such a heavy-tailed distribution is from a standard distribution, we include in the figure a plot of the complement-to-one of the cumulative distribution of a normal distribution, using simulation data and a mean value of 2. We present curves for two different values of the standard deviation: $\sigma = 1$ (left-most curve) and $\sigma = 10^6$ (right-most curve). The key property to observe is the sharp, faster-than-linear decay of the normal distribution in the log–log plot, which is consistent with the exponential decay in the tail of the distribution. We included a normal distribution with $\sigma = 10^6$ to show that the drop-off in the tail remains sharp even when the normal distribution has a large standard deviation. (The normal distribution is symmetrical; the figure gives only the right-hand side.)

3.2 Tree Search Model

Intuitively, heavy-tailed behavior in backtrack-style search arises from the fact that wrong branching decisions may lead the procedure to explore an exponentially large subtree of the search space that contains no solutions. Depending on the number of such "bad" branching choices, one can expect considerable variation in the time to find a solution from one run to another. Heavy-tailed behavior has been shown not to be inherent to backtrack search in general, but rather to depend on the structure of the underlying tree search and on the *pruning* power of the heuristics (Chen et al., 2001).

In our analysis, we contrast two different tree-search models: a balanced model that does not exhibit heavy-tailed behavior, and an *imbalanced* model.

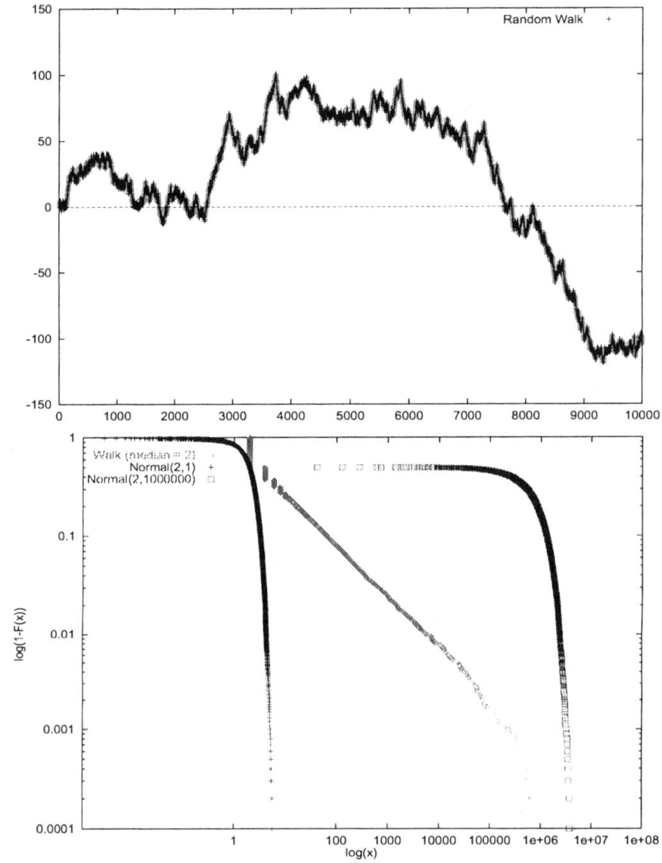

Figure 8.2. Top panel: symmetric random walk (10,000 steps). Bottom panel: the heavy-tailed nature of a random walk contrasted with the exponential decay of the normal distribution.

A key component of the imbalanced model is that it allows for highly irregular and imbalanced trees, which are radically different from run to run. For the imbalanced tree-search model, we formally show that the runtime of a randomized backtrack search method is heavy tailed for a range of values of the model parameter p, which characterizes the effectiveness of the branching heuristics and pruning techniques. The heavy-tailedness leads to a runtime distribution with an infinite variance, and sometimes an infinite mean. The imbalanced tree model is depicted in the top panel in figure 8.3.

The imbalanced tree model assumes that the probability that the branching heuristic will make a wrong decision is p, hence that, with probability $(1-p)$ the search is guided directly to a solution. With probability $p(1-p)$, a search space of size b, with $b \geq 2$, needs to be explored. In general, with probability $p^i(1-p)$, a search space of b^i nodes needs to be explored. Intuitively, p is the probability

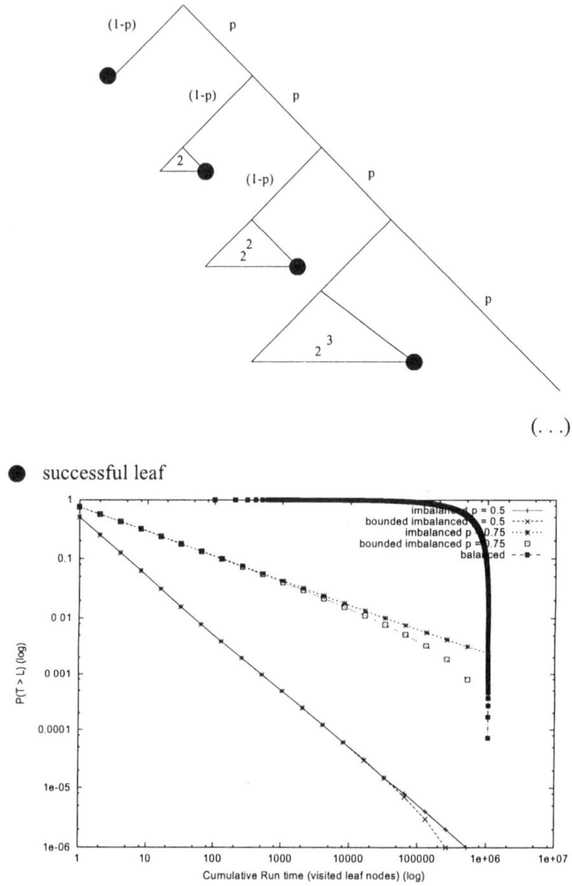

Figure 8.3. Top panel: imbalanced tree model; p is the probability that the branching heuristic will make a wrong decision, and b is the branching factor (in this case, $b = 2$). This tree model has a finite mean and an infinite variance when $1/b^2 < p < 1/b$; both the mean and the variance are infinite when $p \geq 1/b$. Bottom panel: example distributions the imbalanced and bounded imbalanced models, contrasted with the balanced model. Parameters: $b = 2$, $n = 20$, $p = 0.5$ and 0.75 (n is the number of variables).

that the overall amount of backtracking increases geometrically by a factor of b. This increase in backtracking is modeled as a global phenomenon. The larger b and p are, the "heavier" the tail. Indeed, when b and p are sufficiently large, so that $bp \geq 1$, the expected value of T (where T is the runtime of the algorithm)[1] is infinite: $E[T] = \infty$. If, however, $bp < 1$ ("better search control"), we obtain a finite mean of of $E[T] = (1-p)/(1-pb)$. Similarly, if $b^2p > 1$ the variance of T is infinite; otherwise, it is finite.

Bounded Heavy-Tailed Behavior for Finite Distributions. The imbalanced tree-search model does not put an *a priori* bound on the size of the search space. In practice, however, the runtime of a backtrack search procedure is bounded above by some exponential function of the size of the instance. We can adjust our model by considering heavy-tailed distributions with bounded support — "bounded heavy-tailed distributions", for short (see *e.g.*, (Harchol-Balter et al., 1998)). Our analysis of the bounded case shows that the main properties of the runtime distribution observed for the unbounded, imbalanced search model have natural analogues when dealing with finite but exponential-sized search spaces. Heavy-tailed distributions have infinitely long tails with power-law decay, while bounded heavy-tailed distributions have exponentially long tails with power-law decay. Also, the concept of an infinite mean in the context of a heavy-tailed distribution translates into a mean that is exponential in the size of the input, when considering bounded heavy-tailed distributions.

The bottom panel in figure 8.3 illustrates and contrasts the distributions for the various models. We use a log-log plot of the tails of the various distributions to highlight the differences between them. The linear behavior over several orders of magnitude for the imbalanced models is characteristic of heavy-tailed behavior (see section 4.1.2). The drop-off at the end of the tail of the distribution for the bounded case illustrates the effect of the boundedness of the search space. However, given the relatively small deviation from the unbounded model (except at the very end of the distribution), we see that the effect of bounding is relatively minor. In the plot we also contrast the behavior of the imbalanced model that of a balanced model (for details, see (Chen et al., 2001)). The sharp drop-off for the balanced model indicates the absence of heavy-tailedness.

The heavy-tailedness of both of the imbalanced tree-search models occurs as a consequence of the competition between two critical factors: an exponentially increasing penalty in the size of the space to be searched as the number of "mistakes" caused by the branching heuristic increases, and an exponentially decreasing probability that a series of branching mistakes will be made.

4. Heavy and Fat-Tailed Distributions

We first identified the phenomenon of heavy tails in combinatorial search in a study of the quasigroup (or Latin-square) completion problem (QCP) (Gomes et al., 1997; Gomes et al., 2000). A Latin square, which is an abstract structure from finite algebra, is an NxN table on N symbols in which each symbol occurs exactly once in each row and column. A Latin square with N rows corresponds to the multiplication table of a quasigroup of order N. QCP consists in determining whether a partially filled Latin square can be completed in such a way that we obtain a full Latin square (see figure8.4). QCP is NP-complete (Colbourn, 1984). QCP was introduced as a benchmark problem for evaluating combinatorial search methods, since its structure is similar to that found in real-world problems such as scheduling, timetabling, routing, and design of statistical experiments. In fact, QCP directly maps into the problem of fiber-optics routing, which explains the use of the term *Latin Routers* (Laywine and Mullen, 1998; Kumar et al., 1999).

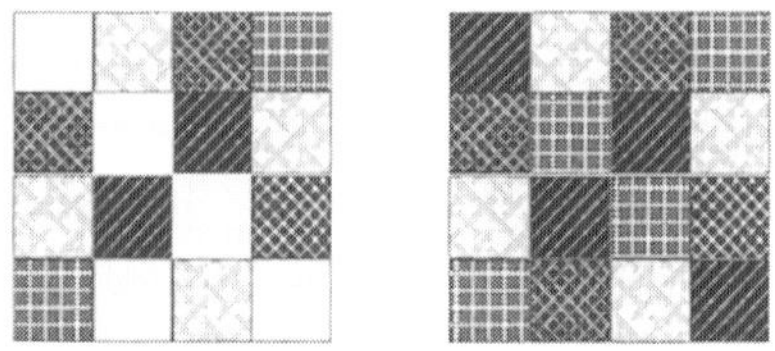

Figure 8.4. Quasigroup Completion Problem (QCP).

In our study, we encoded QCP as a constraint satisfaction problem (CSP), associating a different variable with each cell of the Latin square (Gomes and Selman, 1997b). Given a partial Latin square of order N, QCP can be expressed as follows:

$$x_{i,j} \in \{1, \ldots, N\} \quad \forall i, j$$

$$x_{i,j} = k \quad \forall i, j \text{ such that } \mathrm{QCP}_{ij} = k$$

$$\texttt{alldiff}\ (x_{i,1}, x_{i,2}, \ldots, x_{i,n}), \quad i = 1, 2, \ldots, n$$

$$\texttt{alldiff}\ (x_{1,j}, x_{2,j}, \ldots, x_{n,j}), \quad j = 1, 2, \ldots, n.$$

$x_{i,j}$ denotes the symbol in cell i, j, and the statement "$\mathrm{QCP}_{ij} = k$" denotes that symbol k is pre-assigned to cell i, j.

The `alldiff` constraint states that all the variables involved in the constraint have to have different values.

For our experiments we used a randomized backtrack search procedure, with the so-called first-fail heuristic combined with forward checking. In the first-fail heuristic, the next variable to branch on is the one with the smallest remaining domain, *i.e.*, the search procedure chooses to branch on the variable with the fewest possible options left to explore, thereby leading to the smallest branching factor. In case of a tie, the standard approach is to break the tie using lexicographical order. Since we were interested in analyzing the variance of the backtrack search procedure on a particular instance, *isolating it from the variation that results from considering different instances*, we simply replaced the standard (lexicographical) order for tie-breaking with random tie-breaking. With this change, we obtained a *randomized* backtrack search procedure. In other words, each run of our randomized backtrack search algorithm on the *same* instance may vary in the order in which choices are made —- and, potentially, in solution time as well. Again, we note that introducing randomness into the branching-variable/value selection does not affect the completeness of the backtrack search.

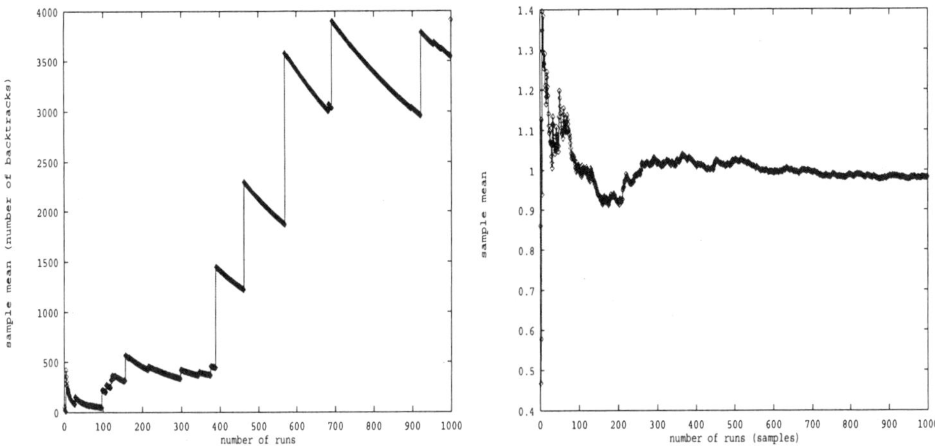

(a) (b)

Figure 8.5. Erratic behavior of mean cost of completing a quasigroup (order 11, 30% pre-assignment) vs. stabilized behavior of mean for a standard distribution (gamma)

In order to obtain *uncensored* empirical runtime distributions for our backtrack search procedure, *i.e.*, without having to interrupt any run, we considered relatively easy QCP instances, of order 11. The phase transition in solvability of QCP, for instances of order 11, occurs when approximately 42% of the cells of the Latin square are prefilled, with a corresponding peak in complexity (Gomes and Selman, 1997b). In our experiments, we sampled instances from the under-constrained area (30%). For each empirical runtime distribution, we typically performed at least $1,000$ runs of the randomized backtrack search procedure on the same instance.

Figure 8.5(a) shows the mean cost of our randomized backtrack-style search procedure on our QCP instance of order 11 with 30% pre-assignment, calculated over an increasing number of runs. The figure illustrates some of the puzzling features of randomized complete search procedures, such as extreme variability in (and a seemingly wandering value of) the sample mean. In other words, the mean of the distribution exhibits highly erratic behavior that does not stabilize with increasing sample size. Again, we note that this instance was relatively *easy*. In fact, 60% of the runs took no more than *one* backtrack. (The median, not shown in the figure, is 1, and it stabilizes rather quickly.) Contrast this behavior with that of the mean of the standard probability distribution[2] given in figure 8.5(b). In this plot, which depicts a gamma distribution, we see that the sample mean converges rapidly to a constant value as the sample size increases.

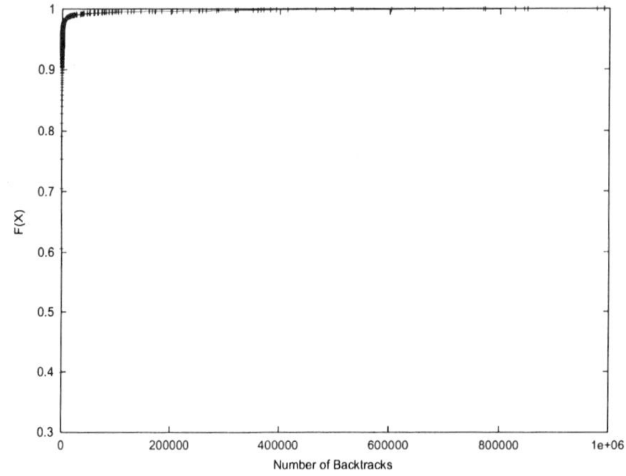

Figure 8.6. The phenomenon of long tails with randomized complete search procedures (QCP instance, order 11, 30% pre-assignment)

Figure 8.6 provides some insight into the wandering-mean phenomenon. The figure shows the surprisingly long "tail" of the distribution. The curve gives the empirical cumulative distribution of the search cost (measured in number of backtracks). It was produced by running the randomized backtrack search procedure 1,000 times on the same instance, namely, the one shown in figure 8.5 (a) (order 11, 30% pre-assignment). Even though 60% of the runs were solved with no more than one backtrack, 0.5% of them were still not solved after $100,000$ backtracks, and some runs took over $1,000,000$ backtracks.

In order to model the long tails of runtime distributions of randomized backtrack search we consider *heavy-tailed* distributions.

4.1 Heavy-Tailed Distributions

Heavy-tailed distributions were first introduced by the Italian-born Swiss economist Vilfredo Pareto in the context of income distribution. They were extensively studied mathematically by Paul Lévy in the period between the world wars. Lévy worked on a class of probability distributions with heavy tails, which he called *stable* distributions. At the time, however, these distributions were largely considered probabilistic curiosities or pathological cases, mainly used as counterexamples. This situation changed dramatically with Mandelbrot's work on fractals. In particular, two seminal papers by Mandelbrot were instrumental in establishing the use of stable distributions for modeling real-world phenomena (Mandelbrot, 1960; Mandelbrot, 1963).

Recently, heavy-tailed distributions have been used to model phenomena in areas as diverse as economics, statistical physics, and geophysics. In particular, they have been applied to stock-market analysis, Brownian motion, weather forecasting, and earthquake prediction — and even for modeling of time delays on the World Wide Web (see *e.g.*, (Adler et al., 1998; Mandelbrot, 1983; Samorodnitsky and Taqqu, 1994)).

4.1.1 Definition. Heavy-tailed distributions have tails that are asymptotically of the Pareto-Lévy form:

$$\Pr\{X > x\} \sim Cx^{-\alpha}, \quad x > 0. \tag{8.1}$$

where α is a positive constant. These are distributions whose tails exhibit a *hyperbolic* (i.e., slower than exponential) decay. In the case treated here, it suffices to consider this tail behavior for only positive values of the random variable X, so in what follows we will assume that the distribution is supported on only the positive line (hence that $\Pr\{0 \leq X < \infty\} = 1$).

The constant α is called the *index of stability* of the distribution, because *which* moments of X (if any) are finite is completely determined by the tail behavior. Note that $\alpha = \inf\{r > 0 : E[X^r] = \infty\}$; hence all the moments of X which are of order strictly *less than* α are finite, while those of higher order ($\geq \alpha$) are infinite. For example, when $\alpha = 1.5$, the distribution has a finite mean but not a finite variance. With $\alpha = 0.6$, the distribution has neither a finite mean nor a finite variance.

(Mandelbrot, 1983), provides an excellent introduction to heavy-tailed distributions, with a discussion of their inherently self-similar or fractal nature. Many aspects of random walks involve heavy-tailed distributions (see *e.g.*, (Feller, 1968; Feller, 1971)). (Adler et al., 1998), provide a collection of essays concerning techniques and approaches for the analysis of heavy-tailed distributions, as well as applications of heavy-tailed modeling (*e.g.*, in the areas of telecommunications, Web, and finance). Heavy-tailed distributions are closely related to stable distributions. For a complete treatment of stable distributions, see either (Zolotarev, 1986), or the more modern approach of (Samorodnitsky and Taqqu, 1994).

4.1.2 Visualization of Heavy-Tailed Behavior. In order to check for the existence of heavy tails in our distributions, we start by plotting and analyzing the behavior of the tails.

If a Pareto-Lévy tail is observed, then the rate of decrease of the estimated probability density of a distribution is hyperbolic — *i.e.*, slower than exponential. The complement-to-one of the cumulative distribution $1 - F(x)$ ($= \Pr\{X > x\}$), also displays hyperbolic decay. By equation 8.1, we have $1 - F(x) \sim Cx^{-\alpha}$.

Given the hyperbolic decay of the complement-to-one of the cumulative distribution of a heavy-tailed random variable, the tail of a log–log plot of its empirical distribution should show an *approximately linear decrease*; moreover, the slope of the observed linear decrease provides an estimate of the index α. This is in contrast to a distribution that decays exponentially, where the tail of a log–log plot should show a faster-than-linear decrease. Figure 8.7(a) displays

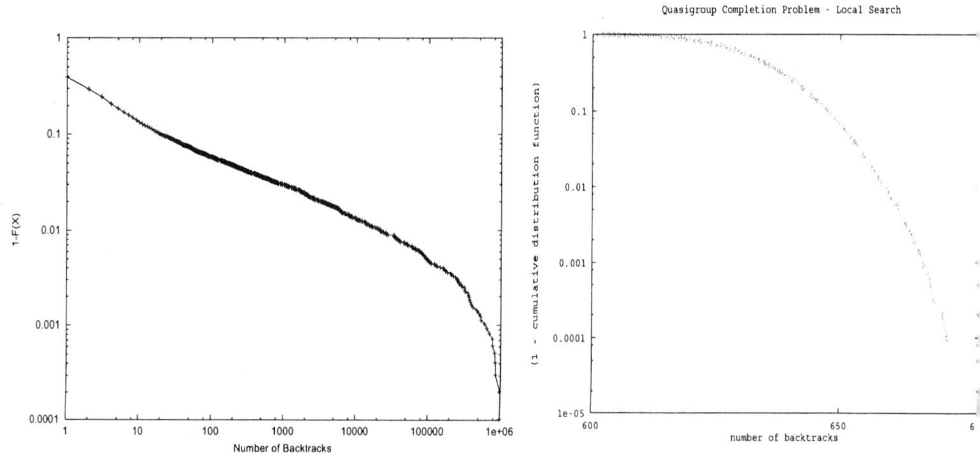

(a) (b)

Figure 8.7. Left panel: log–log plot of heavy-tailed behavior (QCP instance, order 11, 30% pre-assignment; uncensored data). Right panel: no heavy-tailed behavior.

a log–log plot of the complement-to-one of the cumulative distribution for our QCP data (order 11, 30% pre-assignment). The linear nature of the tail of this log–log plot extends over several orders of magnitude, which strongly indicates that the tail is of the Pareto-Lévy type. We were able to obtain the data for this plot by running our backtrack search procedure without a *cutoff* in the number of backtracks — *i.e.*, the data were *uncensored*. (The small drop-off at the end of the tail corresponds to less than 0.2% of the data, and thus is not statistically significant; see also section 3.2.)

By way of comparison, in figure 8.7(b) we show a log–log plot of the complement-to-one of the cumulative distribution of a randomized *local* search procedure on a harder quasigroup instance. (Local search procedures solve our original instance — of order 11, with 30% pre-assignment — trivially.) It is clear from the plot that the distribution for the harder instance does not exhibit heavy-tailed behavior, given the faster-than-linear decay of the tail. (See also section 5.)

4.1.3 Estimation of Index of Stability (α). To supplement the visual check of heavy-tailed behavior, we estimate the index of stability, α.

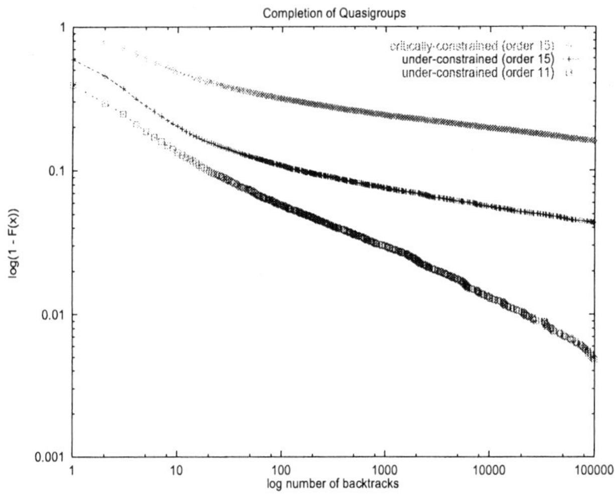

Figure 8.8. Log-log plot of heavy-tailed behavior.

There is a standard estimator, called the Hill estimator (Hill, 1975; Hall, 1982), for determining the index of stability. Quite often, however, we cannot use this analytical tool, since the final outcomes of some of our runs are not observable: In practice, quite often we have to run the experiments with a certain *high cutoff* in the number of backtracks (*i.e.,* we have to resort to the use of *censored* data), to avoid literally getting stuck in the most extreme outliers in the tail of the distribution during our data collection. The selection of the cutoff depends on the problem instance and the algorithm. Obviously, the higher the better. In general, we try to gather data over a fairly sizable range in the value of the random variable, ideally over at least 5 to 6 orders of magnitude (i.e., with the number of backtracks in the range 10^5–10^6), in which case we see some indication of stabilization of the tail behavior. Ideally, we would like not to exclude a significant portion of the tail, but sometimes this is just too costly from a computational point of view.

Figure 8.8 depicts the tails of one critically constrained QCP instance (order 15, 40% pre-assignment) and two solvable under-constrained QCP instances (one being of order 15, with 30% pre-assignment, and the other being the same instance of order 11, with 30% pre-assignment, that was used in figures 8.5(a), 8.6, and 8.7). The computational cost for the instances of order 15 was so high that we were forced to censor the runs, with a cutoff of 10^5 backtracks.

We adapted the Hill estimator for application to a truncated set of data by deriving the following maximum-likelihood estimator for the index of stability

Cases	k	u	α
Quasigroup (order 11, 30%)	10,000	0	0.466 (0.009)
Quasigroup (order 15, 30%)	10,000	431	0.319 (0.006)
Quasigroup (order 15, 40%)	10,000	1933	0.153 (0.003)

Table 8.1. Estimates of the index of stability (α). k is the sample size, and u is the number of runs that failed to terminate before reaching the cutoff in the number of backtracks. The values within parentheses are the estimated asymptotic standard deviations.

α (for details, see (Gomes et al., 2000)):

$$\hat{\alpha}_{r,u} = \left(\frac{1}{r} \sum_{j=1}^{r-1} \ln X_{n,n-r+j} + \frac{u+1}{r} \ln X_{n,n} - \frac{u+r}{r} \ln X_{n,n-r} \right)^{-1} \tag{8.2}$$

where n is the number of runs that terminate within the maximum number of backtracks allowed in the experiment, $n + u$ is the total number of runs, and $X_{n,1} \leq X_{n,2} \leq \cdots \leq X_{n,n} \leq X_{n,n+1} \leq \cdots \leq X_{n,n+u}$ is a numerical ordering of the observations in the sample (*i.e.,* the observed numbers of backtracks for different runs of our procedure). Because of the imposition of the cutoff, the highly extreme values $X_{n,n+1}, X_{n,n+2}, \ldots, X_{n,n+u}$ are not observable. The parameter r (which is less than n) translates into a lower-bound cutoff on search cost, which enables us to focus only on the tail end of the distribution (note that only the observations $X_{n,n-r} \leq X_{n,n-r+1} \leq \cdots \leq X_{n,n}$ enter into the estimation of α). If all the runs terminate (the $u = 0$ case), the above estimator reduces to the Hill estimator.

Table 8.1 displays the maximum-likelihood estimates of the index of stability (the values of α) for the QCP instances in figure 8.8. Note that for each of the instances shown in the table, the estimated value of α is consistent with the hypothesis of an infinite mean and an infinite variance, since $\alpha \leq 1$.

4.2 Fat *vs.* Heavy-Tailed Distributions

The attribution of the name "heavy tails" to this class of distributions definitely captures their main feature, namely, the large proportion of the total "probability mass" which is concentrated in the tail. This high concentration of probability mass is reflected in the slow decay of the tail and is responsible for the fact that *all the moments of a heavy-tailed distribution from some order on are infinite.* The heavier the tail, the lower the order at which the onset of the infinite moments occurs. Distributions with very heavy tails have an infinite mean (first moment), while those with "lighter" tails have a finite mean—and some have finite moments for one or more of the higher orders as well

Figure 8.9 contrasts three distributions: standard normal, Cauchy, and Lévy. The key property to observe is the dramatic difference in the decay of the

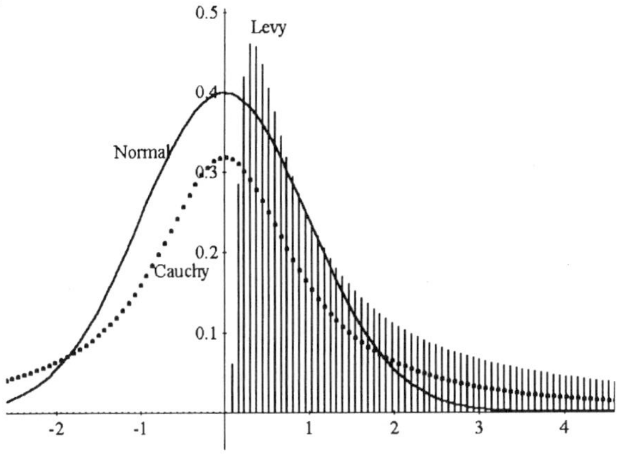

Figure 8.9. Standardized normal, Cauchy, and Lévy densities

c	Normal	Cauchy	Lévy
0	0.5000	0.5000	1.0000
1	0.1587	0.2500	0.6827
2	0.0228	0.1476	0.5205
3	0.001347	0.1024	0.4363
4	0.00003167	0.0780	0.3829
5	0.0000002866	0.0628	0.3453

Table 8.2. Comparison of tail probabilities, $\Pr\{X > c\}$, for standard normal, Cauchy, and Lévy distributions

tails of the distributions. In table 8.2, we present the proportion of the total probability mass in the tail of each of these distributions for various values of the random variable. It is clear that this quantity quickly becomes negligible for the standard normal distribution, whereas the other two distributions have a significant proportion of the total probability mass in their tails. The lower the value of α (the *index of stability* of the distribution), the heavier the tail. For example, the Cauchy distribution has $\alpha = 1.0$, and the Lévy distribution has $\alpha = 0.5$.

Related to *heavy*-tailedness is the notion of *fat*-tailedness, which is defined in terms of a property of a distribution known as the *kurtosis*.

The *kurtosis* of a distribution is the quantity μ_4/μ_2^2, where μ_2 and μ_4 are the second and fourth moments about the mean, respectively (note that μ_2 is the variance). The kurtosis is independent of the location and scale parameters of

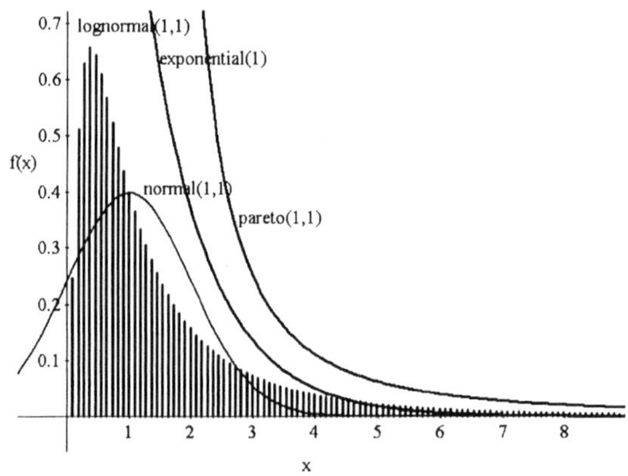

Figure 8.10. Comparison of the *fatness* of tails: normal (1,1), lognormal(1,1), exponential(1), and pareto(1,1).

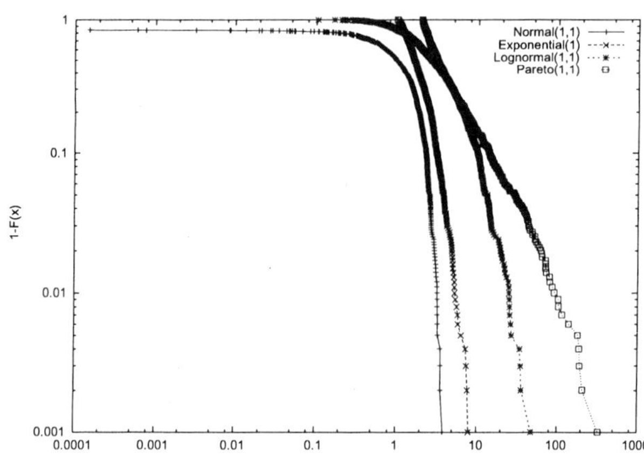

Figure 8.11. Comparison of the tail *fatness*: normal (1,1), lognormal(1,1), exponential(1), and pareto(1,1).

a distribution. If a distribution has a high central peak and long tails, then in general the kurtosis is large.

The kurtosis of the standard normal distribution is 3. A distribution with a kurtosis larger than 3 is said to be *fat-tailed* or *leptokurtic*. Like a *heavy*-tailed distribution, a *fat*-tailed distribution has a long tail and a considerable concentration of mass in its tail. Nevertheless, the tail of a *fat*-tailed distribution decays at a faster rate than that of a heavy-tailed distribution. In fact, all the moments of a fat-tailed distribution are finite, which is in stark contrast to the moments of a heavy-tailed distribution.

Examples of distributions that are characterized by fat tails are the exponential, lognormal, and Weibull distributions. In figure 8.10 we compare the tails of four distributions: normal (1,1), lognormal (1,1), exponential (1), and Pareto (1,1). Clearly, the tail of the Pareto distribution, a heavy-tailed distribution with $\alpha = 1$ (and therefore an infinite mean), is the heaviest. The tail of the lognormal distribution is fatter than that of the exponential, and both of them have fat tails —fatter, in fact, than the tail of the normal distribution.

Figure 8.11 shows a log–log plot of the tails of the same four distributions. (For each one, we generated 1000 data points from the corresponding (theoretical) distribution.) The relative fatness of the tails becomes more apparent in this plot. Note the qualitatively different (and approximately linear) behavior of the drop-off in the Pareto tail, which is suggestive of the heavy-tailed nature of the Pareto distribution.

As a final note, we should mention that there are a number of different parametric models for performing extreme-value analysis, each of which is suited to a certain class of distributions. We have already mentioned fat-tailed and heavy-tailed distributions, such as lognormal and sum-stable distributions, respectively. Extreme-value (EV) and generalized Pareto (GP) models are also central to the statistical analysis and modeling of extremes. A treatment of such models is beyond the scope of this book. (For further information on this topic, see, *e.g.,* (Adler et al., 1998; Mandelbrot, 1983; Reiss and Thomas, 2001; Samorodnitsky and Taqqu, 1994).)

5. Heavy and Fat-Tailed Distributions in Backtrack Search

The phenomena of *heavy* and *fat* tails in randomized backtrack search have been observed in several domains other than QCP, and in considering different search paradigms, namely, general constraint satisfaction problem (CSP) formulations and mixed integer programming (MIP) formulations. They have also been observed when considering more specific problem formulations such as Boolean satisfiability (SAT), graph coloring (GC), and theorem proving (TP). Theorem proving is beyond the scope of this book. (See *e.g.*, (Meier et al., 2001), for a description of results on heavy-tailed behavior in proof planning, a particular area of theorem proving.)

5.1 CSP Formulations

For the QCP experiments, we incorporated randomization into a CSP approach. Specifically, we used the Ilog constraint-solver engine, which provides a powerful C++ constraint programming library (Ilog, 2001b). As described in section 4, we randomized the first-fail heuristic. Not only did we consider several variants of this heuristic, by randomizing variable and/or value selection, but we considered a popular extension of it — namely, the Brelaz heuristic (Brelaz, 1979), which was originally introduced for graph coloring procedures.

The Brelaz heuristic specifies a way for breaking ties that are encountered on applying the first-fail rule: If two variables have equally small remaining domains, the Brelaz heuristic chooses the variable that shares constraints with the largest number of the remaining unassigned variables. Any ties that remain after the Brelaz rule are resolved randomly. As with the first-fail heuristic, we studied several variants of the Brelaz heuristic, by randomizing variable and/or value selection. Though the variants of the Brelaz heuristic proved more powerful than those of the first-fail heuristic, the runtime distributions were qualitatively similar. One significant difference between the two was that in using the Brelaz heuristic, we observed heavy tails, similar to those in figure 8.8.

Sports scheduling is another domain in which heavy-tailed behavior has been observed, using a CP paradigm (Gomes et al., 1998b). There has been an increasing interest in applying CP techniques to sports-scheduling problems, given their highly combinatorial nature (Easton et al., 2001; McAloon and Tretkoff, 1997; Nemhauser and Trick, 1998; Regin, 1999). In sports scheduling problems, one of the issues is timetabling, where one has to take into consideration constraints on how the competing teams can be paired, as well as how each team's games are distributed over the entire schedule. In particular, we consider the timetabling problem for a "round-robin" schedule: Every team must play every other team exactly once. The *global* nature of the pairing constraints makes this a particularly hard combinatorial search problem. Typically, a game will be scheduled on a certain field or court, at a certain time, etc. This kind of combination will be called a slot. These slots can vary in desirability with respect to such factors as lateness in the day, location, and condition of the field. The problem is to schedule the games in such a way that the different periods are assigned to the teams in an equitable manner over the course of the season.

The round-robin sports-scheduling problem considered here is something of a "classic" in the operations research community, because it presents a very tough challenge for integer programming methods. An instance of this problem of size 10, encoded as an integer programming problem, is included in the OR-Library (Beasley, 1990).

In order to study the runtime distributions of randomized backtrack search on this benchmark problem, we considered a problem instance of size 12, even though at the time of these experiments the best algorithm could solve instances of size up to 14 teams.[3] Such an instance would have required a tremendous amount of computational resources, making it difficult for us to obtain a good empirical distribution without eliminating a large portion of the tail. In fact, even for 12 teams, we had to use a cutoff of 250,000 backtracks per run. We randomized the backtrack solver, randomly breaking ties on variable/value selection.

Figure 8.12 shows a log–log plot of the complement-to-one of the cumulative distribution, 1-F(x), for our round-robin sports-scheduling problem ($N = 12$). Plot (a) gives the full range of the distribution, while plot (b) shows only the tail ($X > 10,000$). The linear nature of the tail, which is magnified in plot (b), strongly suggests that it is of the Pareto-Lévy type. (Our cutoff truncated only about 0.2% of the tail.)

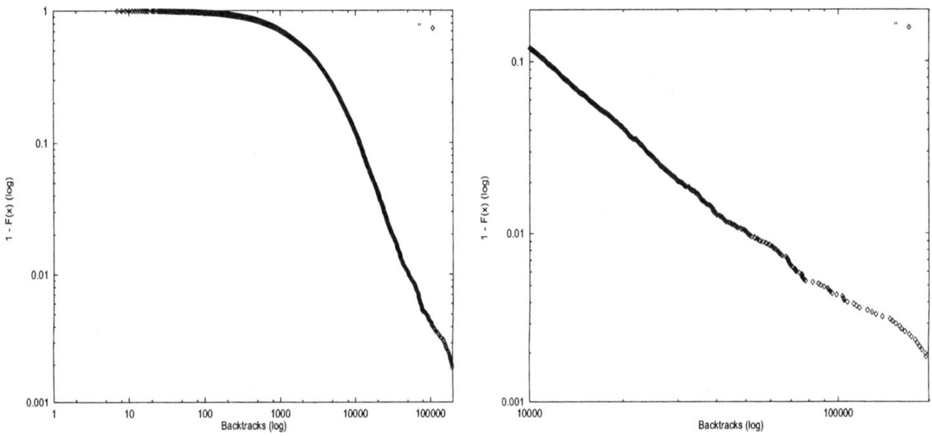

(a) (b)

Figure 8.12. Log-log plot of the complement to one of the cumulative distribution of the runtime for the round-robin sports scheduling (12 teams; CSP formulation); (a) full distribution; (b) tail ($X > 10,000$).

5.2 Mixed Integer Programming Formulations

The standard approach used by the OR community to solve mixed integer programming problems (MIP) is branch-and-bound search. First, a linear program (LP) relaxation of the problem instance is considered. In such a relaxation, all variables of the problem are treated as continuous variables. If the solution to the LP relaxation problem has non-integer values for some of the integer variables, we have to branch on one of those variables. This way we create two new subproblems (nodes of the search tree), one with the floor of the fractional value and one with the ceiling. (For the case of binary (0/1) variables, we create one instance with the variable set to 0 and another with the variable set to 1.) The standard heuristic for deciding which variable to branch on is based on the degree of infeasibility of the variables ("max-infeasibility variable selection"): We select the variable whose non-integer part in the solution of the LP relaxation is closest to 0.5. Informally, we pick the variable whose value is "furthest away from an integer".

Following the strategy of repeatedly fixing integer variables to integer values will lead at some point to a subproblem with an overall integer solution, provided we are dealing with a feasible problem instance. (We call any solution where all the integer variables have integer values an "integer solution.") In practice, it often happens that the solution of the LP relaxation of a subproblem already *is* an integer solution, in which case we do not have to branch further from this node.

Once we have found an integer solution, its objective function value can be used to prune other nodes in the tree, namely, those whose relaxations have worse values. This is because the LP relaxation bounds the optimal solution of the problem. For example, for a minimization problem, the LP relaxation of a node provides a lower bound on the best possible integer solution.

A critical issue that determines the performance of branch-and-bound is the way in which the next node to be expanded is selected. In OR, The standard approach is to use a best-bound selection strategy. That is, from the list of nodes (subproblems) to be considered, we select the one with the best LP bound. (This approach is analogous to an $A^{\star}$-style search. The LP relaxation provides an admissible search heuristic.)

The best-bound node-selection strategy is particularly well-suited for reaching an optimal solution (because of the greedy guidance), which has been the traditional focus of much of the research in OR. However, one significant drawback of this approach is that it may take a long time before the procedure finds an integer solution, because of the breadth-first flavor of the search. Also, the approach has serious memory requirements because the full fringe of the tree has to be stored.

Given problems that have a difficult feasibility part, the best-bound approach may take too long before reaching an integer solution. (Note that an integer solution is required before any nodes can be pruned.) In our experiments, therefore, we also considered a depth-first node-selection strategy. Such a strategy often quickly reaches an integer solution but may take longer to produce an overall optimal value.

In our experiments, we used a state-of-the-art MIP programming package called Cplex (Ilog, 2001a), which provides a set of libraries that allows one to customize the branch-and-bound search strategy. For example, one can vary node selection, variable-selection, variable setting strategies, or the LP solver. We used the default settings for the LP solver, which is primal simplex for the first node and dual simplex for subsequent nodes. We modified the search strategies to include some level of randomization. We randomized the variable-selection strategy by introducing noise into the ranking of the variables, based on maximum infeasibility. (Note that even in this scenario, the completeness of the search method is maintained.) We experimented with several other randomization strategies. For example, in Cplex one can assign an *a priori* variable ranking, which is fixed throughout the branch-and-bound process. We experimented with randomizing this *a priori* ranking, and we found that the dynamic randomized variable-selection strategy described above is more effective (Gomes and Selman, 1999).

In our experiments we were interested in problems that combine a hard combinatorial component with numerical information. Integrating numerical information into standard AI formalism is becoming of increasing importance. In planning, for example, one would like to incorporate resource constraints or a measure of overall plan quality. We considered examples that are based on logistics-planning problems but are formulated as mixed integer programming problems. These formulations extend the traditional AI planning approach by combining the hard constraints of the planning operators (the initial state and goal state) with a series of soft constraints that capture resource utilization. Such formulations have been shown to be very promising for modeling AI planning problems (Kautz and Walser, 1999; Vossen et al., 1999).

Experimentation with the Cplex MIP solver showed that these problem instances are characterized by a non-trivial feasibility component.[4]

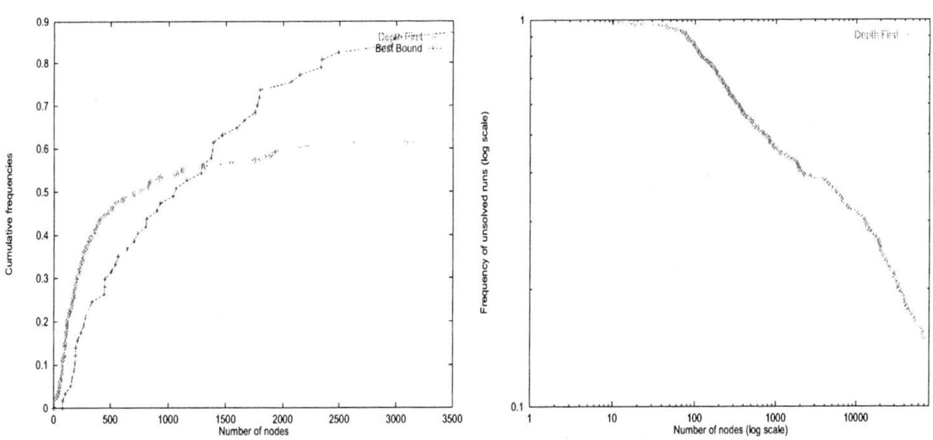

(a) (b)

Figure 8.13. (a) Cost profiles for a logistics planning problems for depth-first and best-bound search strategies. (b) Hevay-tailed behavior of depth-first search.

In figure 8.13(a), we compare the runtime profile of a depth-first strategy with that of a best-bound strategy to solve a hard feasibility problem in the logistics domain, formulated as a mixed integer programming problem. The search is terminated when an optimal or near-optimal (within 10% of optimal) solution is found, but without the requirement of *proving* optimality. The figure shows the cumulative distribution of the solution time (in number of expanded nodes). For example, with 500 or fewer nodes, the depth-first search finds a solution on approximately 50% of the runs. Each run had a time limit of 5000 seconds. As we see from the figure, depth-first search initially outperforms best-bound search. After more than 1500 node expansions, however, the best-bound approach becomes more effective. For example, best-bound search finds a solution on approximately 75% of the runs with 2000 or fewer node expansions. In contrast, depth-first search could find a solution on only 55% of the runs with that many node expansions. These data are consistent with the observation above that the best-bound method may take more time to find an intial integer solution. Once such an initial integer solution is found, however, best-bound search becomes more effective.

We now look at the runtime distributions more closely. figure 8.13(b) gives a log–log plot of the complement-to-one of the cumulative distribution for the depth-first procedure. For example, from this plot we see that after 10,000 nodes, approximately 30% of the runs have not yet found a solution. The figure shows near-linear behavior over several orders of magnitude, an indication of heavy-tailedness. The curve for best-bound search also appears to exhibit heavy-tailed behavior, but less dramatically than that for depth-first search.

5.3 Boolean Satisfiability Formulations

We have also analyzed the runtime distributions of problems encoded as Boolean satisfiability (SAT) problems. The domains considered were AI planning, code optimization, and timetabling (Gomes et al., 1998a). We randomized

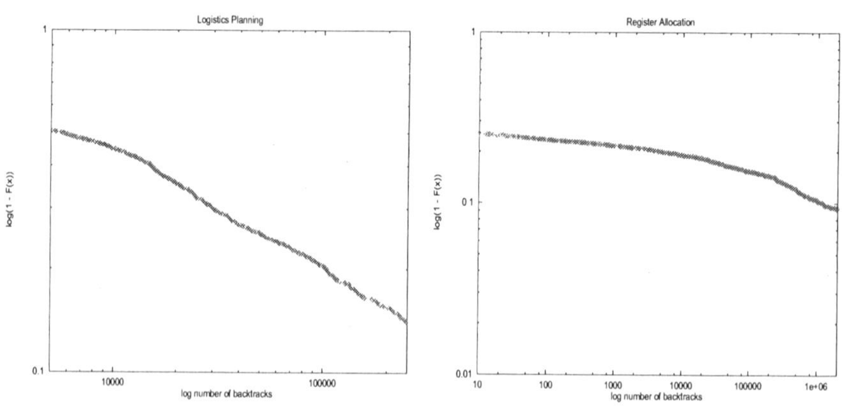

(a) (b)

Figure 8.14. Log-log plot of heavy-tailed behavior for SAT formulations of (a) logistics planning , and (b) register allocation

a complete SAT solver, Satz, by (Li and Anbulagan, 1997). Satz is a version of the Davis–Putnam–Logemann-Loveland (DPLL) procedure (Davis and Putnam, 1960; Davis et al., 1979), with a highly tuned heuristic which is based on choosing a branching variable that, when set positively or negatively, maximizes a certain function of the number of unit propagations performed. Because Satz's heuristic usually chooses a single branching variable without ties, randomization of Satz was implemented using the uniform H-percent window rule (see section 2). (Kautz and Selman, 1996), showed that propositional SAT encodings of difficult STRIPS-style AI planning problems could be efficiently solved by SAT engines. We considered the particular domain of "logistics planning," which involves moving packages on trucks and airplanes between different locations in different cities (Veloso, 1992)).

A second set of problems involving SAT encodings comprised instances from the Dimacs Challenge benchmark (Johnson and Trick, 1996): a code-optimization problem, involving register allocation ("mulsol" instance), and circuit-synthesis problems ("adder" instances). For the circuit-synthesis problems, (Kamath et al., 1990), developed a technique for expressing the problem of synthesizing a programmable logic array (PLA) as a SAT problem. The statement of the problem includes a table specifying the function to be computed and an upper bound on the number of gates that may appear in the circuit. In general, these problems become more difficult to solve as the number of gates is reduced, until the limit is reached where the instance becomes unsatisfiable. These problems are quite hard to solve with complete SAT procedures, and have been used as part of the test beds for numerous SAT competitions and research studies. The problems considered were the "3bit-adder-32" and "3bit-adder-31", based on synthesizing a 3-bit adder using 32 and 31 gates, respectively.

Figure 8.14 displays log–log plots of the complement-to-one of the cumulative distributions for our SAT domains: the logistics-planning problem and the register-allocation problem. The approximately linear nature of the tails for several orders of magnitude indicates heavy-tailed behavior.

Cases	k	α
Logistics Planning (Satz)	1,442	0.360 (0.017)
Register Allocation	10,000	0.102 (0.002)

Table 8.3. Estimates of the index of stability (α). k is the sample size. The values within parentheses are the estimated asymptotic standard deviations (SAT encoding).

Table 8.3 displays the maximum-likelihood estimates of the indices of stability (the values of α) for the instances encoded as SAT problems. Note that for all the instances shown in the table, the estimates of α are consistent with the hypothesis of an infinite mean and an infinite variance, since $\alpha < 1$. We should point out, however, that given the computational difficulty of these problems, we had to truncate a larger portion of the tails of these distributions than we did in the previous examples, which renders our estimates of the indices of stability for the SAT-encoded problems less well-grounded in comparison (see also section 5.5).

5.4 Graph Coloring Formulations

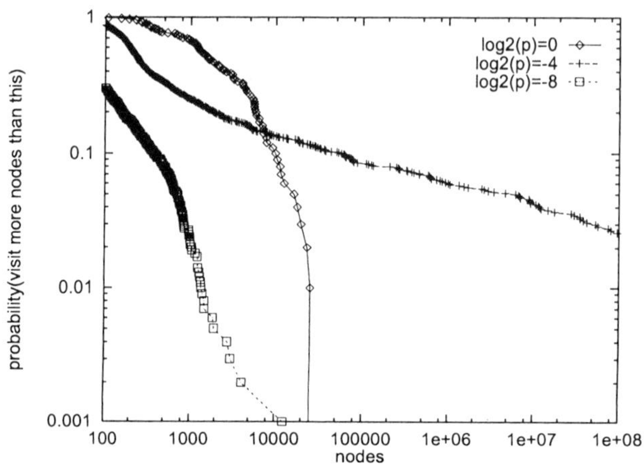

Figure 8.15. Log–log plot of the tail of the runtime distributions of an exact backtrack search coloring algorithm on three different graph topologies: random ($log_2(p) = 0$); small-world ($log_2(p) = -4$); and more structured ($log_2(p) = -8$). p is the probability of randomly rewiring an edge of a regular graph. Size of the regular graph: 100 nodes and degree 8 (Walsh, 1999).

In a study of the coloring of graphs with so-called small-world properties, (Walsh, 1999),[5] also identifies heavy-tailed behavior. In a small-world topology, nodes are highly clustered, yet the path length between them is small (Walsh, 1999; Watts and Strogatz, 1998). In contrast, random graphs tend to have short path lengths but little clustering, while regular graphs tend to have high clustering but large path lengths.

Walsh used random reordering of the variables to "randomize" a deterministic exact coloring algorithm based on DSATUR, by (Trick, 1996) (see also section 2). Figure 8.15(a) shows a log–log plot of the tails of the runtime distributions of the complete backtrack search algorithm on three different graph structures: random, small-world, and a more structured topology. The approximately linear behavior of the tail of the distribution that corresponds to the small-world topology, *over 8 orders of magnitude*, is a clear indication of heavy-tailedness. (About 0.3% of the runs were truncated.) In sharp contrast

Cases	k	α
School Timetabling	10000	0.219 (0.004)

Table 8.4. Estimate of the index of stability (α) for the school-timetabling problem. k is the sample size. The value within parentheses is the estimated asymptotic standard deviation

to that case, the tails of the distributions corresponding to the random topology and the more structured topology have a very fast drop-off, indicating much lighter tails. Walsh conjectures that a small-world topology induces heavy-tailed behavior.

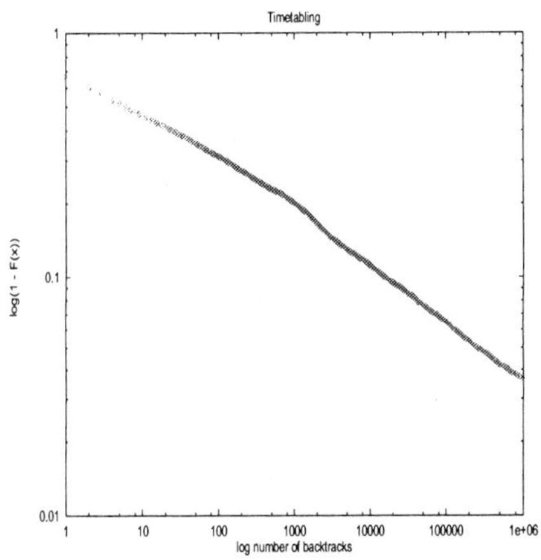

Figure 8.16. Log–log plot of heavy-tailed behavior for the timetabling problem

We also studied the runtime distributions of Dimacs color instances for school timetabling (Johnson and Trick, 1996). Figure 8.16 gives the tail of the distribution for the school timetabling instance, *school1_nsh.col*. The figure displays a log–log plot of the complement-to-one of the cumulative distribution. Again, the approximately linear nature of the tail over several orders of magnitude indicates heavy-tailed behavior. (About 3% of the runs were truncated.) The estimate for the index of stability ($\alpha = 0.219$; see table 8.4) provides further evidence of the heavy-tailedness of the distribution.

5.5 Discussion

Finite Search Space. In general, the computational cost of a complete backtrack algorithm has a finite upper bound.[6] Technically speaking, within the realm of finite bounds it is not rigorous to speak of distributions with infinite moments. Nevertheless, the heavy-tailed model provides a good approximation for the tail behavior of the search cost when dealing with the degree of complexity inherent in NP-complete problems. The heavy-tailed model emphasizes the persistence of power-law decay in the tail of the distribution over several orders of magnitude. Also, since the upper bound of the search space is exponential in the size of the problem, it is generally not reached in practice. As mentioned in section 3.2, one can also speak of *bounded* heavy-tailed behavior in distributions that exhibit power-law decay over an extensive portion of the tail but, given the finiteness of the search space, show a sudden drop-off further out in the tail. Thus, when allowing for a very large number of backtracks, so that one can explore most of the complete search tree, one will observe boundedness effects in the tail of the distribution.

Tail Truncation. As mentioned in section 4.1.3, in practice, we often have to impose a cutoff on the number of backtracks allowed, especially when studying the runtime distributions of hard instances. In other words, we have to *censor* the data, to avoid literally getting stuck in the most extreme outliers in the tail of the distribution during our data collection. As a consequence, the tail gets *truncated*. The selection of the *cutoff* depends on the problem instance and algorithm. Obviously, the higher the better. In general, we try to get a fairly large range, typically over at least 5 to 6 orders of magnitude in the value of the random variable, so that we can see some indication of stabilization in the tail behavior. Ideally, we would like to refrain from truncating a significant portion of the tail, but sometimes this is prohibitive from a computational stand point.

The more severe the truncation, the more difficult it becomes to differentiate between *heavy*-tailedness and *fat*-tailedness. From a practical point of view, this is not a problem, as long we have a clear indication that the tail exhibits behavior of one of these two types. Heavy-tailedness is not a necessary condition for speed-ups in randomized search methods. In fact, one can also speed up the performance of randomized backtrack search methods in the presence of *fat* tails. Nevertheless, the heavier the tail, the higher the frequency of very long runs — and as a result, the greater the advantage we stand to gain from randomization and restarts, so the more dramatic the speed-ups.

Mixtures of Distributions. When modeling real-world phenomena, one often has to combine "mixtures" of distributions. For example, it is not uncommon to model income distributions with a mixture of a lognormal model, for the body of the distribution, and a Pareto model for the tail. One can also combine mixtures of the same family of distributions, *e.g.*, mixtures of exponential distributions. (In fact, it has formally been proved that any distribution can be approximated by a mixture of exponential distributions.) In the study of computational phenomena, we often observe that the behavior of randomized backtrack search methods is best modeled by mixtures of distributions. For example, in figure 8.12 (round-robin sports scheduling problem) we can see that a mixture of distributions (for example, one for the body ($X \leq 10,000$) and another to model a Pareto-Lévy tail), would be a good approach to take. (See also (Hoos, 2000) in which mixtures of exponential distributions are proposed to model certain runtime distributions.)

Stronger Search Methods. Clearly, the existence and nature of heavy- or fat-tailed behavior depends largely on the particular randomized backtrack search algorithm used. Sometimes, just varying the randomization method causes the tails to change dramatically.

Let us consider the effect of more computation-intensive propagation methods on heavy-tailed behavior. In the experiments on QCP described earlier, for example, we used a simple backtrack search method with forward checking. When using stronger propagation techniques, such as general arc consistency, one can solve substantially larger problems (Régin, 1994; Shaw et al., 1998; Gomes et al., 2000). We maintain general arc consistency on the N-ary "alldifferent" constraints using the algorithm of (Régin, 1994). This method uses a matching algorithm to ensure that, at each point during the search, in each row the set of remaining colors available for the yet-to-be-assigned cells is such that there is a different color available for each cell, and similarly for the columns. Using an efficient CSP-based implementation, we can solve instances of order 30 in under five minutes using uncensored data, which allows us to gather runtime information within a reasonable time frame.

By using general arc consistency for QCP, we observed that the tails became considerably *lighter* for quasigroups of low order. The reason for this is that those instances are solved mostly by propagation, with pratically no search. As we increased the order of the quasigroups, however, the heavy-tail phenomenon reappeared. The left panel in figure 8.17 shows the heavy tail for an instance of order 30 with 55% pre-assignment when applying the alldifferent constraints and using uncensored data (55% is close to the phase transition for order 30; (Gomes et al., 2000)).

The right panel in figure 8.17 shows a comparison between two different SAT procedures applied to the logistics-planning problem. We compare Satz (Li and Anbulagan, 1997) with Relsat (Bayardo and Schrag, 1997). As mentioned earlier, both procedures use sophisticated propagation rules and heuristics to guide their search. However, the Relsat backtracking strategy also includes a look-back technique, based on conflict-directed backjumping and relevance-bounded learning (Prosser, 1993; Bayardo and Miranker, 1996). We present this comparison to demonstrate the possible influence of such look-back strategies on heavy-tailed behavior. From the figure, we see that Relsat results in the *lighter* of the two tails, which is reflected in the steeper slope of the log–log plot. This means that the index of stability for Relsat is higher (see table 8.5). This

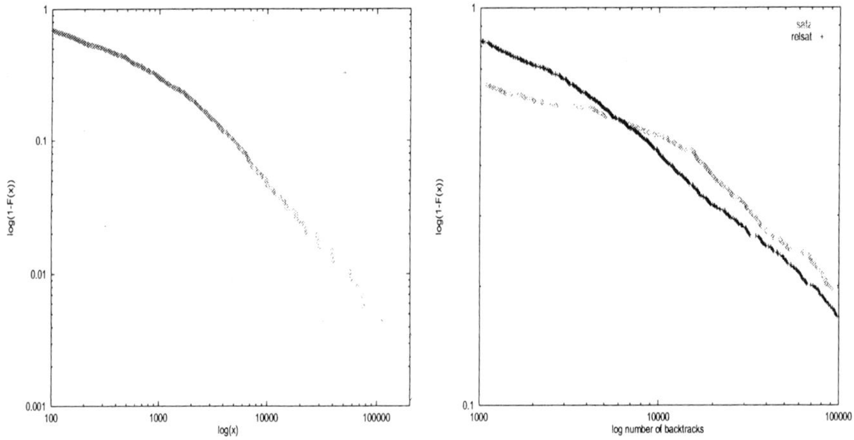

Figure 8.17. Left panel: heavy tail for quasigroup completion with extensive propagation (CSP formulation, order 30, with 55% pre-asssignment; uncensored data). Right panel: comparison of Satz to Relsat on logistics planning (SAT formulations).

Cases	k	α
Logistics Planning (Satz)	1442	0.360 (0.017)
Logistics Planning (Relsat)	1000	0.670 (0.009)

Table 8.5. Index of stability: impact of look-back strategies on heavy-tailed behavior (SAT encodings).

observation is consistent with related work in which look-back strategies were shown to be effective in solving so-called exceptionally hard SAT instances (Bayardo and Schrag, 1997). In effect, the look-back approach reduces the degree of variability in the search, though it does not necessarily eliminate it altogether.

Non–Heavy-Tailed Behavior. Clearly, the phenomenon of heavy-tailedness results from the interactions between the randomized algorithm and a particular instance. In fact, in general, for a given randomized backtrack-style algorithm and a problem domain, one should not expect to see heavy-tailed behavior on all instances. Typically, we observe the following pattern: On extremely easy instances, the tail of the runtime distribution decays very fast. As the level of difficulty of the instance increases, the tail becomes heavier and heavier and eventually goes through a region where heavy-tailedness is clearly in evidence. Beyond this heavy-tailed region, the instances become so hard for the given algorithm that the runtime distribution shifts to the right, and all the runs become very long and terminate after roughly the same number of steps. As a conse-

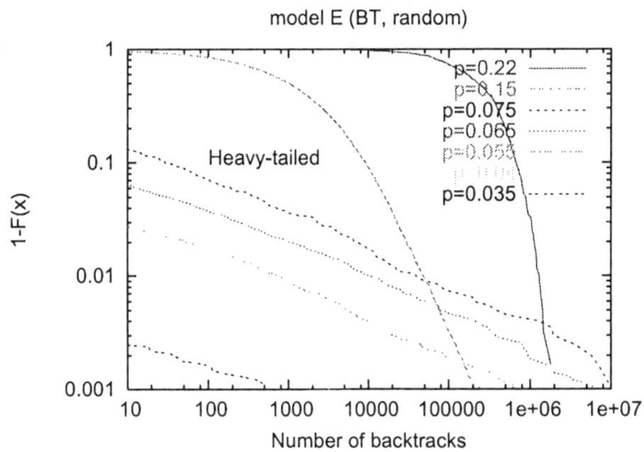

Figure 8.18. Survival function $(1 - F(x))$ of the number of backtracks needed to solve different instances of model E with 20 variables and a domain size of 10. The parameter p captures the constrainedness of the instances; it is the probability of choosing one constraint (out of the total number of possible constraints). Two different regimes in the decay of the survival function can be identified: a heavy-tailed regime (curves with linear behavior) and a non–heavy-tailed regime.

quence, the variance of the runtime distribution decreases, the heavy-tailedness disappears, and the heavy tail is replaced by a fat tail. As the degree of difficulty of the instances increases even further, the variance of the runtime distribution continues to decrease.

Figure 8.18 illustrates this phenomenon for a simple backtrack search algorithm (no look-ahead and no look-back) using a random variable-ordering heuristic in conjunction with random value selection. We studied random instances — generated using model E of random binary CSP (Achlioptas et al., 1997) — with different values of the constrainedess parameter p. In figure 8.18, the curves corresponding to instances with $p \leq 0.075$ exhibit Pareto-like behavior. The instances with p right around 0.075 are clearly heavy tailed. To some extent, in fact, the boundary of this region constitutes a threshold for this particular backtrack search algorithm. Beyond this region, the instances become very hard and all the runs become homogeneously long, so that the variance of runtime distribution decreases and the tail of its "survival function" (which is just another name for the complement-to-one of the cumulative distribution) decays exponentially. For example, the curve corresponding to the instance with $p = 0.22$ exhibits exponential decay, which is much faster than linear decay.

Figure 8.19 shows log–log plots for two unsolvable instances from the quasigroup completion domain. One is a rare unsolvable instance in the underconstrained area (best fit: a gamma distribution); the other is in the critically

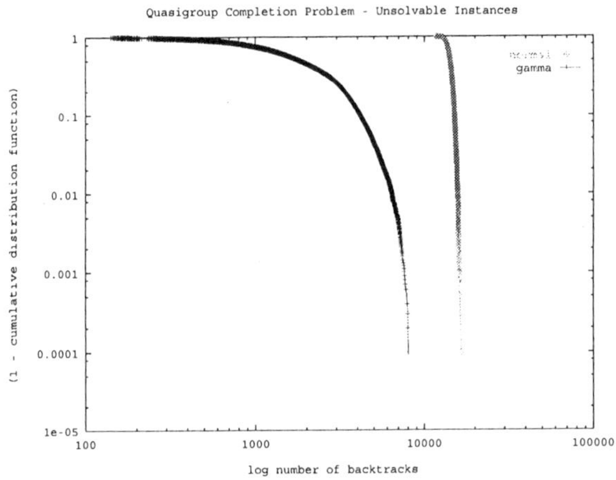

Figure 8.19. Absence of heavy-tailed behavior for unsolvable QCP instances (CSP formulation)

constrained region (best fit: normal distribution). We see that both curves exhibit a sharp, rounded drop-off, which is indicative of the absence of heavy-tailed behavior.

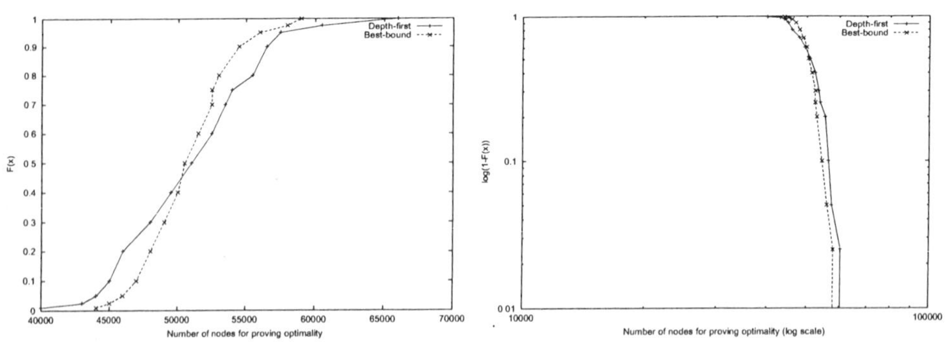

Figure 8.20. Left panel: comparison of runtime profiles for proving optimality of `misc07` from MIPLIB. Right panel: absence of heavy-tailed behavior in proving optimality of `misc07`

Figure 8.20 shows the distributions for proving optimality of the `misc07` problem from the MIPLIB library. It is apparent that the runtime for this instance exhibits relatively little variance in both the best-bound and depth-first searches. The sharp drop-off in the log–log plot in the left panel in the figure is a clear indication of the absence of heavy tails. Our experiments in proving optimality on other instances from the MIPLIB library also did not reveal heavy-tailed behavior.

We conjecture that methods for proving optimality and inconsistency may produce *fat* tails, but that it is less likely that they will produce *heavy*-tails, since the entire search space has to be explored and therefore the variance in the runtime would tend not to be very large. This is consistent with the results by (Frost et al., 1997), on proving inconsistency in constraint satisfaction problems. They showed that fat-tailed distributions, such as the Weibull and lognormal distributions, underlie the cost of proving inconsistency. Given the presence of fat tails in such cases, the use of randomization, in combination with restart strategies can still be effective in boosting backtrack search methods. In fact, as mentioned earlier, use of randomized strategies, together with restarts and nogood learning has been extremely successful in software and hardware verification, where unsatisfiability must be proved. Whether backtrack-style methods can produce heavy-tails in cases where optimality or inconsistency is to be proved is still an open question.

In the next section we show how the large variance in search methods, as characterized by heavy and fat-tailed behavior, can be exploited to boost the performance of randomized backtrack search methods by using restart and portfolio strategies.

6. Restart Strategies

We have shown that randomized backtrack search methods are characterized by *fat* (and, in some cases, even *heavy*) tails. The intuition is that, more often than one would expect, the heuristics guiding the search procedure make "wrong turns," causing the solver to be "stuck" in portions of the search space that contain no competitive solution — or no solution at all.

Given such a phenomenon, a randomized backtrack procedure is, in a sense, most effective early on in a search, which suggests that a sequence of short runs may be a more effective use of computational resources than a single long run. In this section and the next, we show how *restart* and *portfolio* strategies can boost performance of randomized search methods.

6.1 Elimination of Heavy-Tails

Figure 8.21(a) shows the result of applying a strategy of fixed-length short runs — *rapid randomized restarts* of a complete randomized search procedure — to a QCP instance of order 20 with 5% pre-assignment. The figure dislays a log–log plot of the tail of the distribution. From the figure, we see that without restarts and given a total of 50 backtracks, we have a failure rate of about 70%. Using restarts (once after every 4 backtracks), this failure rate drops to about 10%. With an overall limit of only 150 backtracks, the restart strategy nearly always solves the instance, whereas the original procedure still has a failure rate of about 70%. Such a dramatic improvement due to restarts is typical for heavy-tailed distributions; in particular, we get similar results on critically constrained instances. The fact that the curve for the experiment with restarts

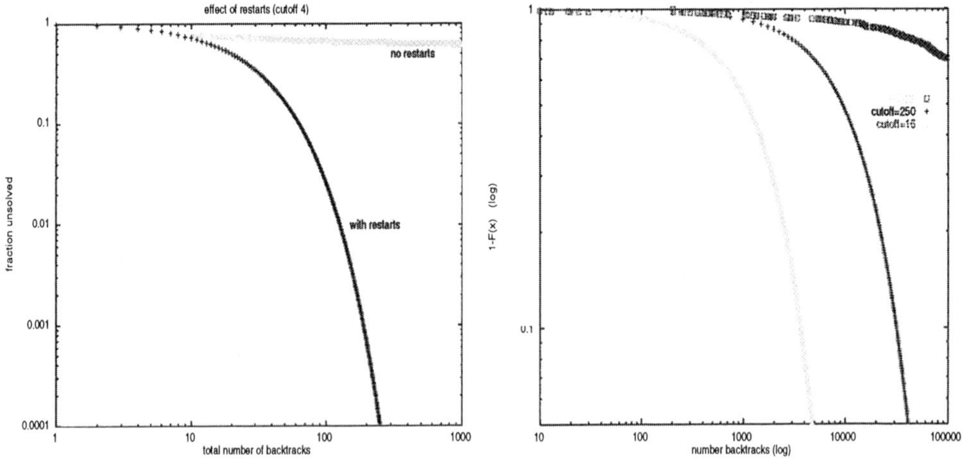

(a) (b)

Figure 8.21. Restarts: (a) Tail (of $(1 - F(x))$ as a function of the total number of backtracks for a QCP instance (CSP formulation), log–log scale; (b) same as (a), but for a logistics instance (SAT formulation). In (b) the left-most curve is for a cutoff value of 16, the middle curve is for a cutoff of 250, and the right-most curve is without restarts.

takes a definite downward turn is a clear indication that the heavy-tailed nature of the original cost distribution has disappeared.

Figure 8.21(b) shows the effect of restarts on the logistics-planning problem (for the same instance as in figure 8.14(a)). The figure gives the distributions resulting from running randomized Satz with cutoffs (in the number of backtracks before each restart) of 16 (near optimal) and 250. The sharp drop-off in the log–log plot shows the disappearance of the heavy tail. We see, for example, that with restarts and using a cutoff value of 16, after a total of about 5,000 backtracks, we obtain a failure rate of approximately 5%. *Without restarts*, on the other hand, even after 100,000 backtracks the failure rate is still up around 70%.

The different plots in figure 8.21 clearly show that restarts shorten the tails of the distributions. Random restarts therefore provide an effective mechanism for dealing with *fat*- and *heavy*-tailed distributions. Of course, the cutoff parameter limits the size of the space that can be searched exhaustively between restarts. In practice, we gradually increase the cutoff, to allow us to maintain completeness (Gomes et al., 1998a).

We have proved formally that the underlying distribution of a restart strategy with a fixed cutoff eliminates heavy-tailed behavior, hence that all the moments of the distribution are finite. (The proof can be found in (Gomes et al., 2000).)

cutoff	success rate	mean cost ($\times 10^6$)
200	0.0001	2.2
5,000	0.003	1.5
10,000	0.009	1.1
50,000	0.07	0.7
100,000	0.06	1.6
250,000	0.21	1.2
1,000,000	0.39	2.5

(a)

cutoff	success rate	mean cost
2	0.0	>300,000
4	0.00003	147,816
8	0.0016	5,509
16	0.009	1,861
32	0.014	2,405
250	0.018	13,456
128000	0.32	307,550

(b)

Table 8.6. Solving (a) a 16-team round-robin scheduling problem (CSP formulation) and (b) the logistics.d instance (SAT formulation) for a range of cutoff values.

6.2 Cutoff Value for Restart Strategy

Different cutoff values will result in different overall mean solution times. This can be seen in table 8.6, where mean solution times (based on 100 runs) for the round-robin scheduling problem (part (a), 16 teams; see also section 5.1) and the logistics-planning problem (part (b)) are considered for a range of cutoff values. The deterministic version of our CSP backtrack search was not able to solve the round robin for 16 teams.[7] The two sets of results in the table show the same overall pattern, revealing a clear optimal range of cutoff values. For the round-robin problem, the mean cost is minimized with a cutoff of about 50,000, and for the planning problem the optimal cutoff value is near 16. With a higher than optimal cutoff value, the heavy-tailed behavior to the right of the median begins to dominate the overall mean, whereas, for cutoffs below the optimal range the success rate is too low, requiring too many restarts to give a good overall value for the mean cost. These characteristics are apparent from table 8.6.

Figure 8.22 gives the data from table 8.6(b) in graphical form, with interpolation between data points. From the plot we see that the optimal cutoff value is around 12. The logarithmic vertical scale indicates that one can shift the performance of the procedure by several orders of magnitude by tuning the cutoff parameter.

In the previous examples, we used a *fixed cutoff* value f for the restart strategy, which was calculated on the basis of the underlying empirical distribution. For the logistics-planning problem, the "optimal" cutoff value for the restart strategy is around 12, as illustrated in figure 8.22. In other words, we computed the fixed cutoff value f that minimizes the expected runtime if the search procedure is

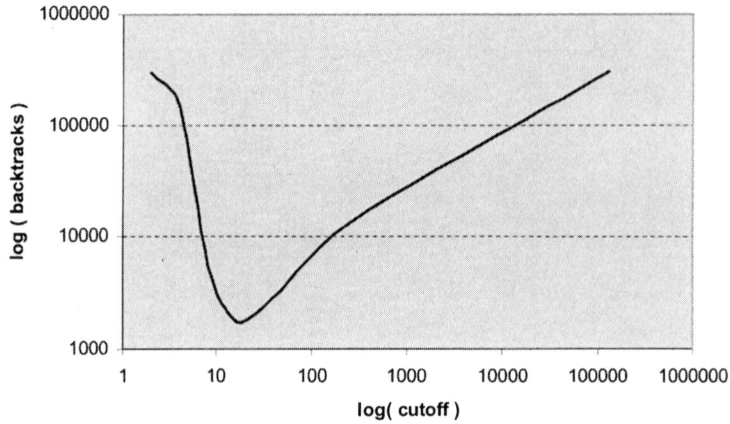

Figure 8.22. The effect of random restarts on solution cost for the logistics.d planning problem (SAT formulation)

Cutoff	Num. Runs	Succ. Run Backtracks	Succ. Run Time
50000	100		
		23420	629.89
250000	100		
		130613	990.8
		165859	1034.39
500000	100		
		43620	653.52
1000000	100	N.A.	N.A.

Table 8.7. Restarts using a hybrid CSP/LP randomized backtrack search on a balanced QCP instance of order 50, at the phase transition.

restarted after every f backtracks, until the solution is found, assuming that the runs are mutually independent.

(Walsh, 1999), introduced a novel restart strategy, inspired by the analysis carried out by Luby *et al.*, in which the cutoff value increases geometrically, after each restart. The advantage of such a strategy is that it is less sensitive to the details of the underlying distribution.

As mentioned in the Introduction, state-of-the-art SAT solvers now incorporate restarts. In practice, the solvers use a default cutoff value, which is increased, linearly, after a specified number of restarts, guaranteeing the completeness of the solver in the limit (Moskewicz et al., 2001).

6.3 Formal Results on Restarts

The idea of a fixed-cutoff restart strategy is based on theoretical results by (Luby et al., 1993), which describe provably optimal restart policies. In cases where the underlying runtime distribution of a randomized procedure is fully known, they showed that the optimal policy is the restart strategy with a fixed cutoff f, that minimizes the expected runtime. In our case, therefore, where we assume that the empirically determined distribution of our search procedure is a good approximation of the real distribution, the "optimal" strategy is just a sequence of runs of fixed length.

Luby *et al.* also provide a strategy for minimizing the expected cost of randomized procedures even in cases where there is no *a priori* knowledge about the distribution—a *universal* strategy that is suitable for all distributions. The universal strategy consists of a sequence of runs each of whose lengths is some power of two. Every time a pair of runs of a given length has been completed, a run of twice that length is immediately executed. The universal strategy is of the form: $1, 1, 2, 1, 1, 2, 4, 1, 1, 2, 4, 8 \cdots$.

Although the universal strategy of Luby *et al.* is provably within a constant log factor (in the *limit*) of the strategy that results from using a fixed cutoff (the optimal strategy to be used when the distribution is known), we found that in practice the schedule often converges too slowly. The work behind Luby's universal strategies was motivated by Ertel's observation of potentially long runs of theorem proving methods and their effect on parallel strategies (Ertel and Luby, 1994).

In (Williams et al., 2003a; Williams et al., 2003b) the connections between so-called backdoors, restarts, and the phenomenon of heavy-tailedness are described. Backdoors are small sets of variables that capture the overall combinatorics of a problem instance. More precisely, a backdoor is a set of variables for which there is a value assignment such that the simplified problem can be solved by a poly-time algorithm, called the "subsolver." The subsolver captures any form of poly-time simplification procedure as used in current CSP solvers. Of course, the set of all variables in a problem forms a trivial backdoor set, but many interesting practical problem instances possess much smaller backdoors and strong backdoors. When given a set of backdoor variables of a problem instance, one can restrict the combinatorial search by branching only on the backdoor variables and thus search a drastically reduced space. Interestingly, (Williams et al., 2003a) observed that structured problem instances can have suprisingly small sets of backdoor variables.

The notion of backdoor came about in the context of the study of heavy-tailed behavior as observed in backtrack-style search. Heavy-tailed distributions provide a justification of why restarts are effective, namely to prevent the search procedure from getting stuck in unproductive portions of the search space that do not contain solutions. Such distributions also imply the existence of a wide range of solution times, often including short runs. This is where backdoors enter the picture: Intuitively, a small backdoor explains how a backtrack search can get "lucky" on certain runs, since the backdoor variables are identified early on in the search and set in the right way.

(Williams et al., 2003a; Williams et al., 2003b) provide a detailed mathematical model that explains heavy-tailed behavior (Pareto-like tails) in backtrack search as a function of the size of a minimal backdoor set. They show, furthermore, that even though finding a small set of backdoor variables is computationally hard, the very existence of a small backdoor in a problem provides a concrete computational advantage to its solution. They consider three scenarios. First, a deterministic scenario is considered with an exhaustive search of backdoor sets. In this scenario, one obtains provably better search complexity when the backdoor contains up to a certain fraction of all variables. They then show that a randomized search technique, which in effect repeatedly guesses backdoor sets, provably outperforms a deterministic search. Finally, in the third scenario the availability of a variable selection heuristic is considered, which provides guidance in looking for a backdoor set. This strategy can reduce the search space even further. By exploiting restart strategies, for example, one can obtain a polynomially solvable case when the backdoor contains at most $\log(n)$ variables, where n is the total number of variables in the problem instance. We believe that this final scenario is closest to the behavior of the effective randomized backtrack SAT and constraint solvers that are currently available.

6.4 Restart Results on a Range of Problem Instances

Table8.7 shows the results of randomization and restarts on a *balanced* QCP instance of size 50, and right at the *phase transition*—a very hard instance, not previously solved by any other complete search method. A balanced instance is one in which the numbers of pre-assigned cells in the various rows and columns are nearly equal to one another. Balanced QCP instances are considerably harder than instances in which the numbers of pre-assigned cells are randomly distributed (Kautz et al., 2001). We used a complete backtrack search method that combines CSP propagation techniques with linear programming (LP) randomized rounding (Gomes and Shmoys, 2002). The LP randomized rounding is based on the relaxation of a 0-1 integer programming formulation of the QCP problem. In the LP randomized rounding the optimal variable values may be fractional, between 0 and 1. The optimal value of each such variable is interpreted as the probability that the variable will be assigned the value 1. The LP randomized rounding is used within the backtrack procedure as a heuristic for selecting the next variable/value to be assigned. (See (Gomes and Shmoys, 2002), for details.)

From the table we can see that we could only solve the instance in only one out of 100 runs when using a cutoff of $50,000$ backtracks, in only out of 100, when using a cutoff of $250,000$ backtracks, and in only one out of 100, when using a cutoff of $500,000$ backtracks. We also ran the backtrack search procedure with a cutoff of $1,000,000$ backtracks, but with such a high cutoff we were not able to solve the instance at all in 100 runs. In general, complete backtrack search methods that incorporate randomization and restarts allow us to solve considerably harder QCP instances than the standard, deterministic backtrack procedures.

Problem	Solver	Deterministic soln. time	RRR mean soln. time
logistics.d	Satz	108 min	95 sec
3bit-adder-32	Satz	$>$ 24 hrs	165 sec
3bit-adder-31	Satz	$>$ 24 hrs	17 min
round-robin 14	Ilog	411 sec	250 sec
round-robin 16	Ilog	$>$ 24 hrs	1.4 hrs
round-robin 18	Ilog	$>$ 48 hrs	$\approx$ 22 hrs
blocks-world.d	Satz	30 min	23 min

Table 8.8. Randomized rapid restarts (RRR) versus deterministic versions of backtrack search procedures (Satz solver used on SAT encodings; Ilog solver on CSP encodings).

In table 8.8 we give the mean solution times of randomized backtrack procedures with rapid restarts (RRR), for a range of problem instances. In each case, the runtimes were averaged over 100 runs. For comparison, we include the runtime of the original deterministic procedure (SAT for the logistics, adder, and blocks-world instances; CSP for the round-robin problems).[8]

Our deterministic CSP procedure on the round-robin scheduling problem gives us a solution for the 14 team problem in 411 seconds; randomization improves this to 250 seconds. We could not find solutions for the 16 and 18 team problem with the deterministic version. Apparently, the subtle interaction between global and local constraints makes the search for a globally consistent solution surprisingly difficult. These problem instances are too hard to obtain a full cost distribution, making it difficult to calculate an optimal cutoff value.[9] For example, in the 16 team case, running with a cutoff of 1,000,000 backtracks gives a success rate of less than 40%, so, we do not even reach the median point of the distribution. (We estimate that the median value is around 2,000,000 backtracks.) Each run takes about 2 hours to complete.

In order to find a good cutoff value for very hard problem instances, one possible strategy is to start at relatively low cutoff values, since the optimal cutoff for these problems tends to lie far below the median value of the distribution, and incrementally increase the cutoff in order to guarantee the completeness of the algorithm. Using such a strategy with RRR, we were able to solve the 16 team instance in 1.4 hours and the 18 team in approximately 22 hours.

For the SAT encodings of the 3bit-adder problems, which are examples of Boolean-circuit synthesis problems from the Dimacs benchmark (Johnson and Trick, 1996), the RRR solution times are—to the best of our knowledge—the first successful runs of a backtrack search procedure (DPLL) on these instances, though they were previously solved with local search methods, (Selman and Kautz, 1993).

The results given in tables 8.7 and 8.8 show that introducing a stochastic element into a backtrack-style SAT or CSP procedure, combined with an appropriate restart strategy, can significantly enhance the procedure's performance. In fact, as we see here, it allows us to solve several previously unsolved problem instances.

6.5 Learning Dynamic Restart Strategies

In section 6.2 we described a fixed-cutoff restart strategy based on the work of Luby *et al.*. Though these results are quite important, it is often mistakenly assumed that they solved the issue of restart strategies *completely*. It should be noted that the analysis of Luby *et al.* of their universal strategy is based on a few key assumptions, namely that (1) no information about the prior run-time distribution of the solver on the given instance, or complete information (2) the only information we have about the solver is the runtime; (3) we have unlimited time and resources to solve the problem.

In general, these assumptions turn out to be invalid when it comes to real-world scenarios—in fact, quite often we have *partial* knowledge about the runtime distribution of a given instance. For example, we know that the median run time for solving a QCP instance in the under-constrained area is considerably shorter than for an instance at or near the phase transition. It can be shown that if one has prior knowledge about the problem instance (such as experience in solving instances of the same problem class), or if one has some time constraints (*e.g.*, a need to solve the instance within, say, two days of cpu time), then we can improve upon the fixed-cutoff restart strategy.

Motivated by these considerations, (Horvitz et al., 2001), introduced a Bayesian framework for learning predictive models of randomized backtrack solvers. The basic idea is to predict the length of a run based on several features of the runs, captured during the initial execution of the runs. The features considered were both domain dependent and domain independent, such as: depth of backtrack points, number of unit propagations, size of the remaining search tree.

Extending that work, (Kautz et al., 2002), considered restart policies that can factor in information about a solver's behavior which is based on real-time observations. In particular, they introduce an *optimal* policy for dynamic restarts that takes such observations into account. To demonstrate the efficacy of their restart policy, they carried out empirical studies of two randomized backtrack solvers, one for CSP formulations and the other for SAT formulations. (Ruan et al., 2002), further extended this work by considering the case of dependency among runs, when the runs have the same underlying distribution, given two different *a priori* distributions. They show how, in such a scenario, an offline dynamic programming approach can be used to generate the optimal restart strategy, and how to combine the resulting policy with real-time observations, in order to boost the performance of backtrack search methods.

The study of dynamic restart strategies based on learning models about the behavior of the solvers is a new are of research, offering a new class of procedures for tackling hard computational problems. Clearly, much research is needed in this new emerging area.

6.6 Variants of Restart Strategies

(Lynce et al., 2001), propose "randomized backtracking", a technique for randomizing the *target* decision points of a backtrack solver, *i.e.*, the points to which the solver backtracks during the search. Such randomized target points are decided based on randomly picking a decision variable from learned conflict clauses. This backtracking strategy contrasts with the standard, non-chronological backtracking strategy, in which the most recent decision variable is selected as the target backtrack point.

Several variants of this "randomized backtracking" protocol can be considered, namely a *destructive* strategy that undoes all the decisions performed after the *target* decision point, or a *non-destructive* strategy that undoes only the decision associated with the target point itself, leaving all the other assignments unaffected. The former strategy is more drastic, since it basically erases a portion of the search tree previously searched. Non-destructive backtracking has the flavor of a local search procedure, since the current assignment is only locally modified.

Lynce *et al.* propose a parameter K that controls the frequency with which to apply randomized backtracking. When K is set to 1, "randomized backtracking" is performed every time a conflict is detected; otherwise, randomized backtracking is performed every K conflicts. In the latter case, either just the most recent conflict or all the interim conflicts can be considered when picking the *target* point.

Both *destructive* and *non-destructive* procedures can lead to *unstable* algorithms. Lynce *et al.* discuss ways of dealing with such situations in order to guarantee the completeness of the algorithm (*e.g.*, by preventing clause deletion). Even though the number of recorded clauses can grow exponentially in the worst-case, in practice (based on experiments with real-world instances) this does not seem to pose a problem (Lynce et al., 2001).

An alternative to "randomized backtracking," proposed by Lynce *et al.* is *unrestricted backtracking*, a form of "deterministic randomization", that uses a deterministic formula (specifically, a function of the clauses learned during search) to select the target point. This strategy is guaranteed to be complete, and the system will still behave (deterministically) randomly (*i.e.*, it is very unlikely that it will pick a *target point* previously selected). Analogously to randomized backtracking, unrestricted backtracking can also be applied every K conflicts.

(Zhang, 2002), introduces a "random jump" strategy, which can be viewed as a variant of the complete restart strategy. Like restarts, the main motivation behind the strategy proposed by Zhang is to allow the solver to escape unproductive portions of the search tree. However, rather than restarting from scratch, the backtrack solver randomly jumps to unexplored portions of the search space, ensuring that no portion of the search space is visited twice. The jumping decisions are based on analyzing, at given *checkpoints*, to what extent the remaining search space is too large to be fully explored in the remaining time allotted for the search. If the percentage of the remaining space is larger than that of the remaining allotted time, the remaining space is considered *sufficiently large*, and so the solver skips some open right branches of the search tree along the way. The *checkpoints* of Zhang's strategy are analogous to the cutoffs for the fixed restart strategy

Zhang's "random jump" strategy is implemented in SATO, an efficient implementation of the DPLL backtrack search procedure for Boolean satisfiability. The strategy is quite effective, and it has solved a dozen previously open problems in the area of finite algebra. As reported in (Zhang, 2002), SATO, could not solve those open problems considered— even with a full week of runtime allotted per problem—by using its standard approach. With the "random jump" strategy, SATO was able to solve each of them in an overnight run.

As a final point, we mention "probing." Probing is extensively used in CSP and MIP formulations (see *e.g.*, (Crawford and Baker, 1994) and (Savelsbergh, 1994)). In a "probing" strategy one repeatedly goes down a random branch of the search tree, with no backtracking, which can be seen as a restart with a cutoff of 1.

In the next section we discuss portfolio strategies, another technique for combatting the high variance of search methods.

7. Portfolio Strategies

In the previous sections we showed how the performance of backtrack search methods can vary significantly from instance to instance—and, when using different random seeds, even on the same instance. One can take advantage of such high variance by combining several algorithms into a portfolio, and either running them in parallel or interleaving them on a single processor. In this section we describe portfolio strategies and identify conditions under which they can have a dramatic computational advantage over the best traditional methods.

7.1 Portfolio Design

A *portfolio of algorithms* is a collection of different algorithms and/or different copies of the same algorithm running on different processors (Gomes and Selman, 1997b; Huberman et al., 1993).

We consider randomized algorithms—Las Vegas and Monte Carlo algorithms. Therefore, the computational cost associated with a portfolio is a random variable. The expected computational cost of the portfolio is simply the expected value of the random variable associated with the portfolio, and its standard deviation is a measure of the "dispersion" of the computational cost obtained when using the portfolio of algorithms. In this sense, the standard deviation is a measure of the risk inherent in using the portfolio.

The main motivation to combine different algorithms into a portfolio is to improve on the performance of the component algorithms, mainly in terms of expected computational cost but also in terms of the overall risk. As we will show, some portfolios are strictly preferable to others, in the sense that they provide a lower risk *and* a lower expected computational cost. However, in some cases, we cannot identify any portfolio (within a set that of portfolios) that is best, in terms of both expected value and risk. Such a set of portfolios corresponds to the *efficient set* or *efficient frontier*, following terminology used in the theory of mathematical finance (Martin et al., 1988). Within an efficient set, one has to accept deterioration in the expected value in order to minimize the risk; conversely, one has to assume greater risk in order to improve the expected value of the portfolio.

In this context, where we characterize a portfolio in terms of its mean and variance, combining different algorithms into a portfolio only makes sense if they exhibit different probability profiles and none of them *dominates* the others over the whole spectrum of problem instances. An algorithm A dominates algorithm B if the cumulative frequency distribution function for algorithm A lies above the corresponding value for algorithm B at every point, that is for every value of the random varaible.

7.2 Portfolio Results

In section 5.2 we have shown that there are several interesting trade-offs between depth-first branch-and-bound and best-bound branch-and-bound. In

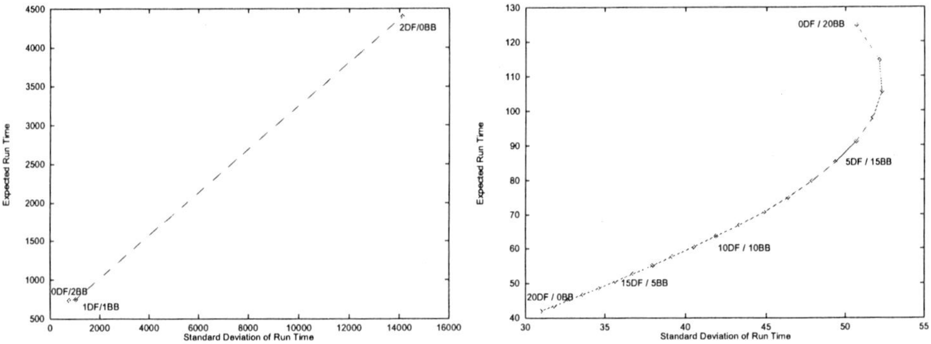

Figure 8.23. Portfolio results for logistics planning: left panel, 2 processors; right panel, 20 processors

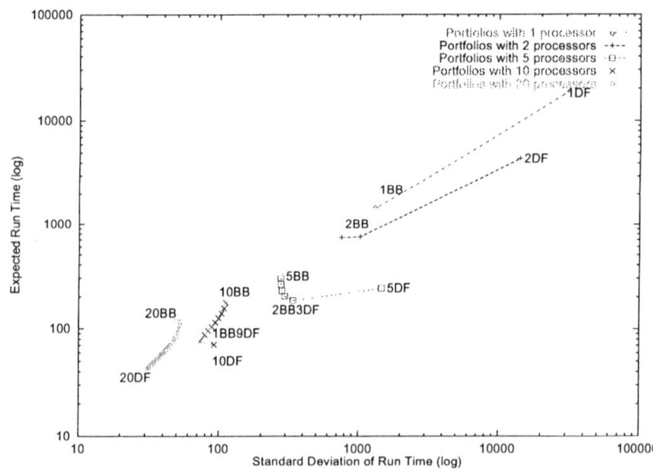

Figure 8.24. A range of portfolios for the MIP formulation of logistics planning (expected runtime).

particular, depth-first search performs better early on in the search, whereas best-bound is better on longer runs. We also identified *fat* and *heavy*-tailed behavior in the cost profiles of the MIP algorithms.

These results indicate that choosing a single search strategy — which is the standard current practice — may not be the most effective approach when dealing with these types of MIP problems. In fact, as we will show below, one can obtain better results by combining strategies, either by interleaving them or by running them in parallel.

In this section we illustrate the applicability of the portfolio framework by using the results from section 5.2 for the general paradigm of mixed integer programming. The portfolio approach is general and can therefore be applied to

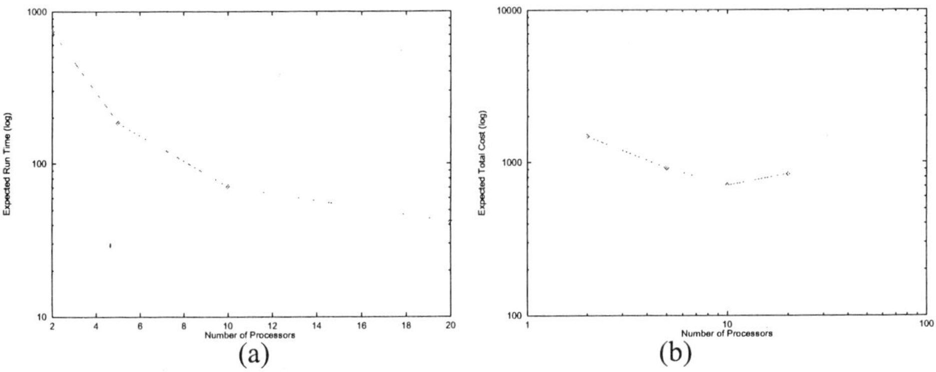

Figure 8.25. Expected runtime (a) and expected total cost (b) of optimal portfolio for different numbers of processors.

exploit different types of algorithms, including mixtures of complete backtrack-style algorithms and local search algorithms.

In the left panel in figure 8.23, we present our results for the case of using two processors to solve the logistics problem (the same instance as the one to which figure 8.13 applies). The plot gives the expected runtime and standard deviation for different ways of combining a branch-and-bound search procedure using depth-first search and a branch-and-bound search procedure using best-bound search, assuming two processors. These values can be computed analytically from the cost profiles in figure 8.13, by using the portfolio probability distribution (see (Gomes and Selman, 2001) for details). From the left panel in figure 8.23 we see that in the case of two processors, the best choice in terms of minimizing the expected runtime and standard deviation consists of running branch-and-bound with best-bound on both processors (i.e., 0DF/2BB). The expected runtime of such a strategy is approximately 700 nodes, with a standard deviation of about 750. Note the contrast between these values, and the much higher values corresponding to the strategy of running branch-and-bound with depth-first on both processors (i.e., 2DF/0BB, with an expected runtime of 4397 and a standard deviation 14,112).

From the right panel in figure 8.23 we see that in the case of 20 processors, the best strategy is to use only depth-first search (*i.e.*, 20DF/0BB). Note that this is quite counterintuitive, especially in the view that is exactly the opposite situation from that of the one- or two- processor scenario, where using only best-bound strategies is optimal. So, we see that in certain cases, *optimal behavior emerges from running many copies of a single suboptimal strategy.*

In figure 8.24, we present data for a range of portfolios that we used in solving our logistics problem. Each of these portfolios consists of anywhere from one to twenty processors.

The figure gives the expected runtime and standard deviation for different ways of combining branch-and-bound search procedures using depth-first search with those that use best-bound search. From this plot we see that the mixing strategy changes as we increase the number of processors or the amount of interleaving. In the case of one processor, the best choice in terms of minimizing the expected runtime and standard deviation consists of running branch-and-bound with best-bound. Best-bound remains the best strategy when we scale up two processors (*i.e.*, the best portfolio consists of running branch-and-bound with best-bound on both processors, as mentioned earlier). In the experiments with five and ten processors, the best strategy is a combination of depth-first search and best-bound search. In the case of ten processors, for example, the best strategies are 9DF/1BB and 10DF/0BB. Neither of those two portfolios dominates; taken together, therefore, they constitute an efficient set. There is a tradeoff between the expected runtime and the standard deviation : The 10DF/0BB portfolio minimizes the expected runtime, while the 9DF/1BB portfolio minimizes the risk (standard deviation).

Figure 8.25(a) shows that once the number of processors exceeds 10, the expected runtime decreases at a very slow rate. In figure 8.25(b), we show the expected total cost of solving the chosen problem instance. It is clear from this figure that when total resource usage is factored in the most cost-effective solution[10] is obtained with a 10-processor portfolio.

7.3 Variants of Portfolios

(Baptista and Marques-Silva, 2000), consider algorithm portfolios based on restarts: Each time the search is restarted, a *different* algorithm is selected from a set of algorithms, $\{A_1, A_2, \ldots, A_k\}$. Each algorithm A_i has probably p_i of being selected. A key aspect of their approach is that restarts are not independently performed: A database of learned clauses is maintained from run to run, allowing the system to reuse information about the search space learned during each run. In their approach, they select different configurations of the same algorithm (GRASP) for the portfolio mix, rather than using radically different algorithms; they vary parameters such as the branching heuristics and the amount of randomization. Baptista and Marques-Silva report very promising results with their portfolio approach: They were able to prove, for example, that several hard superscalar-processor verification instances were *unsatisfiable*.

(Davenport, 2000), uses portfolios of a cooperative problem solving team of heuristics that evolve algorithms for a given problem instance. They solve difficult instances of a bicriteria sparse multiple knapsack problem based on real-world inventory instances.

(Leyton-Brown et al., 2002) proposed a promising methodology for building portfolio of algorithms. Their approach uses machine learning techniques to identify key features of instances with respect to different algorithms, and was initially motivated by the combinatorial auction winner determination problem.

8. Conclusions

Researchers have informally observed that the performance of *complete* backtrack search methods can vary considerably from instance to instance. Early formal results showed that randomization and restarts can increase the robustness of such search methods. Nevertheless, until recently, such techniques were not believed to provide significant *practical* benefits in a *complete* search setting: Up until about five years ago, state-of-the-art complete backtrack search solvers did not exploit randomization and restart strategies.

Advances in the understanding of heavy-tailedness of certain runtime distributions of backtrack search methods has led to a dramatically different attitude toward randomization, restart, and portfolio strategies, for *complete* or *exact* methods. Restarts and portfolios can significantly reduce the variance in runtime and the probability of failure of such search procedures, resulting in search methods that are both more robust and more efficient overall. Randomization and restart strategies are now becoming an integral part of state-of-the-art complete backtrack solvers. We hope that this overview of the study and design of randomized complete backtrack search methods will stimulate additional research efforts in this new and exciting research area.

Acknowledgments

I thank all my colleagues with whom I have collaborated and exchanged ideas on the study and design of randomized complete search methods over the past five years. In particular, I would like to thank Dimitris Achlioptas, Ramon Béjar, Christian Bessière, Hubie Chen, Nuno Crato, César Fernández, Eric Horvitz, Henry Kautz, Felip Manya, João Marques-Silva, Ken McAloon, Bart Selman, David Shmoys, Carol Tretkoff, Toby Walsh, Steve Wicker, and Ryan Williams. I thank Toby Walsh for providing me with the plot in figure 8.15, and Henry Kautz and Joachim Walser for the MIP formulations of the logistics-planning problems.

This research was partially supported by AFRL grants F30602-99-1-0005 and F30602-99-1-0006, AFOSR grant F49620-01-1-0076 (Intelligent Information Systems Institute) and F49620-01-1-0361 (MURI grant on Cooperative Control of Distributed Autonomous Vehicles in Adversarial Environments), and DARPA F30602-00-2-0530 (Controlling Computational Cost: Structure, Phase Transitions, and Randomization) and F30602-00-2-0558 (Configuring Wireless Transmission and Decentralized Data Processing for Generic Sensor

Networks). The views and conclusions contained herein are those of the author and should not be interpreted as necessarily representing the official policies or endorsements, either expressed or implied, of the U.S. Government.

Notes

1. T is the number of leaf nodes visited, up to and including the successful leaf node.

2. By a standard distribution we mean a distribution that has all its moments defined and finite.

3. CP techniques can now solve the round-robin sports-scheduling problem up to 40 teams (Regin, 2002).

4. We thank Henry Kautz and Joachim Walser for providing us with MIP formulations of the logistics-planning problems.

5. We thank Toby Walsh for providing us with the plot for the study of coloring graphs with different structures.

6. There are some applications that are characterized by unbounded search spaces —for example, in the area of theorem proving.

7. CP techniques can now solve the round-robin sports-scheduling problem up to 40 teams (Regin, 2002).

8. The deterministic runs can also be viewed as single runs of the randomized procedure with an infinite cutoff value. Note that, of course, one might be "lucky" on any given instance and have the deterministic procedure branch in just the right way. On the harder instances, however, we often need several hundred restarts to find a solution. On those instances, it becomes quite unlikely that a single deterministic run would succeed.

9. For problems for which we can empirically determine the overall cost profile, we can use the empirical distribution to calculate the cutoff value that minimizes the expected cost of finding a solution.

10. The expected the total cost is the product of the time to find a solution and the number of processors in the portfolio.

References

Achlioptas, D., Kirousis, L. M., Kranakis, E., Krizanc, D., Molloy, M. S. O., and Stamatiou, Y. C. (1997). Random constraint satisfaction: A more accurate picture. In *Principles and Practice of Constraint Programming*, pages 107–120.

Adler, R., Feldman, R., and Taqqu, M. (1998). *A practical guide to heavy tails*. Birkhäuser.

Agrawal, M., Kayal, N., and Saxena, N. (2002). Primes in P. (www.cse.iitk.ac.in/news/primality.html).

Baptista, L. and Marques-Silva, J. (2000). Using randomization and learning to solve hard real-world instances of satisfiability. In *Proceedings of the 6th International Conference on Principles and Practice of Constraint Programming (CP00)*.

Bayardo, R. and Miranker, D. (1996). A complexity analysis of space-bounded learning algorithms for the constraint satisfaction problem. In *Proceedings of the Thirteenth National Conference on Artificial Intelligence (AAAI-96)*, pages 558–562, Portland, OR. AAAI Press.

Bayardo, R. and Schrag, R. (1997). Using csp look-back techniques to solve real-world sat instances. In *Proceedings of the Fourteenth National Conference on Artificial Intelligence (AAAI-97)*, pages 203–208, New Providence, RI. AAAI Press.

Beasley, J. E. (1990). OR-Library: distributing test problems by electronic mail. *Journal of the Operational Research Society*, 41(11):1069–1072.

Brelaz, D. (1979). New methods to color the vertices of a graph. *Communications of the ACM*, 22(4):251–256.

Chen, H., Gomes, C., and Selman, B. (2001). Formal models of heavy-tailed behavior in combinatorial search. In *Proceedings of 7th Intl. Conference on the Principles and Practice of Constraint Programming (CP-2001), Lecture Notes in Computer Science, Vol. 2239, Springer-Verlag*, pages 408–422.

Colbourn, C. (1984). The complexity of completing partial latin squares. *Discrete Applied Mathematics*, (8):25–30.

Cook, S. and Mitchell, D. (1998). Finding hard instances of the satisfiability problem: a survey. In Du, Gu, and Pardalos, editors, *Satisfiability Problem:*

Theory and Applications. Dimacs Series in Discrete Mathematics and Theoretical Computer Science, Vol. 35.

Crawford, J. and Baker, A. (1994). Experimental Results on the Application of Satisfiability Algorithms to Scheduling Problems. In *Proceedings of The Tenth National Conference on Artificial Intelligence.*

Davenport, A. (2000). Cooperative strategies for solving bicriteria sparse multiple knapsack problems. In *Proceedings of the AAAI Workshop on Leveraging Probability and Uncertainty in Computation.*

Davis, M., Logemann, G., and Loveland, D. (1979). A machine program for theorem proving. *Communications of the ACM*, 5:394–397.

Davis, M. and Putnam, H. (1960). A computing procedure for quantification theory. *Journal of the ACM*, 7:201–215.

Easton, K., Nemhauser, G., and Trick, M. (2001). The Traveling Tournament Problem: Description and Benchmarks. In *Proceedings of the 7th International Conference on Principles and Practice of Constraint Programming (CP01)*, Cyprus.

Ertel, W. and Luby, M. (1994). Optimal Parallelization of Las Vegas Algorithms. In *Symp. on Theoretical Aspects of Computer Science, Lect. Notes in Computer Science 775*, pages 463–475. Springer Verlag.

Feller, W. (1968). *An introduction to probability theory and its applications Vol. I.* John Wiley & Sons, New York.

Feller, W. (1971). *An introduction to probability theory and its applications Vol. II.* John Wiley & Sons, New York.

Frost, D., Rish, I., and Vila, L. (1997). Summarizing CSP hardness with continuous probability distributions. In *Proceedings of the Fourteenth National Conference on Artificial Intelligence (AAAI-97)*, New Providence, RI. AAAI Press.

Gent, I. and Walsh, T. (1993). Easy Problems are Sometimes Hard. *Artificial Intelligence*, 70:335–345.

Goldberg, E. and Novikov, Y. (2002). Berkmin: A fast and robust sat solver. In *Proc. of Design Automation and Test in Europe (DATE-2002)*, pages 142–149.

Goldwasser, S. and Kilian, J. (1986). Almost all primes can be quickly certified. In *Proceedings of Annual IEEE Symposium on Foundations of Computer Science*, pages 316–329.

Gomes, C. and Selman, B. (2001). Algorithm portfolios. *Artificial Intelligence*, 126:43–62.

Gomes, C. P. and Selman, B. (1997a). Algorithm Portfolio Design: Theory vs. Practice. In *Proceedings of the Thirteenth Conference On Uncertainty in Artificial Intelligence (UAI-97)*, Linz, Austria. Morgan Kaufman.

Gomes, C. P. and Selman, B. (1997b). Problem Structure in the Presence of Perturbations. In *Proceedings of the Fourteenth National Conference on Artificial Intelligence (AAAI-97)*, New Providence, RI. AAAI Press.

Gomes, C. P. and Selman, B. (1999). Search Strategies for Hybrid Search Spaces. In *Proceedings of the Eleventh International Conference on Tools with Artificial Intelligence (ICTAI-99)*, Chicago, IL. IEEE.

Gomes, C. P., Selman, B., and Crato, N. (1997). Heavy-tailed Distributions in Combinatorial Search. In *Proceedings of the Third International Conference of Constraint Programming (CP-97)*, Linz, Austria. Springer-Verlag.

Gomes, C. P., Selman, B., Crato, N., and Kautz, H. (2000). Heavy-tailed phenomena in satisfiability and constraint satisfaction problems. *J. of Automated Reasoning*, 24(1–2):67–100.

Gomes, C. P., Selman, B., and Kautz, H. (1998a). Boosting Combinatorial Search Through Randomization. In *Proceedings of the Fifteenth National Conference on Artificial Intelligence (AAAI-98)*, New Providence, RI. AAAI Press.

Gomes, C. P., Selman, B., McAloon, K., and Tretkoff, C. (1998b). Randomization in Backtrack Search: Exploiting Heavy-Tailed Profiles for Solving Hard Scheduling Problems. In *Proceedings of Fourth International Conference on Artificial Intelligence Planning Systems 1998*, Pittsburgh, PA, USA.

Gomes, C. P. and Shmoys, D. (2002). The promise of LP to boost CSP techniques for combinatorial problems. In Jussien, N. and Laburthe, F., editors, *Proceedings of the Fourth International Workshop on Integration of AI and OR Techniques in Constraint Programming for Combinatorial Optimisation Problems (CP-AI-OR'02)*, pages 291–305, Le Croisic, France.

Hall, P. (1982). On some simple estimates of an exponent of regular variation. *Journal of the Royal Statistical Society*, 44:37–42.

Harchol-Balter, M., Crovella, M., and Murta, C. (1998). On choosing a task assignment policy for a distributed server system. In *Proceedings of Performance Tools '98*, pages 231–242. Springer-Verlag.

Hill, B. (1975). A simple general approach to inference about the tail of a distribution. *Annals of Statistics*, 3:1163–1174.

Hogg, T., Huberman, B., and Williams, C. (1996). Phase Transitions and Complexity. *Artificial Intelligence*, 81.

Hoos, H. (1998). PhD Thesis,TU Darmstadt.

Hoos, H. (2000). Personal communications.

Horvitz, E., Ruan, Y., Gomes, C., Kautz, H., Selman, B., and Chickering, M. (2001). A Bayesian Approach to Tackling Hard Computational Problems. In *Proceedings of the Seventeenth Conference On Uncertainty in Artificial Intelligence (UAI-01)*.

Huberman, B., Lukose, R., and Hogg, T. (1993). An economics approach to hard computational problems. *Science*, (265):51–54.

Ilog (2001a). Ilog cplex 7.1. user's manual.

Ilog (2001b). Ilog solver 5.1. user's manual.

Johnson, D. and Trick, M. (1996). Cliques, coloring, and satisfiability: Second dimacs implementation challenge. In *Dimacs Series in Discrete Mathematics and Theoretical Computer Science, Vol. 36.*

Kamath, A., Karmarkar, N., Ramakrishnan, K., and Resende, M. (1990). Computational experience with an interior point algorithm on the satisfiability problem. In *Proceedings of Integer Programming and Combinatorial Optimization*, pages 333–349, Waterloo, Canada. Mathematical Programming Society.

Kautz, H., Horvitz, E., Ruan, Y., Gomes, C., and Selman, B. (2002). Dynamic Restart Policies. In *Proceedings of the Eighteenth National Conference on Artificial Intelligence (AAAI-02)*, Edmonton, Canada. AAAI Press.

Kautz, H., Ruan, Y., Achlioptas, D., Gomes, C., and Selman, B. (2001). Balance and Filtering in Structured Satisfiable Problems. In *Proceedings of the Seventeenth Inernational Joint Conference on Artificial Intelligence (IJCAI-01)*, Seattle, WA.

Kautz, H. and Selman, B. (1996). Pushing the envelope: planning, propositional logic, and stochastic search. In *Proceedings of the Fourteenth National Conference on Artificial Intelligence (AAAI-97).* AAAI Press.

Kautz, H. and Walser, J. (1999). State-space planning by integer optimization. In *Proceedings of the Sixteenth National Conference on Artificial Intelligence (AAAI-99)*, Orlando, FA. AAAI Press.

Kirkpatrick, S., Gelatt, C., and Vecchi, M. (1983). Optimization by simulated annealing. *Science*, (220):671–680.

Kirkpatrick, S. and Selman, B. (1994). Critical behavior in the satisfiability of random Boolean expressions. *Science*, 264:1297–1301.

Kumar, S. R., Russell, A., and Sundaram, R. (1999). Approximating latin square extensions. *Algorithmica*, 24:128–138.

Kwan, A. (1995). Validity of normality assumption in csp research. In *Proceedings of the Pacific Rim International Conference on Artificial Intelligence*, pages 459–465.

Laywine, C. and Mullen, G. (1998). *Discrete Mathematics using Latin Squares.* Wiley-Interscience Series in Discrete mathematics and Optimization.

Leyton-Brown, K., Nudelman, E., and Shoham, Y. (2002). Learning the empirical hardness of optimization problems: the case of combinatorial auctions. In *Principles and Practice of Constraint Programming CP-2002.*

Li, C. and Anbulagan (1997). Heuristics based on unit propagation for satisfiability problems. In *Proceedings of the International Joint Conference on Artificial Intelligence*, Nagoya, Japan.

Li, C. M. (1999). A constrained-based approach to narrow search trees for satisfiability. *Information processing letters*, 71:75–80.

Luby, M., Sinclair, A., and Zuckerman, D. (1993). Optimal Speedup of Las Vegas Algorithms. *Information Process. Lett.*, pages 173–180.

Lynce, I., Baptista, L., and Marques-Silva, J. (2001). Stochastic systematic search algorithms for satisfiability. In *LICS Workshop on Theory and Applications of Satisfiability Testing (LICS-SAT)*.

Lynce, I. and Marques-Silva, J. (2002a). Building state-of-the-art sat solvers. In *Proceedings of the European Conference on Artificial Intelligence (ECAI)*.

Lynce, I. and Marques-Silva, J. (2002b). Complete unrestricted backtracking algorithms for satisfiability. In *Fifth International Symposium on the Theory and Applications of Satisfiability Testing (SAT'02)*.

Lynce, I. and Marques-Silva, J. (2002c). The effect of nogood recording in mac-cbj sat algorithms. In *ERCIM/CologNet Workshop on Constraint Solving and Constraint Logic Programming, June 2002*.

Mandelbrot, B. (1960). The Pareto-Lévy law and the distribution of income. *International Economic Review*, 1:79–106.

Mandelbrot, B. (1963). The variation of certain speculative prices. *Journal of Business*, 36:394–419.

Mandelbrot, B. (1983). *The fractal geometry of nature*. Freeman: New York.

Marques-Silva, J. P. and Sakallah, K. A. (1999). Grasp - a search algorithm for propositional satisfiability. *IEEE Transactions on Computers*, 48(5):506–521.

Martin, J., Cox, S., and MacMinn, R. (1988). *The Theory of Finance: Evidence and Applications*. The Dryden Press.

McAloon, K. and Tretkoff, C. (1997). Sports League Scheduling. In *Proceedings of the Third ILOG Conference*, Paris, France.

Meier, A., Gomes, C., and Mellis, E. (2001). An Application of Randomization and Restarts to Proof Planning. In *Proceedings of the Sixth European Conference On Planning (ECP-01)*.

Mitchell, D. and Levesque, H. (1996). Some pitfalls for experimenters with random SAT. *Artificial Intelligence*, 81(1–2):111–125.

Moskewicz, M. W., Madigan, C. F., Zhao, Y., Zhang, L., and Malik, S. (2001). Chaff: Engineering an efficient SAT solver. In *Design Automation Conference*, pages 530–535.

Motwani, R. and Raghavan, P. (1995). *Randomized algorithms*. Cambridge University Press.

Nemhauser, G. and Trick, M. (1998). Scheduling A Major College Basketball Conference. *Operations Research*, 46(1):1–8.

Prosser, P. (1993). Hybrid algorithms for the constraint satisfaction problem. *Computational Intelligence*, 9(3):268–299.

Rabin, M. (1980). Probabilistic algorithm for testing primality. *J. Number Theory*, 12:128–138.

Régin, J.-C. (1994). A filtering algorithm for constraints of difference in CSPs. In *Proceedings of The Twelfth National Conference on Artificial Intelligence*, pages 362–367.

Regin, J. C. (1999). Constraint Programming and Sports Scheduling Problems. Informs.

Regin, J. C. (2002). Personal communications.

Reiss, R. and Thomas, M. (2001). *Statistical Analysis of Extreme Values with Applications to Insurance, Finance, Hydrology and Other Fields*. Birkhäuser.

Rish, I. and Frost, D. (1997). Statistical analysis of backtracking on inconsistent csps. In *Proceedings of Third International Conference on Principles and Practices of Constraint Programming*, pages 150–162.

Ruan, Y., Kautz, H., and Horvitz, E. (2002). Restart Policies with Dependence among Runs: a Dynamic Programming Approach. In *Proceedings of the 8th International Conference on Principles and Practice of Constraint Programming (CP02)*, Ithaca, USA.

Samorodnitsky, G. and Taqqu, M. (1994). *Stable Non-Gaussian Random Processes: Stochastic Models with Infinite Variance*. Chapman and Hall.

Savelsbergh, M. W. P. (1994). Preprocessing and probing for mixed integer programming problems. *ORSA Journal on Computing*, (6):445–454.

Selman, B. and Kautz, H. (1993). Local Search Strategies for Satisfiability Testing. In *Proceedings of DIMACS Workshop on Maximum Clique, Graph Coloring, and Satisfiability*.

Selman, B. and Kirkpatrick, S. (1996). Finite-Size Scaling of the Computational Cost of Systematic Search. *Artificial Intelligence*, 81(1–2):273–295.

Shaw, P., Stergiou, K., and Walsh, T. (1998). Arc consistency and quasigroup completion. In *Proceedings of ECAI-98, workshop on binary constraints*.

Smith, B. and Grant, S. (1995). Sparse constraint graphs and exceptionally hard problems. In *Proceedings of the International Joint Conference on Artificial Intelligence*, pages 646–651. AAAI Pess.

Solovay, R. and Strassen, V. (1977). A fast monte carlo test for primality. *SIAM Journal on Computing*, 6:84–86.

Trick, M. (1996). http://mat.gsia.cmu.edu/color/solvers/trick.c. Source code for the implementation of an exact deterministic algorithm for graph coloring based on DSATUR.

Vandengriend and Culberson (1999). The G_{nm} Phase Transition is Not Hard for the Hamiltonian Cycle Problem. *Journal of Artificial Intelligence Research*.

Veloso, M. (1992). Learning by analogical reasoning in general problem so lving. Ph.D. Thesis, CMU, CS Techn. Report CMU-CS-92-174.

Vossen, T., Ball, M., Lotem, A., and Nau, D. (1999). Integer Programming Models in AI Planning. Technical report, Workshop on Planning as Combinatorial Search, held in conjunction with AIPS-98, Pittsburgh, PA.

Walsh, T. (1999). Search in a Small World. In *Proceedings of the International Joint Conference on Artificial Intelligence*, Stockholm, Sweden.

Watts, D. J. and Strogatz, S. (1998). Collective Dynamics of *Small World* Networks. *Nature*, 393:440–442.

Williams, R., Gomes, C., and Selman, B. (2003a). Backdoors to typical case complexity. In *To appear in Proceedings of the Eighteenth International Joint Conference on Artificial Intelligence (IJCAI-03).*

Williams, R., Gomes, C., and Selman, B. (2003b). On the connections between backdoors, restarts, and heavy-tailedness in combinatorial search. In *Proceedings of Sixth International Conference on Theory and Applications of Satisfiability Testing (SAT-03).*

Wolfram, S. (2002). *A New Kind of Science*. Stephen Wolfram.

Zhang, H. (2002). A random jump strategy for combinatorial search. In *Proc. of International Symposium on AI and Math.*

Zolotarev, V. (1986). One-dimensional stable distributions. Vol. 65 of "Translations of mathematical monographs", American Mathematical Society. Translation from the original 1983 Russian Ed. *ORSA Journal on Computing*, (6):445–454.

Chapter 9

LOCAL SEARCH AND CONSTRAINT PROGRAMMING

LS and CP illustrated on a transportation problem[1]

Filippo Focacci
ILOG S.A.
9, rue de Verdun
BP 85, 94253 Gentilly, France
ffocacci@ilog.fr

Francois Laburthe
BOUYGUES S.A.
1, avenue E. Freyssinet
78061 St-Quentin-en-Yvelines Cedex, France
flaburthe@bouygues.com

Andrea Lodi
D.E.I.S., University of Bologna
Viale Risorgimento 2, 40136 Bologna, Italy
alodi@deis.unibo.it

Abstract Real-world combinatorial optimization problems have two main characteristics which make them difficult: they are usually large, and they are not *pure*, i.e., they involve a heterogeneous set of side constraints. Hence, in most cases, exact approaches cannot be applied to solve real-world problems, whereas incomplete algorithms, and among them Local Search and Metaheuristic methods, have proved to obtain very good results in practice. Moreover, real-world applications typically lead to frequent update/addition of constraints, thus the algorithmic approach requires flexibility, and this flexibility can be guaranteed by Constraint Programming.

In this chapter we review hybrid algorithms combining Local Search and Constraint Programming using a didactic transportation problem to illustrate the techniques.

Keywords: Local Search, Metaheuristics, Incomplete Algorithms

1. Introduction

Real-world combinatorial optimization problems have two main characteristics which make them difficult: they are usually large (see, for example, (Caprara et al., 1997), which describes real-world crew scheduling applications), and they are not *pure*, i.e., they involve a heterogeneous set of side constraints (see, e.g., union contract regulations for crew scheduling and rostering again in (Caprara et al., 1997)).

Hence, in most cases, exact approaches cannot be applied to solve real-world problems, whereas incomplete algorithms, and among them *Local Search* (LS) and *Metaheuristic* methods, have proved to obtain very good results in practice.

LS techniques are based on a simple and general idea. Let P be the combinatorial optimization problem we want to solve, and s a current solution which, for the moment, we assume to be feasible for P, and to have value $z(s)$. A *neighborhood* is defined for s with respect to a *move type* $\mathcal{N}$, i.e., a *function* mapping s (actually, any feasible solution of P) in a subset $\mathcal{N}(s)$ of the overall solution space. In other words, $\mathcal{N}(s)$ contains all the feasible solutions of P which can be reached from s by means of a move of type $\mathcal{N}$. Roughly speaking, the move is a manipulation of s whose effect is the transition to another solution $x \in \mathcal{N}(s)$. The LS framework explores the neighborhood by searching for the solution $x^* \in \mathcal{N}(s)$ such that $\delta z = z(s) - z(x^*)$ is maximized (for minimization problems). If $\delta z > 0$, then an improved solution has been found, and the process is iterated by considering x^* as new current solution. Otherwise, a local optimum has been reached, and several very effective techniques can be applied to escape from it.

These escaping methods characterize the so-called Metaheuristic techniques. Roughly speaking, the idea is to use a higher level heuristic (actually, metaheuristic) framework so as to continue the search if a local optimum has been reached. This higher level heuristic can consist of a trivial random jump to another solution $\widetilde{s}$ to restart the search with, or can be much more sophisticated by using the memory of previous moves as escaping mechanism. The reader is referred to (Aarts and Lenstra, 1997) for a complete description of metaheuristic techniques.

If problem P presents relevant feasibility issues, i.e., it is not easy in practice to find a feasible initial solution, the same framework can be applied anyway by using as current solution an infeasible one. In order to drive the local search process towards feasible solutions, the cost function needs to be modified to measure the infeasibility of the candidate solution. In this sense, the evaluation of a move can be more complex (penalty functions are usually necessary), but the framework does not change substantially.

Three important issues must be taken into account when dealing with real-world problems.

1 Huge problems require large neighborhoods whose exploration can be computationally expensive.

2 When dealing with problems involving many heterogeneous side constraints it is often preferable to consider them as hard constraints rather than transforming them into penalty functions. In these cases, the neighborhood may contain few feasible solutions, and again large neighborhoods are required in order to avoid getting trapped into local minima too often.

3 Real-world applications typically lead to frequent update/addition of constraints (recall again union contract regulations), thus the algorithmic approach requires flexibility.

In the last decade Constraint Programming (CP) has shown its effectiveness in modeling and solving real-world combinatorial optimization problems. CP is a programming paradigm exploiting Constraint Satisfaction techniques (Mackworth, 1977), and in the following we restrict our attention to CP on *Finite Domains* (CP(FD)) which is the case of all constraint tools for discrete optimization. A good introduction to CP is given in Chapter 1.

CP systems can be very useful in the context of real-world combinatorial optimization problems by adding flexibility to the algorithmic approaches. Indeed, CP supports the design of declarative, compact and flexible models where the addition of new constraints is straightforward and does not affect the previous model. Indeed, the propagation of the previous constraints remain unchanged (since they locally model parts of the overall problem), and the previous constraints simply interact with the new ones through shared variables.

Hybrid methods that combine principles from LS and CP can be partitioned in two main sets.

A first set of such hybrids belongs to the family of local search methods and use CP as a way to efficiently explore large neighborhoods with side constraints. It is often referred to as *Constrained LS*.

A second set belongs to the family of global search (tree search) methods and use LS as a way to improve some of the nodes of the search tree or to explore a set of paths close to the path selected by a greedy algorithm in the search tree. It is often referred to as *Incomplete Global Search*.

There is to date no proven recipe for finding the appropriate hybrid algorithm solving a particular optimization problem. The study in this paper is nevertheless inspired by our own practice, which could be summarized as follows.

Starting from a simple model, we first investigate greedy heuristics (methods constructing a solution by assigning variables one after another). Two situations may occur: Either, finding a reasonable greedy heuristic (i.e. constructing solutions with decent objective values) seems within reach, in which case, a variety of techniques will be used to improve its performance. Or, finding a reasonable greedy heuristic seems out of reach, as the problem is too loosely connected (exhibits too little structure) to give enough information to distringuish good decisions from bad ones. In that second case, one will resort to local search, by reducing the problem to a basic optimization problem (for which neighborhoods are already known) enriched with side constraints.

In the first case, the following steps can be followed to improve the original naive greedy heuristic:

- The model is enriched with redundant modeling to increase the effect of propagation and diminish the search space. Redundant constraints involving the cost variable (such as constraints implementing relaxations) are of particular interest, as the objective variables are often little reduced by simple constraint models.
- Constructive heuristics can be significantly tailored to a precise problem; however, some general principles can be observed: one should instantiate variables that are key in driving the search towards feasible regions, variables that have a significant impact through propagation and variables who do not have too much impact on lower bounds (in the case of minimization).
- Once a good search heuristic has been crafted, the greedy heuristic can be diversified by turning it into an algorithm exploring small variations over the greedy solution.
- The greedy heuristic can also be randomized.
- Last, once the algorithm has been improved by all these means, a post-optimization can be done by means of local search.

In the second case, a simple local optimization is built, as a variation over a local procedure over the basic problem, checking for all neighbors the validity of the additional side constraints. Such algorithms can be improved using the following techniques:

- Checking side constraints earlier. The loops for exploring the neighborhoods can be rewritten so as to check all side-constraints from the model within the neighborhood iteration.
- Using several neighborhoods. The procedure can use several neighborhood structures as a means of diversifying the search.
- Using CP to explore neighborhoods. As one considers larger neighborhoods, it may be worthwhile to use techniques from global search for

each local move: the search for the next neighbor can be modeled as an optimization problem of its own, and techniques (concerning both propagation and search) from CP can be used to guide this search.

- Using control strategies from metaheuristics.
- Using schemes for building large neighborhoods from smaller ones, such as ejection chains.

In short, LS may use ideas from CP in order to make large neighborhoods more tractable, while CP may use ideas from LS to explore a set of solutions close to the greedy path in a tree search and converge more quickly towards the optimum.

The remainder of the chapter is organized as follows. The transportation problem introduced in (Focacci et al., 2002) as a didactic case study is considered in Section 2. A CP model for the problem and the corresponding propagation scheme are discussed in Section 3. Then, all the techniques to devise either constrained LS algorithms or incomplete global search ones are described making use of this transportation problem. Specifically, constructive algorithms are discussed in Section 4, while LS techniques used as post-optimization steps are presented in Section 5. Metaheuristic methods are described in Section 6, and LS steps within the search tree are proposed in Section 7. Finally, some conclusions are drawn in Section 8.

In almost every section or subsection a brief introduction tries to give the intuition of what the general idea is, by then applying it to the specific problem.

2. A didactic transportation problem

The techniques for combining Local Search and Constraint Programming outlined in the previous section are applied here to the transportation problem already introduced in (Focacci et al., 2002). This is a variant of the classical *Vehicle Routing Problem* (VRP), (see, (Toth and Vigo, 2002)) in which several side constraints are considered modeling real-world requirements.

Informally, we are given a set of *clients* and a depot in which a fixed number of *trucks* are located. Each client produces a given amount of *goods* of a certain type and has to be visited within a *time window*. Early arrivals are allowed, in the sense that a truck can arrive before the time window lower bound, but, in this case, it has to wait until the client is ready for the beginning of the service. The service time, for each client, only depends on the type and quantity of goods to be collected (i.e., it does not depend on the truck), and each truck has two *capacitated bins* which each can contain a unique type of goods. The travel times and costs between each pair of clients, and between each client and the depot, are given.

This VRP variant, referred to as *didactic* Transportation Problem (dTP) in the following, calls for the determination of the set of trucks' routes minimizing the total travel cost, and satisfying the following constraints:

- each route is associated to a truck, and starts/ends at the depot;
- each client is visited exactly once, and within the time window;
- the bins' capacity constraints are respected.

3. A CP approach for dTP

The CP approach to combinatorial optimization problems decomposes the solving process in three logical phases. It enables to declaratively model the problem by means of variables and constraints linking several variables. It provides propagation algorithms that reduce the search space by eliminating infeasible combinations of variable-value assignments. Finally, it provides a framework that can be used to specify the ways the search space is explored.
In a CP model we often recognize decision variables and auxiliary variables. The decision variables are typically instantiated during search as a mean to explore the search space. Once all decision variables are instantiated, all auxiliary variables are also instantiated by propagation and a solution is reached.

3.1 A CP model for dTP

CP models for combinatorial optimization problems consist in the definition of three different sets of objects: an objective function, decision objects, and constraints. Here, we explicitly separate the three sets with the keywords *min*, *on*, and *subject to*. Moreover, most CP languages provide modeling objects such as *Activities*, *Resources*, etc. Using such objects rather than only variables yields more concise models.

The CP model of the transportation problem dTP can be formally stated by using the following notation:

- $i, j \in \{1, \ldots, N\}$ for the locations (0 denotes the depot) and, by extension, for the clients;
- $k \in \{1, \ldots, M\}$ for the trucks, and, by extension, for their routes;
- $h \in \{1, \ldots, 2M\}$ for the bins;
- $\ell \in \{1, \ldots, P\}$ for the types of goods.

For each location i ($i = 1, \ldots, N$) the corresponding goods are denoted by $type_i$ and q_i, where $type_i \in \{1, \ldots, P\}$ indicates the type of goods to be collected from client i, while $q_i > 0$ is its quantity. Moreover, each client i has an associated time window $[a_i, b_i]$ representing the time frame during which the service must start. The fleet of vehicles located at the depot 0 is composed by

M trucks: each truck k has two bins of identical capacity C, and each bin may collect a unique type of goods $\ell \in \{1, \ldots, P\}$. The duration of the service at location i is $d_i > 0$. Finally, for each pair of locations i and j, the travel time $tt_{ij} \geq 0$ and the travel cost c_{ij} is reported.

An example of dTP is given in Figure 9.1 with $N = 10, M = 3, C = 10$. With each vertex is associated a triple: the client identifier, the color specifying the type of goods collected from the client[2] and the quantity of goods to be collected.

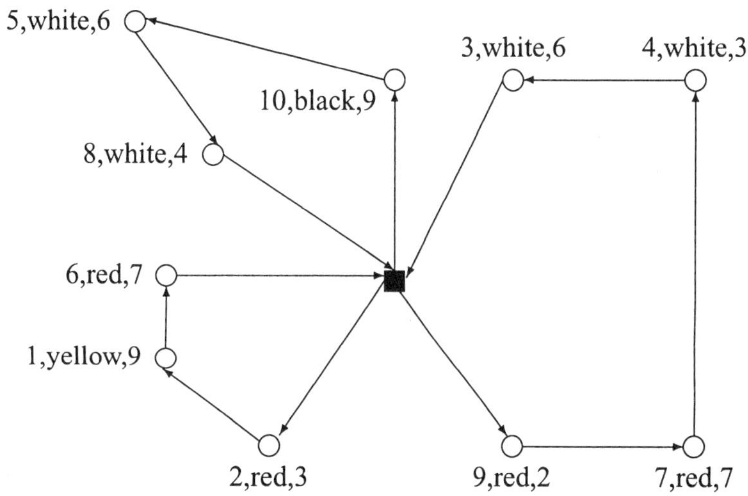

Figure 9.1. Example of dTP with $N = 10, M = 3, C = 10$.

A possible CP model for dTP is the following:

$$\begin{array}{rl}
\min & totCost = \sum_{k=1}^{M} cost_k \\
\text{on} & \\
\forall k \in \{1, \ldots, M\} & cost_k \geq 0, \\
 & truck_k = UnaryResource(tt, c, cost_k) \\
\forall h \in \{1, \ldots, 2M\} & collects_h \in [1..P] \\
\forall i \in \{1, \ldots, N\} & start_i \in [a_i..b_i], \\
 & service_i = Activity(start_i, d_i, i), \\
 & visitedBy_i \in [1..M], \\
 & collectedIn_i \in [1..2M]
\end{array}$$

subject to

$$\forall i \in \{1,\ldots,N\} \qquad service_i \text{ requires } truck[visitedBy_i] \tag{9.1}$$

$$\forall h \in \{1,\ldots,2M\} \qquad \sum_{i \mid collectedIn_i = h} q_i \leq C \tag{9.2}$$

$$\forall i \in \{1,\ldots,N\} \qquad collects[collectedIn_i] = type_i \tag{9.3}$$

$$\forall i \in \{1,\ldots,N\} \qquad visitedBy_i = \left\lceil \frac{collectedIn_i}{2} \right\rceil \tag{9.4}$$

Even if the actual syntax of such objects may vary depending on the specific CP language at hand, we may safely assume that the model proposed can easily be coded using most CP languages.

3.1.1 Basic model. We are given M trucks, $2M$ bins, and N clients that must be visited within a time window by exactly one truck. Each truck is a $UnaryResource$ object containing the information on the travel time and cost among locations (clients), and a variable representing the total travel cost for the truck. The service at each client is an $Activity$ object defined by a variable start-time, a constant duration, and a location. Constraint (9.1) '$service_i$ requires $truck_k$' enforces that from $start_i$ to $start_i + d_i$ the $truck_k$ $UnaryResource$ is used by the $Activity$ $service_i$ without interruption. A $UnaryResource$ cannot be used simultaneously by more than one $Activity$, and, in addition, it is not used during the time needed to move from location to location. Moreover a given time and cost must be considered before the first and after the last activities are executed. In case tt satisfies the *triangle inequality* ($tt_{i_1,i_3} \leq tt_{i_1,i_2} + tt_{i_2,i_3}$), an equivalent model is that for every pair of $Activity$ $service_i$, $service_j$ such that '$service_i$ requires $truck_k$', and '$service_j$ requires $truck_k$', $(start_i \geq start_j + d_j + tt_{ji}) \vee (start_j \geq start_i + d_i + tt_{ij})$. Note that the same objects used to model trucks and visits in dTP could also be used to model machines with maximal capacity equal to 1, and tasks with sequence dependent setup times and cost in scheduling problems. A client $service_i$ needs to be visited by exactly one of the M trucks, say the k-th, thus we model the alternative choice of trucks by defining, for each client, a variable $visitedBy_i$ referring to that index k. The requirement constraint is thus stated on the $truck_k$ array of alternative resources indexed by the variable $visitedBy_i$.

Each client i produces a quantity q_i of goods of type $type_i$. The variable $collectedIn_i$ identifies the bin used to serve client i. The bin capacity constraint (9.2) for each bin h simply states that the sum of all quantities q_i such that q_i is collected in h is less or equal to the maximal capacity C. The variable $collects_h$ identifies the type associated to bin h. The constraint requiring that each bin must contain goods of the same type can be stated using variables $collects_h$ and $collectedIn_i$ by imposing that for each client i the type associated to the bin used by the client i is equal to $type_i$ (9.3).

Bins with index $2k - 1$ and $2k$ are associated to truck k (bins 1 and 2 are placed on truck 1, bins 3 and 4 on truck 2, etc.). The link between bins and trucks is modeled with constraint (9.4).

The *basic model* correctly models dTP. Decision variables are start-time ($start_i$) and bin selection (*collectedIn$_i$*) variables for each client i. Once all variables $start_i$ and *collectedIn$_i$* are instantiated to values satisfying all the constraints of the *basic model*, a solution is reached, since all the other variables of the model ($cost_k$, $visitedBy_i$, etc.) are instantiated by propagation.

Nevertheless, when dealing with routing problems, it may be convenient to explicitly manipulate the sequence of services performed by each truck. For this purpose an extension of the model is considered. This model, based on the abstraction of multiple path, is redundant with respect to the *basic model*, since it is not necessary for correctly modeling dTP.

3.2 Propagation

> CP constraints embed domain propagation algorithms aimed at removing values from variable domains that are proven infeasible with respect to the constraint itself (see, (Mackworth, 1977) and Chapter 1). Several propagation algorithms can be designed for the same CP constraint.

We briefly sketch some algorithms that can be used to implement the propagation of the constraints used in the dTP model.

Concerning notation, given a variable v, in the following we will refer to the minimum (resp. maximum) value in its domain as $\texttt{inf}(v)$ (resp. $\texttt{sup}(v)$). Moreover, the domain itself is referred to as $\texttt{domain}(v)$, whereas once v has been instantiated its value is returned as $\texttt{value}(v)$.

3.2.1 Disjunctive Relations. Consider the $UnaryResource$ object used to model the trucks and the *require* constraints linking objects $Activity$ and $UnaryResource$. As mentioned these constraints state that two activities must not be executed simultaneously on the same unary resource, and a transition time must be granted between two subsequent activities. A simple propagation algorithm updates the start-time variable of activities by looking at all pairs of activities performed on the same resource and deducing precedence constraints. Consider two activities $service_i$, $service_j$, if $(\texttt{inf}(start_j) + d_j + tt_{ji} > \texttt{sup}(start_i)) \wedge (\texttt{value}(visitedBy_i) = \texttt{value}(visitedBy_j))$ then it can immediately be deduced that $service_i$ must precede $service_j$, i.e.,
$\texttt{inf}(start_j) := \max\{\texttt{inf}(start_j), \texttt{inf}(start_i) + d_i + tt_{ij}\}$
and
$\texttt{sup}(start_i) := \min\{\texttt{sup}(start_i) + d_i, \texttt{sup}(start_j) - tt_{ij}\} - d_i$.
This propagation rule finds new time bounds for the activities by considering the current time bounds, the capacity availability, and the resource assignments. Similarly, new possible assignments are deduced by considering current time

bounds, and capacity available: if $(\texttt{inf}(start_j) + d_j + tt_{ji} > \texttt{sup}(start_i)) \wedge (\texttt{inf}(start_i) + d_i + tt_{ij} > \texttt{sup}(start_j)) \wedge (\texttt{value}(visitedBy_i) = k)$ then $visitedBy_j \neq k$. More sophisticated propagation algorithms have been implemented in different CP languages, see for example Edge Finding techniques for unary and discrete resources, and constructive disjunctive techniques for alternative resources (see, (Nuijten, 1994)). These techniques have been successfully applied to Scheduling Problems and Traveling Salesman Problem with Time Windows (TSPTW) (see, (Pesant et al., 1998)). In the proposed model, beside the modeling of capacity availability and transition times, the object $UnaryResource$ is also responsible for maintaining a cost variable corresponding to the resource specific component of the total cost.

3.2.2 Linking trucks and bins.

For its simplicity, it is worth describing in detail the constraint linking the variables used for the choice of trucks and bins, $visitedBy_i = \lceil collectedIn_i/2 \rceil$. The constraint is considered consistent iff the following condition hold:

$$\forall k \in \texttt{domain}(visitedBy_i)\ \exists h \in \texttt{domain}(collectedIn_i)\ |k = \lceil h/2 \rceil$$

and vice-versa

$$\forall h \in \texttt{domain}(collectedIn_i)\ \exists k \in \texttt{domain}(visitedBy_i)\ |k = \lceil h/2 \rceil$$

A simple propagation algorithm iterates on all values k belonging to the domain of $visitedBy_i$ and checks for the existence of a value h in the domain of $collectedIn_i$ that satisfies the relation $k = \lceil h/2 \rceil$. If no value h is found, k is removed from the domain of variable $visitedBy_i$. The propagation algorithm also iterates on all values h belonging to the domain of $collectedIn_i$ and checks for the existence of a value k in the domain of $visitedBy_i$ that satisfy the relation $k = \lceil h/2 \rceil$. If no value k is found, h is removed from the domain of variable $collectedIn_i$.

3.2.3 Propagating costs.

In many cases the calculation of good lower bounds on the objective function variable is especially important. As shown in (Focacci et al., 1999a) good lower bounds on the optimal cost can be used for discarding a priori uninteresting parts of the solution space. Moreover, lower bounds can also be used in greedy algorithms as described in Section 4.2 for guiding towards more promising parts of the solution space. In CP the link between the objective variable and the decision variables is maintained through a constraint.

For the specific dTP case, several important substructures belonging to the classical literature of *packing* and *routing* can be recognized and exploited, such as the Linear Assignment Problem and many others.

Concerning propagation, in summary, three things should be stressed. First, whenever a simple or global constraint acts on several variables the modification of any of the variables triggers propagation algorithms that propagate the modification on the other variables involved. Second, for a given constraint several propagation algorithms can enforce different degrees of consistency. Finally, even without giving complexity results of the propagation algorithms, it is clear that some of them can be computationally more expensive than others. For example, checking a constraint for the feasibility on a set of instantiated variables is usually much easier than eliminating all values that can be proved to be infeasible on a set of yet uninstantiated ones. Therefore, the propagation algorithms to use may vary depending on the size of the problem, and on the type of solving procedure used.

3.3 A redundant routing model

Given a CP model M for a combinatorial optimization problem, a redundant model M' adds to M one or more constraints and possibly variables linked to the decision variables of M.
The main advantage of using redundant models in CP consists in better filtering out inconsistent values. Indeed, despite their redundancy (the search phase would be able to prune these inconsistent values), these constraints generally perform additional propagation since they implement a different view of the problem. Then, the interaction among constraints, through shared variables, allows to effectively reduce the search phase.
Moreover, another purpose of redundant models is to introduce decision variables so as to enable each one of several search methods to use the most appropriate variables.
However, redundant models should be used with care. Depending on the problem size, the use of several linked models could eventually be computationally penalizing. In order to show a wide range of CP-based local search techniques, we used all redundant models simultaneously. Performance issues arising with very large instances could be addressed by using lighter models or other techniques such as decomposition.

Starting from the basic model of Section 3.1.1, a redundant model for dTP can be devised. In this specific case, many neighborhood structures can be easily defined by means of the additional variables ($next_i$, $succ_i$) introduced by the redundant models.

One possible redundant *routing* model is the following:

$$\forall k \in \{1, \ldots, M\} \quad first_k \in [1..N]$$
$$\forall i \in \{1, \ldots, N\} \quad next_i \in [1..N + M], \quad succ_i \in [\{\}..\{1, \ldots, N\}]$$

$$multiPath(first, next, succ, visitedBy) \tag{9.5}$$
$$costPaths(first, next, succ, c, totCost) \tag{9.6}$$
$$\forall i, j \in \{1, \ldots, N\} \quad j \in succ_i \Leftrightarrow (visitedBy_i = visitedBy_j \land start_j > start_i) \tag{9.7}$$

Let $G = (V, A)$ be a digraph, and a partition of V be defined by a set S of M *start* nodes, a set I of N *internal* nodes, and a set E of M *end* nodes. A multiple path constraint *multiPath* enforces that M paths will exist starting from a start node, ending in an end node and covering all internal nodes exactly once. All internal nodes have one direct predecessor (previous node) one direct successor (next node).

Internal nodes are labeled $1, 2, \ldots, N$, end nodes are labeled $N+1, N+2, \ldots, N+M$, while start nodes are labeled $N+M+1, N+M+2, \ldots, N+2M$. The multiple path constraint makes use of four arrays of variables. Variable $first_k$ is associated with the start node $N+M+k$, and identifies its next node in the path. Variables $next_i$, $succ_i$, and $visitedBy_i$ are associated with the internal node i. Variable $next_i$ identifies the node next of i; variable $succ_i$ identifies the set of internal nodes occurring after i in the path containing i; finally, variable $visitedBy_i$ identifies the path containing node i.

The *multiPath* constraint can be used to model a subproblem of dTP: each internal node represents a client, the set of start and end nodes represents copies of the depot. The path k starting from the start node $N+M+k$, covering a set of internal nodes, and ending at an end node represents the route of $truck_k$. The *redundant routing model* and the *basic model* are linked by the variables $visitedBy_i$ used in both models, and by the constraint (9.7).

Note that the cost on each truck only depends on the sequence of services performed, therefore the multiple path structure can also be used to define the objective function. Since the variable $next_i$, and $succ_i$ will be extensively used as decision variables for dTP, it is indeed useful to explicitly link the multiple path model to the cost variables $cost_k$, as done with the constraint (9.6).

3.3.1 Propagation. Since the multi path model will be extensively used in the discussion of the local search methods, it is important to describe it in details. A model equivalent to the $multiPath$ constraint based on only simple arithmetical and logical constraints is the following:

$$\forall i, j \in \{1, \ldots, N\}, i > j \qquad next_i \neq next_j \tag{9.8}$$

$$\forall k_1, k_2 \in \{1, \ldots, M\}, k_1 > k_2 \qquad first_{k_1} \neq first_{k_2} \tag{9.9}$$

$$\forall i \in \{1, \ldots, N\}, \forall k \in \{1, \ldots, M\} \qquad first_k \neq next_i \tag{9.10}$$

$$\forall i, j \in \{1, \ldots, N\} \qquad next_i = j \Rightarrow succ_i = \{j\} \cup succ_j \tag{9.11}$$

$$\forall i, j \in \{1, \ldots, N\} \qquad j \in succ_i \Rightarrow visitedBy_i = visitedBy_j \tag{9.12}$$

$$\forall k \in \{1, \ldots, M\}, \forall i \in \{1, \ldots, N\} \qquad first_k = i \Rightarrow visitedBy_i = k \tag{9.13}$$

$$\forall i \in \{1, \ldots, N\} \qquad i \notin succ_i \tag{9.14}$$

As mentioned in the model description, the set variables $succ_i$ are used to enforce precedence relations among nodes. An integer set variable is a variable whose domain contains sets of integer values. In analogy with the definition of lower and upper bounds on integer variables, lower and upper bounds can also be defined for an integer set variable: the lower bound, called *required set*, is the intersection of all sets belonging to the domain of the variable; the upper bound, called *possible set*, is the union of all sets belonging to the domain of the variable. The variable is considered instantiated when the domain contains a single set of integer values, thus the required set is equal to the possible set. Required set, possible set, and value of the set variable v will be referred to as $\texttt{req}(v)$, $\texttt{poss}(v)$, and $\texttt{value}(v)$.

Constraints (9.8)-(9.10) force all internal nodes $\{1, \dots, N\}$ to have exactly one incoming, and one outgoing arc, while all start nodes are forced to have exactly one outgoing arc belonging to the set of internal nodes. Constraints (9.11) and (9.12) link $succ$ and $next$, and $succ$ and $visitedBy$ variables; constraint (9.13) links the $visitedBy$ variable of the first internal node in a path starting from node $N + M + k$ to the value k. Finally the last constraint prevents cycles in the graph.

From the above discussion follows that the global constraint *multiPath* could be imposed through simple constraints (9.8)-(9.14). However, more effective and efficient propagation algorithms can be designed enforcing the global constraint itself. More precisely, basic propagation on the $next$ variables enforces that, whenever a variable $next_i$ is instantiated to a value t, the value t is removed from the domain of all variables $next_j, j \neq i$. A more powerful propagation can be obtained by using flow algorithms (see, (Régin, 1994)). Cycles can be avoided either by simple cycle removal algorithms, or by performing a more sophisticated propagation using strong connected components or isthmus detection. Some of the propagations that can be performed on the variables $succ_i$ exploit the transitivity of the $succ$ array: $i_2 \in \texttt{req}(succ_{i_1}) \wedge i_3 \in \texttt{req}(succ_{i_2}) \Rightarrow \texttt{req}(succ_{i_1}) := \texttt{req}(succ_{i_1}) \cup \{i_3\}$. This propagation rule can be read as follows: if i_2 is necessarily after i_1, and i_3 is necessarily after i_2, then i_3 is necessarily after i_1.

Beside the simple link between $next$ and $succ$ variables in (9.11), more effective propagation can be performed. For example, if given two nodes i and j, the set of nodes that necessarily follows i has a non-empty intersection with the set of nodes that necessarily precedes j, then at least a node must exist in any path from i to j, therefore j cannot be the next of i. More formally, $\texttt{req}(succ_i) \cap (\{1, \dots, N\} \setminus \texttt{poss}(succ_j)) \neq \emptyset \Rightarrow next_i \neq j$. Note that this last propagation rule is not part of the arithmetic model of the $multiPath$ constraint (9.8)-(9.14); this is a simple example of the extra propagation that can be done by a global constraint.

4. Constructive Algorithms

The model presented in Section 3.1 could be solved as such by complete global search. However, such a solution is impractical in terms of computing time for realistic instances (recall that the maximal size of plain VRPs that can currently be solved to optimality within minutes ranges from a few tens of clients for branch-and-bound methods to 80 for branch-and-cut methods, see (Toth and Vigo, 2002)). This section discusses constructive methods which are realistic for solving dTP instances with hundreds of clients.

4.1 Insertion algorithms

> A global search algorithm produces a solution by taking decisions and backtracking on failure. The decisions taken in a branch amount to adding a constraint to the problem. A general branching scheme, e.g., the first-fail criterion (see, (Haralick and Elliott, 1980)) selects any variable from the model (that with the smallest number of values in its domain) and instantiate it. In general cases, it is often difficult to interpret the state of the system before a solution has been reached, while for problem specific branching schemes, the states after a sequence of decisions may indeed be interpreted as relevant partial solutions making the word "constructive" perfectly clear.

In the dTP model, when all CP variables in the arrays $visitedBy$ and $succ$ are instantiated, then the solution is fully known. Indeed, variables arrays $first$, $next$ and $cost$ can be instantiated from $succ$. When all these variables are instantiated, the objective function is instantiated although the $collectedIn$ and $start$ variables may be left with an interval domain. A simple greedy algorithm can instantiate all these remaining variables. For example one could iteratively choose each client i and instantiate $collectedIn_i$ and $start_i$ to its lower bound[3].

We propose a constructive algorithm based on this dominance property using $(visitedBy_i)$ and $(succ_i)$ as CP decision variables for dTP. Clients are considered one after the other: for each client i, the algorithm instantiates the variable $visitedBy_i$, and for all j such that $visitedBy_j$ has already been instantiated to the same value as $visitedBy_i$, the $succ$ variables are reduced by enforcing one of the two decisions $j \in succ_i$ and $j \notin succ_i$.

Such an instantiation scheme implements a constraint-based version of the standard insertion heuristics for vehicle routing (see, (Golden and Assad, 1988)). Indeed, each time a client i is assigned to a truck, the ordering between i and all others clients assigned to that same route is settled, yielding a total order on all clients of the route. These ordering decisions implement the insertion of i between two consecutive clients in a route. Compared to standard insertion algorithms in Operations Research textbooks, this CP description adds two notions.

- The routing problem can be enriched with side constraints, such as time windows, bin assignments and bin capacity. When selecting the best possibility for insertion, feasibility is ensured. Rather than performing the insertion and checking for the satisfaction of all constraints, we rely on propagation to discard some of the infeasible assignments. Infeasible route assignments are removed from the domain of $visitedBy_i$, and infeasible orderings are discarded by reducing the domain of $succ_i$.

- The overall insertion process is seen as a monotonic process where variable domains are reduced. Indeed, the $next_i$ variables may remain uninstantiated until the very end of the construction process (until the routes are full). Until then, the routes are only partially sequenced with precedence decisions: the relative order among clients already assigned to a given route is settled ($\forall i, j,\ i \in succ_j \ \vee \ j \in succ_i$), but there is room to insert new clients between such i and j. This description of insertion within "open" routes supports the evaluation of lower bounds on the route costs: at any point in the algorithm, each route k has been assigned a subset of clients $i_1, \ldots, i_p$ ($visitedBy_{i_1} = visitedBy_{i_2} = \cdots = visitedBy_{i_p} = k$) which have been ordered: $i_2 \in succ_{i_1}, \ldots, i_p \in succ_{i_{p-1}}$. When the costs c_{ij} satisfy the triangle inequality, the cost of the route $cost_k$ can be bounded with $cost_k \geq c_{0,i_1} + c_{i_1,i_2} + \cdots + c_{i_{n-1},i_n} + c_{i_n,0}$.

The general framework for construction algorithms can thus be described formally as follows:

```
algorithm: INSERTION
 for all clients i
     for each possible value k in domain(visitedBy_i)
         branch on
             try the assignment visitedBy_i = k
             for all j such that domain(visitedBy_j)= {k}
                 branch on try j ∈ succ_i
                        or try j ∉ succ_i
     if some feasible insertions were found
        commit to one among them
     else return(failure)
 return(success)
```

In the end of the algorithm, either a feasible routing plan that services all clients (success) or no solution (failure) is obtained[4].

4.2 Greedy insertion

The search in a constructive algorithm is guided by a heuristic[5]: at each choice point, a function h is evaluated for all possible choices and the choice are ranked by increasing values of h: the choice that minimizes h is considered the preferred decision. In a greedy constructive algorithm, the preferred branch is systematically followed and no backtracking takes place.

The general scheme of the insertion algorithm in the previous section can be refined into a greedy algorithm.

- In order to help the algorithm finding a routing plan covering all clients, it is wise to consider first the clients that are "difficult" to insert into a route (those for which the insertion into a partial route is most likely to be infeasible). We therefore sort all clients and use as priority criterion a function that returns higher values for clients far away from routes already built and with tight time windows over clients close to current routes and with a loose time window. The latter have indeed little propagation impact when assigned to a route while the former drastically augment the lower bound on the cost of the route and significantly reduce the possibilities of further assignments through time window constraints. An example of such a static criterion returning high values for nodes that should be considered first would be $priority(i) = tt_{0,i}/(b_i - a_i)$.
- Since all trucks are similar, each assignment of a client to a (still) empty route amounts to a route creation and is strictly equivalent. Therefore, when there exist several k such that for no client j, domain($visitedBy_j$)= $\{k\}$, then, the tentative assignment ($visitedBy_i = k$) is considered for only one k.
- When several possible insertions are found to be feasible for a client i, rather than selecting any of them, we can choose the assignment $visitedBy_i = k$ and the ordering (reductions on $succ_i$) that causes the lower bound of the overall objective z to increase the least: this turns the constructive method into a greedy algorithm.

The greedy algorithm is then as follows:

```
procedure: BEST_INS(i, k)
 oldInf := inf(cost_k)
 try visitedBy_i = k
 for all j such that domain(visitedBy_j)= {k}
     branch on try j ∈ succ_i
            or try j ∉ succ_i
 over all solutions, minimize Δ :=inf(cost_k) - oldInf

algorithm: GREEDY CONSTRUCTION
 sort all clients i by decreasing values of priority(i)
 for all clients i
     for k ∈ domain(visitedBy_i) | ∃j ≠ i, domain(visitedBy_j) = {k}
         try BEST_INS(i, k)
```

```
    if ∃k₀ ∈ domain(visitedBy_i) | ∀j ≠ i, domain(visitedBy_j) ≠ {k₀}
        try BEST_INS(i, k₀)
    if some feasible insertion was found
        commit to the one that yielded the smallest Δ
    otherwise return(failure)
return(success)
```

Adapting the greedy algorithm to a dynamic priority criterion is straightforward.

4.3 Restricted Candidate Lists and GRASP

Any greedy heuristic can be transformed into a search algorithm through the addition of *backtracking*. By adding a limited amount of backtracking, the algorithm explores a more substantial portion of the solution space at the price of an increased complexity. At each choice point, the heuristic provides a preferred branch, as well as an indication of the quality of the other branches. The idea of Restricted Candidate Lists (RCL, see (Feo and Resende, 1995) and (Glover, 1995)) is to retain only the good branches and to discard the bad ones. The GRASP method (Feo and Resende, 1995) uses RCL within a randomized version of the greedy algorithm: a randomized construction algorithm is run many times. At each choice point of the construction algorithm, one branch is selected at random among the RCL and according to some probability distribution (with higher probability for the most promising branches). GRASP methods are gaining increasing interest in the scientific community, and we refer to Chapter 8 for a detailed description of randomized search.

The greedy insertion algorithm in Section 4.2 can easily be randomized by only considering those insertions that generate small increase in the cost lower bound, and randomly choosing among them. More precisely, let Δ^* be the minimal increase in the lower bound over all possible insertions. Only those insertions k having $\Delta_k \leq \beta\Delta^*$ for some $\beta \geq 1$ are considered. Among all possible insertions k, one is randomly chosen for branching. In the GRASP variant of dTP, the insertion algorithm is run several times. A limited number of backtracks (or a limited amount of time) is granted to each run, and a cost improvement cut is imposed to each following run every time a feasible solution is found.

The resulting GRASP algorithm is then as follows:

```
procedure: SELECT_RANDOM(i)
  bestSet := ∅
  Δ* := ∞
  for k ∈ domain(visitedBy_i) | ∃j ≠ i, domain(visitedBy_j) = {k}
    oldInf := inf(cost_k)
    try visitedBy_i = k
    for all j such that domain(visitedBy_j)= {k}
       try j ∈ succ_i
```

Δ_j :=inf($cost_k$) - oldInf
Δ^* :=min(Δ^*,Δ_j)
if $\exists k_0 \in$ domain($visitedBy_i$) $\mid \forall j \neq i$, domain($visitedBy_j$) $\neq \{k_0\}$
try $visitedBy_i = k_0$
Δ^{k_0} :=inf($cost_{k_0}$)
Δ^* :=min(Δ^*,Δ^{k_0})
for all j such that $\Delta_j \leq \beta\Delta^*$ add the branch $j \in succ_i$ in bestSe
if $\Delta^{k_0} \leq \beta\Delta^*$ add the branch $visitedBy_i = k_0$ in bestSet
return a random value in bestSet

procedure: RANDOMIZED_INSERTION(i)
random branch k^* := SELECT_RANDOM(i)
branch on k^*
call RANDOMIZED_INSERTION($i+1$)
alternate branch $\neq k^*$
call RANDOMIZED_INSERTION(i)
stop if fail limit is reached
if some feasible solution was found
return(success)
otherwise return(failure)

algorithm: GRASP
sort all clients i by decreasing values of $priority(i)$
while global stopping condition is not reached
RANDOMIZED_INSERTION(1)
if solution is found reduce the cost upper bound

4.4 Discrepancy-based search

The previous section introduced the notion of a subtree defined by RCL for a given heuristic. This subtree can either be explored by means of randomized construction procedures (GRASP) or by systematic search. Discrepancy-based Search (see, (Harvey, 1995)) defines a sub-tree by counting the number of times the branch suggested by the heuristic is chosen and follows such a branch always but a limited number of times (the "discrepancies"). More precisely, a discrepancy is introduced each time the preferred branch is not followed, and pure *Limited* Discrepancy Search LDS(K) (Harvey and Ginsberg, 1995) would allow for discrepancies at any choice point, granted that all solutions are generated with less that K discrepancies. LDS explores the search tree for increasing values of K with the aim of considering first the portions of the tree which are more likely to contain "good" solutions (it trusts the heuristic). Obviously, if the search is stopped at a maximum value of K, then the algorithm is transformed into a heuristic one.

For dTP, not only do we have a ranking of insertion choices, but also an evaluation on each branch (the value of Δ for each possibility of insertion), we may precisely control the occurrences of the discrepancies. Thus, the general idea can be refined in several manners:

- one of the interests of Discrepancy-based Search is to generate solutions that are spread across the overall solution space. It is therefore particularly interesting to branch on options that are radically different at each discrepancy. Insertions at different positions within the same route are considered similar; therefore the alternate choices for insertion that are considered involve different routes;
- it is not worthwhile considering the second best possibility of insertion when its impact on the cost is significantly worse than the preferred choice. Branching on discrepancies should be reserved for situations of near-ties. Thus, a discrepancy is generated only when $\Delta_{k_2} \leq \beta\Delta_{k_1}$ where Δ_{k_1} is the best (least) insertion cost and Δ_{k_2} the second best, for some ratio $\beta \geq 1$;
- the branching scheme compares insertions within routes to which some clients have already been assigned with an insertion into one empty route (route creation). Comparing such options by their impact on the cost lower bound is sometimes unfair to the second option. Indeed, let i be the current client, $\Delta_{k_0} = c_{0,i} + c_{i,0}$ whereas $\Delta_{k_1} = c_{j_1,i} + c_{i,j_2} - c_{j_1,j_2}$ if the best place of insertion for i in k_1 is between clients j_1 and j_2. Therefore, Δ_{k_0} accounts for two transitions between locations (back and forth between the depot, 0, and i), while Δ_{k_1} accounts for only one (replacing the direct transition from j_1 to j_2 by a transition from j_1 to i and one from i to j_2). In order to avoid such a bias against route creations, discrepancies where the alternate branch is a route creation are encouraged through a higher ratio threshold $\beta' > \beta$.

The overall Discrepancy-based Search $DISCREPANCY(K)$, defined below, is a refined version of $LDS(K)$. Its integer parameter, K, represents the maximal number of discrepancies allowed in a solution:

```
algorithm DISCREPANCY(K):
  sort all clients i by decreasing values of priority(i)
  DISC(1,K)

procedure: DISC(i, K):
 if i > N return(success)
 else
   for k ∈ domain(visitedBy_i) |∃j ≠ i, domain(visitedBy_j) = {k}
       try BEST_INS(i, k)
```

```
let k1 and k2 be the two branches with smallest Δk
if ∃k0 ∈ domain(visitedBy_i) | ∀j ≠ i, domain(visitedBy_j) ≠ {k0}
   try BEST_INS(i, k0)
if no feasible solution was found
   return(failure)
else
  branch on
     preferred choice:
       if Δk1 < Δk0 commit to BEST_INS(i, k1)
                      DISC(i + 1, K)
       else commit to BEST_INS(i, k0)
            DISC(i + 1, K)
     if K > 0 alternate branch:
       if Δk0 ≤ Δk1
          if Δk1 ≤ βΔk0 commit to BEST_INS(i, k1)
                          DISC(i + 1, K − 1)
       else if Δk0 ≤ β'Δk1 commit to BEST_INS(i, k0)
                             DISC(i + 1, K − 1)
       else if Δk2 ≤ βΔk1 commit to BEST_INS(i, k2)
                             DISC(i + 1, K − 1)
```

5. LS as Post-Optimization

This section illustrates various techniques for the exploration of constrained neighborhoods in the dTP case. We assume to start with a feasible solution which can, for instance, be built through constructive methods presented in the previous sections. We review four kinds of LS procedures on dTP: two simple adaptations of standard LS with constraint checks either at the end or within the neighborhood exploration, and two descriptions of neighborhoods with CP, specifically, frozen fragments and CP models of the neighborhood.

Although we refer to the solution with the notations introduced in Section 3, it is never assumed that the variables of the model are CP domain variables. More precisely, in the two next sections (5.1 and 5.2), the easiest way to implement the LS methods consists in representing a solution with simple data structures (integers to represent variables modeling integers), whereas the approaches in sections 5.3 and 5.4 require that the variables of the model are implemented as CP variables with domains and active propagation.

5.1 LS + constraint checks

Fast LS algorithms typically use neighborhoods of small size which can be explored with a relatively small computational effort. Even if the neighborhood is in principle quite small, the addition of side constraints can considerably increase the computational effort required to explore it, since it is often necessary to test feasibility. The trivial way of addressing problems with side constraints in LS is: the neighborhood of the pure problem is explored, and for each neighbor, the side constraints are iterated and, for each one of them, its feasibility is tested.

A first idea for improving a solution to dTP consists in using classical VRP neighborhoods, and checking, for each neighbor, the feasibility with respect to side constraints. This method is illustrated on two simple neighborhoods, specifically *3-opt* and *node-transfer*.

3-opt. 3-opt is the straightforward extension of the **2-opt** neighborhood classically used by (Lin and Kernighan, 1973) for the Traveling Salesman Problem (TSP) in which three arcs, instead of two, are removed from the solution, and three others are added to obtain a new solution. Since this exchange is performed within the same route, the feasibility test for the new solution needs to be executed only on that route. The best move is the one with the largest decrease of z (if any). An example of 3-opt move is depicted in Figure 9.2.

More formally, let k be a route, and i_1, i_2, i_3 be three different clients such that $visitedBy_{i_1} = visitedBy_{i_2} = visitedBy_{i_3} = k$, and $i_2 \in succ_{i_1}$ and $i_3 \in succ_{i_2}$. A new solution can be obtained by re-assigning variables $next_{i_1}$, $next_{i_2}$, and $next_{i_3}$, and keeping unchanged all the other $next$ variables. Since dTP incorporates side constraints, testing feasibility can be computationally expensive, but using constraints directly supports extensions to additional side constraints. For such 3-opt, since there is no exchange of clients among the routes, bin type and capacity constraints still hold, and only the time window constraints need to be checked.

The entire exploration of the neighborhood can be implemented as follows:

```
procedure: 3-opt_EXPLORE
 for each k ∈ {1,...,M}
     for each i1 | visitedBy_i1 = k
         for each i2 | visitedBy_i2 = k ∧ i2 ∈ succ_i1
             for each i3 | visitedBy_i3 = k ∧ i3 ∈ succ_i2
                 let j1 := next_i1, j2 := next_i2, j3 := next_i3
                 try next_i1 = j2, next_i2 = j3, next_i3 = j1
                 check feasibility for time windows
                 old := c_i1,j1 + c_i2,j2 + c_i3,j3
                 new := c_i1,j2 + c_i2,j3 + c_i3,j1
 over all neighbors, minimize Δ := old - new
```

Once the feasible solution with smallest Δ is found, the data structure $succ$ must be updated, so as the objective function z. As mentioned at the beginning of Section 5, we do not consider here CP variables, but the constraints (in this case the time window ones) are used off-line to check feasibility and update the $start$ variables.

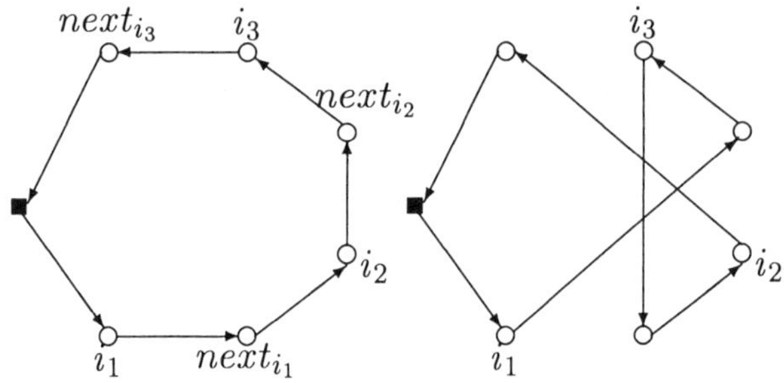

Figure 9.2. Example of 3-opt move.

Node-transfer. Node-transfer moves a single client from a route to another by keeping feasibility of both routes, and decreasing the overall value of z. This move leads to more complicated feasibility issues than 3-opt. Specifically, three constraints need to be checked for the route in which the client is inserted: bin type compatibility, bin capacity and the time window constraints. An example of node-transfer move is depicted in Figure 9.3.

More formally, let k_1 and k_2 be two routes, and let i_1, i_2 and i_3 be three clients such that $visitedBy_{i_1} = visitedBy_{i_2} = visitedBy_{i_3} = k_1$, and $next_{i_1} = i_2$, and $next_{i_2} = i_3$. Moreover, let j be one more client such that $visitedBy_j = k_2$[6]. A new solution is obtained by moving i_2 from k_1 to k_2 at position $next_j$.

A passive way of using constraints within a LS based on a single-client move is to have the following four nested loops:

```
procedure: node-transfer_EXPLORE
 for each k1 ∈ {1,...,M}
     for each i2 | visitedBy_i2 = k1
         for each k2 ∈ {1,...,M} | k2 ≠ k1
             for each j | visitedBy_j = k2
                 try visitedBy_i2 = k2, next_i2 = next_j,
                     next_j = i2, next_i1 = i3
                 check feasibility for all constraints
                 old := c_{i1,i2} + c_{i2,i3} + c_{j,next_j}
                 new := c_{i1,i3} + c_{j,i2} + c_{i2,next_j}
 over all neighbors, minimize Δ := old - new
```

Unlike the 3-opt case, more side constraints need to be checked, since the move changes the client-truck assignments, and again an updating of the data structures is needed.

5.2 Constraint checks within the neighborhood iteration

Checking feasibility only at the very end of the decision process is a passive way of using constraints. Constraints may be used in more active way by factoring

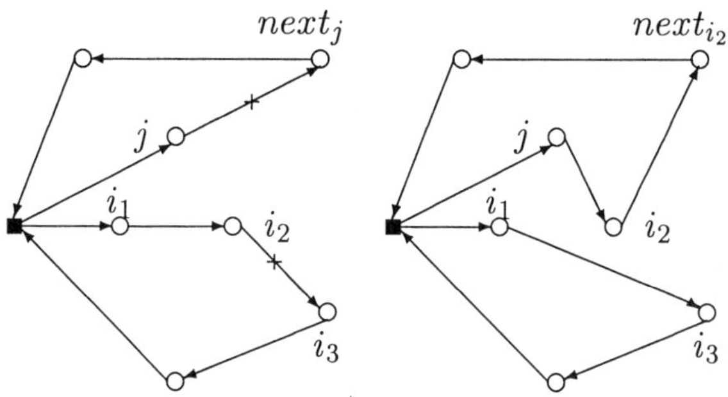

Figure 9.3. Example of node-transfer move.

out some of the constraint checks early on in the iteration. Therefore, single checks may discard (hopefully large) sets of neighbors, thus improving the overall efficiency of the neighborhood exploration.

In the pseudo-code presented for the *node-transfer* neighborhood exploration much of the time is wasted generating trivially infeasible neighbors. For instance, it is useless iterating j over a route k_2 covered by a truck whose bins are not compatible with $type_i$. Therefore, to speed up the exploration, we can anticipate some of the feasibility checks within the nested loop and before the final constraint checks. The infeasibility (or ineffectiveness) of many of the solutions in the neighborhood can be easily detected by testing: (i) if the type of goods of client i_2 is compatible with the types collected by the bins of route k_2, (ii) if the residual capacity of the bins of route k_2 can accommodate the quantity q_{i_2}, and (iii) if the neighbor improves z, i.e., if $c_{i_1,i_3} + c_{j,i_2} + c_{i_2,next_j} - c_{i_1,i_2} - c_{i_2,i_3} - c_{j,next_j} < 0$.

Such an optimized neighborhood exploration can be stated as follows:

```
procedure: node-transfer_EXPLORE_updated
 for each h1 ∈ {1,...,2M}
     for each i2 | collectedIn_{i2} = h1
         for each h2 ∈ {1,...,2M} |
                  h2 ≠ h1 ∧ collects_{h1} = collects_{h2} ∧ RC_{h2} ≥ q_{i2}
             for each j | collectedIn_j = h2
                 old := c_{i1,i2} + c_{i2,i3} + c_{j,next_j}
                 new := c_{i1,i3} + c_{j,i2} + c_{i2,next_j}
                 if old - new < 0
                    try collectedIn_{i2} = h2,
                        next_{i2} = next_j, next_j = i2, next_{i1} = i3,
                        visitedBy_{i2} = visitedBy_j
                    check feasibility for time windows
 over all such neighbors, minimize Δ := old - new
```

where the RC_{h_2} represents the *residual capacity* of bin h_2, i.e.,

$$RC_{h_2} = C - \sum_{i \mid collectedIn_i = h_2} q_i$$

and must be updated (together with $succ$, etc.) once a feasible (improving) move is performed. Not only does this exploration generate less infeasible neighbors, it also needs to check less constraints (only time window ones) on the outcome of the loop.

5.3 Freezing Fragments

> Despite the improvement in the use of the constraints described in the previous section, the main point concerns the effectiveness of small neighborhood for real-world applications. Thus, a possible larger neighborhood can be explored through another kind of move called *shuffle* (see, (Applegate and Cook, 1991)) or *Large Neighborhood Search* (LNS, (Shaw, 1998)). Such a move keeps a fragment of the current solution (freezing the values of variables in that fragment), forgets the value and restores the initial domain of variables outside the fragment. Finding the best completion of such a state into a solution is an optimization subproblem. The search space for this subproblem is smaller than for the original problem because some variables have been fixed. Enforcing part of the assignments from the previous solution is the opposite process of relaxing: the problem is strengthened with additional constraints. The frozen fragments must be selected carefully and should be: (i) large enough so that the subproblem is more tractable than the original one; (ii) not too large so that the subproblem contains enough flexibility to improve over the current solution; and (iii) good enough so that the subproblem consists in good solutions in the overall search space.
> In a way, the fragment plays the role of a strictly followed heuristic while building a solution. However, since the fragment that is kept corresponds to choices that were made at any point in the search tree, and not necessarily at the root of the tree, such a destruction-construction process differs from chronological backtracking. Such LNS algorithms can be easily implemented within CP systems, since the neighborhood exploration consists of solving the original problem while some variables are instantiated and the domain of others are reduced.

Insertion algorithms are natural candidates to LNS. Specifically for the dTP case, clients are partitioned into two sets: a set of clients for which the route assignment and relative sequencing is kept (the fragment from the reference solution that is frozen), and the remainder of the clients, for which all decisions (assignment and sequencing variables) are forgotten.

As described in (Shaw, 1998), the forgotten part of the solution should be of relative significant size and should be made of related clients (clients from the same routes or from a given geographical area).

Several neighborhoods can be proposed for such LNS:

- keeping all decisions concerning a subset of the *routes* and re-optimizing all the remainder of the solution;

- keeping all decisions concerning a subset of the *types of goods* and re-optimizing all the remainder of the solution;
- keeping all decisions concerning a subset of the *clients* and re-optimizing all the remainder of the solution.

Note that such different neighborhoods may be combined through Variable Neighborhood Search (VNS, see, e.g., (Mladenović and Hansen, 1997)).

In the following, a simple example of LNS based algorithm is proposed. The algorithm iteratively freezes a part of the problem and tries to extend the partial solution. The iterative process stops when a time limit is reached.

A given solution s is encoded through three arrays $sNext, sSucc$, and $sVisitedBy$ containing the values of $next, succ$, and $visitedBy$ in s. The algorithm selects randomly a client i^*, and forgets the decisions that were made on the set of clients that are *close to* i^*. The term *close to* is quantitatively defined by the parameter $radium$, and the travel time matrix tt: two clients i, j are considered close if $tt_{ij} \leq radium$. The subproblem obtained is solved by the GREEDY CONSTRUCTION algorithm described in Section 4.2. Whenever an improving solution is found, the arrays encoding the solution must be updated. The overall algorithm is as follows:

```
algorithm: LNS(radium)
 while time is still available
     select at random a client i*
     shuffleSet := {j | tt_{i*,j} ≤ radium}
     restore initial domains for variables not in shuffleSet
     for all clients i|i ∉ shuffleSet
         visitedBy_i = sVisitedBy_i
         if sNext_i ∉ shuffleSet
             next_i = sNext_i
         for all clients j|j ∈ sSucc_i
             if j ∉ shuffleSet
                 j ∈ succ_i
     z < z*
     call algorithm GREEDY CONSTRUCTION
     if an improving solution is found
         for all clients i
             sNext_i := next_i
             sSucc_i := succ_i
             sVisitedBy := visitedBy_i
         z* := z
 if any improving solution was found
     return(success)
```

```
otherwise return(failure)
```

Some interesting modifications of the LNS algorithm can be done.

- The $radium$ parameter can start with small values and increase whenever the greedy algorithm was unable to find improving solutions for a given number of consecutive times. Leading to a Variable Neighborhood Search, this modification will help escaping from local optimal solutions;
- Any constructive algorithm could be used to extend the frozen fragment. The use of Discrepancy-based Search algorithms could indeed increase the likelyhood of finding solutions when the problem becomes harder due to the bounding constraint $z < z^*$;
- The pseudo code presented above forgot the assignments for all clients reachable within a given time from client i^*. This forgotten part of the solution could take into account a more sophisticated distance than mere travel time. For instance, tt_{ij} could be multiplied by a weight depending on the routes, the types of goods, and the time windows of clients i and j. Clients having equal type of goods or overlapping time windows should be considered relatively *closer* than clients with different type of goods or distant time windows.

5.4 CP models for the Neighborhoods

Another way of exploring large neighborhood, introduced in (Pesant and Gendreau, 1996) and (Pesant and Gendreau, 1999), consists of modeling the exploration through CP variables and constraints. Roughly speaking, a CP model of the neighborhood is created such that every feasible solution of the CP problem represents a move that transforms the current solution into a neighbor solution. With such a model, the exploration of the neighborhood by means of explicit enumeration (as an example, two nested loops in a classical *swap* neighborhood) can be replaced by a tree search (such as branch-and-bound) for finding the best move.

In the dTP case a neighborhood, based on ejection chains, is modeled using CP: a set of clients belonging to different routes are chosen and re-located (see Figure 9.4).

A given solution s is encoded through four arrays $sNext$, $sSucc$, $sVisitedBy$, and $sPrev$; the first three arrays contain the values of $next$, $succ$, and $visitedBy$ in solution s; array $sPrev$ contain the inverse of $next$ (if $next_i = j$ in s, then $sNext_i = j, sPrev_j = i$).

The following neighborhood structure can be defined: a move from s chooses $K \leq M$ clients $i_1, i_2, \ldots, i_K$ such that all the visits are performed in different trucks in solution s. Then it removes the edges $(sPrev_{i_j}, i_j)$, $(i_j, sNext_{i_j})$, with $j = 1, \ldots, K$, and re-connects the tours by rotating the nodes $i_1, i_2, \ldots, i_K$.

For all $j = 1, \ldots, K-1$, i_{j+1} replaces i_j; the rotation is closed by replacing i_K with i_1. We may either look for any improving move or for the best move in the neighborhood.

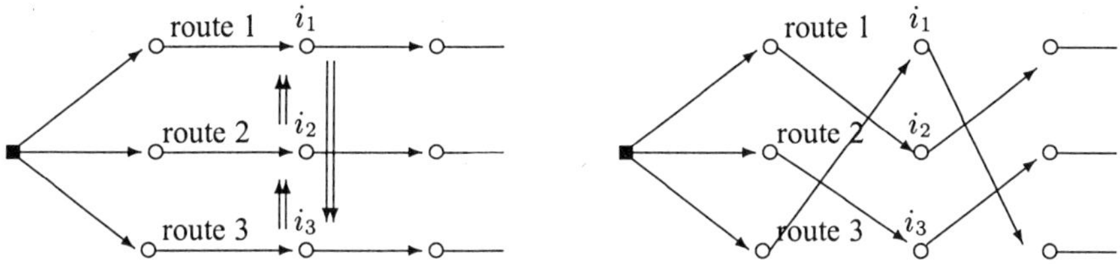

Figure 9.4. Example of ejection chain.

This neighborhood structure can be defined using CP by adding the following model to the existing dTP model:

$$
\begin{array}{rll}
\text{min} & dtCost & \\
\text{on} & & \\
 & dtCost < 0 & \\
\forall j \in \{1, \ldots, M\} & I_j \in [-1, 1..N] & \\
 & size \in [2..M] & \\
\text{subject to} & & \\
\forall i, j \in \{1, \ldots, M\}, i < j & i \leq size \Leftrightarrow & \\
 & \quad sVisitedBy[I_i] \neq sVisitedBy[I_j] & (9.15) \\
\forall j \in \{1, \ldots, M\} & j > size \Leftrightarrow I_j = -1 & (9.16) \\
\forall j \in \{1, \ldots, M\}, i \in \{1, \ldots, N\} & I_j = i \Rightarrow & \\
 & \quad next[sPrev_\ell] = \ell, \ \forall \ell | i \in sSucc_\ell & (9.17) \\
\forall j \in \{1, \ldots, M\}, i \in \{1, \ldots, N\} & I_j = i \Rightarrow & \\
 & \quad next[\ell] = sNext_\ell, \ \forall \ell | \ell \in sSucc_i & (9.18) \\
\forall j \in \{1, \ldots, M-1\} & j < size \Rightarrow next[sPrev[I_j]] = I_{j+1} & (9.19) \\
\forall j \in \{1, \ldots, M-1\} & j < size \Rightarrow next[I_{j+1}] = sNext[I_j] & (9.20) \\
 & next[sPrev[I_{size}]] = I_1 & (9.21) \\
 & next[I_1] = sNext[I_{size}] & (9.22)
\end{array}
$$

$$
\begin{aligned}
& dtCost = \\
& \sum_{i=1}^{size-1} (c[sPrev[I_i], I_{i+1}] + c[I_{i+1}, sNext[I_i]]) + \\
& c[sPrev[I_{size}], I_1] + c[I_1, sNext[I_{size}]] - \\
& (\sum_{i=1}^{size} (c[sPrev[I_i], I_i] + c[I_i, sNext[I_i]]))
\end{aligned} \tag{9.23}
$$

A solution of the neighborhood model defines an ejection chain affecting $size$ clients; since $size$ can take any value between 2 and M, we introduce M variables representing the variable length ejection chain. The first $size$ variables identify the clients changing route while the others are meaningless and are set to -1.

M domain variables $I_1, I_2, \ldots, I_M$, are used to identify the clients. The domain of variable I_j is $[-1, 1..N]$; the first $size$ variables[7] will take values different from -1. Client I_2 goes in place of client I_1, client I_3 goes in place of client I_2, etc. finally client I_1 goes in place of client I_{size}. All variables I_j with $j > size$ take the value -1. The objective function of the neighborhood model is the difference between the cost of the current solution s and the cost of the tentative solution.

Constraints (9.15) and (9.16) define the neighborhood structure: $size$ consecutive I_j variables identify the clients in the ejection chain, and the remaining $M - size$ are constrained to be equal to -1. The clients must be chosen from $size$ different trucks (9.15).

Constraints (9.17)-(9.22) define the interface constraints between the dTP model and the neighborhood model. They are basically of two types: some of them are devoted to restoring of the unchanged part of the solution (9.17)-(9.18), some others are devoted to encoding the move (9.19)-(9.22). Finally, constraint (9.23) links the neighborhood variables to $dtCost$.

Defining the neighborhood structure via a CP model yields a stable neighborhood model that would remain valid if side constraints (such as precedence constraints, etc.) were added to dTP. All side constraints propagate from the problem variables to the neighborhood variables, removing some neighbors. For example, once variable I_1 is instantiated to a value i_1 visited by the first truck, all clients i_2 requiring a capacity exceeding the remaining capacity of the first truck are automatically removed from the domain of variable I_2 by propagation of the interface constraints.

6. Metaheuristics

This section illustrates examples of how to exploit metaheuristic strategies within CP and applications in the dTP context.

6.1 Control strategies from metaheuristics

> Using CP to model neighborhoods that are explored by global search implements by construction the intensification principle. As metaheuristics are based on a careful balance between intensification and diversification, it is crucial to efficiently implement diversification techniques, through constraints, in the model of the neighborhood.

The first possibility of implementing diversification techniques in CP consists in choosing at each iteration between a search for the best neighbor and a search for the first (or any) acceptable neighbor. The best-neighbor strategy is slower but generally proves more efficient towards the end of the walk, while strategies selecting any good neighbor (the first one, a random one, the best found after a given time limit, the best among the first k found, ...) are interesting for generating diversified paths for the start of the local search iterative process.

Another approach to diversification is the control strategy of *Simulated Annealing* (see, e.g., (Aarts and Lenstra, 1997)). A temperature control parameter T is introduced, and moves are accepted according to a probability distribution: a move changing the value of the objective function by δ is always accepted if $\delta < 0$ (strictly improving move) and otherwise (non-improving move), is accepted with probability $e^{\frac{-\delta}{T}}$. The spirit of this strategy can be captured by a constraint: at each iteration of the local search process, an optimization cut threshold is selected at random with a Gaussian probability distribution centered around UB (the value of the best solution found so far). This optimization cut is added to the CP model of the given neighborhood and propagated. This implements a different control strategy from the simulated annealing, as the acceptance threshold is the same for all candidates from the neighborhood, instead of drawing it at random for each candidate. On the other hand, this strategy uses the information of the acceptance threshold early on in the exploration through propagation, pruning subtrees where all candidates can be proven above the acceptance threshold. This control strategy can be applied on the dTP with local search procedures optimizing single routes: the acceptance of k-opt moves can be controlled with an optimization constraint on the cost of the move.

The central diversification tool from *Tabu Search* (see, e.g., (Aarts and Lenstra, 1997)) can also be captured by means of constraints. As noted by (Pesant and Gendreau, 1999), one can forbid tabu moves by adding constraints on the neighborhood variables. For instance, for the TSP, a tabu algorithm using 3-opt moves and a tabu list of vertices can be implemented by adding a constraint that no more than one among the neighborhood variables X_i, X_j and X_k should take a value in the tabu list. The constraint is posted in the model of the neighborhood and is then propagated throughout the neighborhood exploration. This strategy can be useful on dTP with node transfer neighborhoods. In order to avoid cycles in the optimization procedure where a set of nodes would be swapped back and forth among routes, one can consider a list of tabu nodes that must remain in their route. Transfers for such nodes can be forbidden by means of a constraint.

Last, one may also use the principle of *Guided Local Search* (see, (Voudouris and Tsang, 1999)) for diversifying the search. This consists in modifying the coefficients of the objective function by means of penalties. As illustrated in (Kilby et al., 1997), such a modification can very easily be performed in the model of the neighborhood, by changing the constraint defining the objective criterion. In the case of dTP, one can drive the optimization process towards promising regions by analyzing the solution and adding penalties for route to customer, or route to bin type assignment. These penalties are added by changing the coefficients of the sum defining the objective.

Those examples show the versatility of the approach that uses a CP model of the neighborhoods and applies propagation and global search in the search for the next neighbor.

6.2 Managing several neighborhoods

Often, a wide variety of neighborhoods exist for a given combinatorial optimization problem. In order to effectively find good solutions avoiding to get trapped in local minima, a design decision that must be carefully evaluated is which neighborhoods should be used, and how.

Neighborhoods can be combined via VNS (Mladenović and Hansen, 1997). VNS proposes to interleave the exploration of several neighborhoods, modifying the neighborhoods and the solutions once a local minimum is reached. More generally, if a single neighborhood is used we risk to quickly get trapped into a local minimum. If too many neighborhoods are tried one after the other we risk to slow down the improvement rate since some neighborhoods may share a subset of identical solutions which would be evaluated several times.

A clear answer to this problem does not exist; as usual, the user must find a good tradeoff between diversification of the neighborhoods, and speed of the search. We refer to (Mladenović and Hansen, 1997) for a comprehensive description of VNS; here we give some guidelines.

- Fast (small) neighborhoods should be tried before slow (large) neighborhoods.
- Combine neighborhoods exploring disjoint portions of the search space (e.g. inter-route and intra-route neighborhoods).
- Once a local minimum is reached, modify (or enlarge) the neighborhood.
- Switch from a neighborhood to another rather than exhaustively explore it.
- Use more often those neighborhoods that enable to obtain the highest improvement per time unit.

Neighborhoods can be seen as vehicles used to travel in the solution landscape towards the optimal solution. At each point of the trip, we should intelligently choose the vehicle that enables us to move faster towards the target.

In dTP, we could use a two phase process: first, using Constrained Local Search to look for quick improvements of the initial solution, until the process gets stuck in a local minimum. Then, applying Freezing Fragment techniques of Section 5.3 to try to further improve the solution.

7. LS during construction

The previous sections showed how constructive algorithms from greedy to Discrepancy-based Search could yield one or several solutions, and how local moves from various neighborhoods could improve them. This section concerns the use of local moves at the very heart of the construction process. Following (Caseau and Laburthe, 1999), we develop on the idea of Incremental Local Optimization which applies improving local moves after each construction step.

7.1 Single Route Optimization

> This section shows how one can take advantage of a decomposition scheme in order to apply local moves on subproblems during the construction of a solution to the overall problem. Thus, decomposition is used to make the local optimization more incremental (it is applied only on a small subproblem), therefore eligible as a technique applied at each step of a construction process, and not only as a post-optimization technique.

A standard move in routing problems consists in optimizing the sequence of visits for each route. Optimizing a route yields one small constrained TSP per route. This leads to a decomposition of the problem: the partition of clients by routes is kept, and the induced independent TSPs are re-solved. Each neighborhood for the TSP induces a neighborhood for dTP. A solution to dTP may, for instance, be transformed into a neighbor one by applying on any of the routes:

- a feasible 3-opt move (Reinelt, 1994);
- a feasible Lin and Kernighan move (Lin and Kernighan, 1973);
- a sequence of improving feasible 3-opt moves until no more such moves can be found;
- a complete algorithm that sequences each route to optimality.

The last case is of particular interest. Indeed, many routing problems, although involving thousands of clients typically assign less than 15 clients per truck. It is thus easy to solve to optimality the constrained TSP associated to a route by branch and bound with constraint propagation. Such an approach was shown to be viable in (Caseau and Laburthe, 1996) and (Focacci et al., 1999b).

After each insertion, the partial solution can be re-optimized in a state where each route is optimally sequenced. For the decomposition of dTP into a master partitioning sub-problem and an induced sequencing (routing) sub-problem, the sequencing sub-problem is solved to optimality. The complexity of the overall procedure remains tractable because the LS phase that is performed at each insertion step can be done much faster than a LS phase that would be performed in the end, after a complete solution has been built. Indeed, after the insertion of a node i in route k ($visitedBy_i = k$), TSP moves should be applied on route k only. Therefore, it is worthwhile resolving the TSP for route k only. This fast incremental exploration accounts for the overall efficiency of applying local optimization within the construction process.

7.2 Ejection Chains

The neighborhood introduced above is well suited for improving sequencing decisions among clients on the same route but it never questions the partition of clients into routes. As the insertion algorithm inserts clients one at a time, it is likely to make assignments (of clients to routes) that seem good when only a subset of the clients is considered, but that later prove to be suboptimal when further clients are inserted. This section presents a neighborhood whose aim is to repair such partitioning decisions that turn out to be wrong along the insertion process.

Section 5.4 proposed a CP model for an ejection chain neighborhood. The idea of ejection chains, a standard technique for packing problems, is the following. Let $i_1, \ldots, i_r$ be a set of r clients currently assigned to different routes (say, $visitedBy_{i_1} = k_1, \ldots, visitedBy_{i_r} = k_r$) and i_0 be another client not assigned to k_1. The move changes the bin assignment into $visitedBy_{i_0} = k_1$, $visitedBy_{i_1} = k_2, \ldots,$ $visitedBy_{i_{r-1}} = k_r$ and resolves the induced sequencing problems on routes k_1 to k_r. Applied on the dTP, this move amounts to change the route assignment of a few nodes while respecting all constraints (satisfying capacity constraints with the new insertion $visitedBy_{i_j} = k_{j+1}$ is eased by the removal of i_{j+1} from k_{j+1}).

Such moves are particularly interesting in the case of promising infeasible insertions. For instance, one may foresee that the insertion $visitedBy_i = k$ would yield a very small insertion cost Δ_k, but that it is infeasible because of, say, a capacity constraint (there is no more room for client i in the route k). Instead of considering the insertion of i into routes that are geographically farther away (which may cause significant detours, thus higher insertion cost Δ), it is interesting to try to insert i into the route k by removing another client j from route k and transferring it to another route. This leads to the search for ejection chains initiated by the insertion of i in k. Such moves are particularly useful when capacity constraints on the routes are tight. Another situation where ejection chains prove useful is the case of travel optimization under a small number of trucks. In such a case, the routes must be tight in order to cover all clients. What usually happens in the insertion process is that the algorithm is not able to insert some of the clients towards the end of the construction (all tentative insertions turn out to be infeasible). Ejection chains can be used as a way to force these insertions: (Caseau et al., 1999). An alternate chain of insertions and ejections is searched for such that i can be inserted into some route k_1 by transferring another client j from k_1 to k_2, and so on. Such neighborhoods can be explored through Breadth First Search in order to find the least length ejection chain (the one that involves the least number of clients), likely to be less disruptive of the current solution.

Finally, one should be careful about the fact that such ejection chain neighborhoods are very large neighborhoods and that their exploration may take significantly longer than the other neighborhoods presented so far (3-opt, single route direct optimization, etc.). It amounts to a form of LNS and should be used with care. Nevertheless, it complements very efficiently the short-sightness nature of constructive algorithms which build the solution one step at a time by repairing some of the early non-optimal route assignments.

8. Conclusions

This chapter has presented a variety of techniques for using local search methods with constraints. Hybrid combinations have been described for local search methods as well as for constructive global search methods.

Constraints are blended with local search procedures in several ways:

- by expressing the problem as a standard problem with additional side constraints, iterating a neighborhood for the standard problem and, for each neighbor, checking feasibility for all side constraints; or by optimizing the neighborhood exploration procedure with inlined constraint checks. These methods can either be implemented with any programming language following the ideas described in this chapter, or with a CP language using already defined specific data structures (such as constraints, neighborhood objects, and move operators) (De Backer et al., 2000).

- by using global search techniques for exploring the neighborhood defined by a fragment of the current solution; these methods are also described in (Shaw, 1998; Caseau et al., 1999) for routing problem, or in (Nuijten and Le Pape, 1998) for scheduling problems.

- by defining the search for the best neighbor as a constrained optimization problem on its own; originally introduced in (Pesant and Gendreau, 1996; Pesant and Gendreau, 1999), it has been applied mainly to timetabling and routing problems in (Pesant and Gendreau, 1996; Pesant and Gendreau, 1999; Pesant et al., 1997).

Local search techniques can be introduced within a constructive global search algorithm with the following techniques:

- Restricted Candidate Lists to filter the good branches of a choice point (see (Cesta et al., 2002) for a recent application to scheduling problems);

- Discrepancy-based Search to generate near-greedy paths in a search tree;

- Incremental Local Optimization to improve the partial solutions at each step of a construction process such as a greedy insertion algorithm.

In all these cases, CP provides the user with clean formalism to combine technologies: use propagation to reduce the size of neighborhoods, use global search techniques to explore a neighborhood, control the divergence of a construction process from a greedy path, etc. In a sense, CP supports a clean engineering of many "good old tricks" from Operations Research. The first clean integration has been the expression of neighborhoods with CP models. CP models could be introduced for many other algorithm engineering purposes such as objective function combinations for multi-criteria optimization (see, e.g., (Focacci et al., 2000)), or solution fragment combinations for population-based optimization methods. All these are open research topics. The definite impact of CP to Operations Research in general, and to LS in particular, is the introduction of structuring objects that provide a way to easily combine techniques.

Acknowledgments

We warmly thank all the participants to the *CP-AI-OR'02 School on Optimization* for many interesting discussions on the topic. Thanks are also due to an anonymous referee for helpful comments.

Notes

1. This chapter is based on (Focacci et al., 2002).

2. In the example we have $P = 4$, $type_i \in \{1, 2, 3, 4\}$, e.g., red $\equiv 1$, black $\equiv 2$, white $\equiv 3$, and yellow $\equiv 4$.

3. This is true with the current set of constraints while one has to be careful in the case of addition of other constraints.

4. In case of failure, it is always possible to remove from the model the clients which cannot be inserted by the solver and return a feasible routing plan covering only a subset of the clients.

5. Recall that, in the CP context, the word 'heuristic' refers to a function used to compare different branches at a choice point.

6. Since we do not have the variable $next_0$, if i_2 is the first client visited in route k_1, and/or if we want to insert it as first client in route k_2, then it is necessary to consider also variables $first_{k_1}$ and/or $first_{k_2}$.

7. When I_j is used to index an array $array$, and -1 is part of the domain of I_j, the value $(array_i)[-1]$ is conventionally considered equal to -1.

References

Aarts, E. and Lenstra, J. K. (1997). *Local Search in Combinatorial Optimization*. J. Wiley & Sons, Chichester.

Applegate, D. and Cook, W. (1991). A computational study of the job-shop scheduling problem. *ORSA Journal on Computing*, 3:149–156.

Caprara, A., Fischetti, M., Toth, P., Vigo, D., and Guida, P.-L. (1997). Algorithms for railway crew management. *Mathematical Programming*, 79:125–141.

Caseau, Y. and Laburthe, F. (1996). Improving branch and bound for job-shop scheduling with constraint propagation. In Deza, M., Euler, R., and Manoussakis, Y., editors, *Proceedings of Combinatorics and Computer Science - CCS'95*, LNCS 1120. Springer-Verlag.

Caseau, Y. and Laburthe, F. (1999). Heuristics for large constrained routing problems. *Journal of Heuristics*, 5:281–303.

Caseau, Y., Laburthe, F., and Silverstein, G. (1999). A metaheuristic factory for vehicle routing problems. In Jaffar, J., editor, *Principle and Practice of Constraint Programming - CP'99*, LNCS 1713, pages 144–158. Springer-Verlag.

Cesta, A., Oddi, A., and Smith, S. (2002). A constraint-based method for project scheduling with time windows. *Journal of Heuristics*, 8:109–136.

De Backer, B., Furnon, V., Shaw, P., Kilby, P., and Prosser, P. (2000). Solving vehicle routing problems using constraint programming and meta-heuristics. *Journal of Heuristics*, 6:481–500.

Feo, T. and Resende, M. (1995). Greedy randomized adaptive search procedures. *Journal of Global Optimization*, 6:109–133.

Focacci, F., Laborie, P., and Nuijten, W. (2000). Solving scheduling problems with setup times and alternative resources. In *Proceedings of the Fifth International Conference on Artificial Intelligence, Planning Systems - AIPS'00*. AAAI Press.

Focacci, F., Laburthe, F., and Lodi, A. (2002). Local search and constraint programming. In Kochenberger, G. and Glover, F., editors, *Handbook of MetaHeuristics*, pages 369–403. Kluwer Academic Publishers.

Focacci, F., Lodi, A., and Milano, M. (1999a). Cost-based domain filtering. In Jaffar, J., editor, *Principle and Practice of Constraint Programming - CP'99*, LNCS 1713, pages 189–203. Springer-Verlag.

Focacci, F., Lodi, A., and Milano, M. (1999b). Solving TSP with time windows with constraints. In De Schreye, D., editor, *Proceedings of the 1999 International Conference on Logic Programming - ICLP'99*, pages 515–529. The MIT-press.

Glover, F. (1995). Tabu thresholding: Improved search by nonmonotonic trajectories. *ORSA Journal on Computing*, 7:426–442.

Golden, B. and Assad, A. (1988). *Vehicle Routing: Methods and Studies*. North-Holland, Amsterdam.

Haralick, R. M. and Elliott, G. L. (1980). Increasing tree search efficiency for constraint satisfaction problems. *Artificial Intelligence*, 14:263–313.

Harvey, W. (1995). *Nonsystematic Backtracking Search*. PhD thesis, Stanford University.

Harvey, W. and Ginsberg, M. (1995). Limited discrepancy search. In *Proceedings of the 14th IJCAI*, pages 607–615. Morgan Kaufmann.

Kilby, P., Prosser, P., and Shaw, P. (1997). Guided local search for the vehicle routing problem. In Voss, S., Martello, S., Osman, I. H., and Roucairol, C., editors, *Advances and Trends in Local Search Paradigms for Optimization*, pages 473–486. Kluwer Academic Publishers.

Lin, S. and Kernighan, B. W. (1973). An effective heuristic for the traveling salesman problem. *Operations Research*, 21:498–516.

Mackworth, A. (1977). Consistency in networks of relations. *Artificial Intelligence*, 8:99–118.

Mladenović, N. and Hansen, P. (1997). Variable neighborhood search. *Computers & Operations Research*, 24:1097–1100.

Nuijten, W. (1994). *Time and Resource Constrainted Scheduling, a Constraint Satisfaction Approach*. PhD thesis, University of Eindhoven, The Netherlands.

Nuijten, W. and Le Pape, C. (1998). Constraint based job shop scheduling with ILOG scheduler. *Journal of Heuristics*, 3:271–286.

Pesant, G. and Gendreau, M. (1996). A view of local search in constraint programming. In Freuder, E. C., editor, *Principle and Practice of Constraint Programming - CP'96*, LNCS 1118, pages 353–366. Springer-Verlag.

Pesant, G. and Gendreau, M. (1999). A constraint programming framework for local search methods. *Journal of Heuristics*, 5:255–279.

Pesant, G., Gendreau, M., Potvin, J. Y., and Rousseau, J. M. (1998). An exact constraint logic programming algorithm for the travelling salesman problem with time windows. *Transportation Science*, 32:12–29.

Pesant, G., Gendreau, M., and Rousseau, J.-M. (1997). GENIUS-CP: A generic single-vehicle routing algorithm. In Smolka, G., editor, *Principle and Prac-*

tice of Constraint Programming - CP'97, LNCS 1330, pages 420–433. Springer-Verlag.

Régin, J.-C. (1994). A filtering algorithm for constraints of difference in CSPs. In *Proceedings of the Twelfth National Conference on Artificial Intelligence - AAAI'94*, pages 362–367.

Reinelt, G. (1994). *The traveling salesman: computational solutions for TSP applications*. Springer-Verlag, Berlin Heidelberg.

Shaw, P. (1998). Using constraint programming and local search methods to solve vehicle routing problems. In Maher, M. and Puget, J.-F., editors, *Principle and Practice of Constraint Programming - CP'98*, LNCS 1520, pages 417–431. Springer-Verlag, Berlin Heidelberg.

Toth, P. and Vigo, D. (2002). *The Vehicle Routing Problem*. Monographs on Discrete Mathematics and Applications. SIAM.

Voudouris, C. and Tsang, E. (1999). Guided local search and its application to the travelling salesman problem. *European Journal of Operational Research*, 113:469–499.

Chapter 10

OPEN PERSPECTIVES

Mark Wallace
IC-Parc
mgw@icparc.ic.ac.uk

Yves Caseau
Bouygues Telecom
ycaseau@bouyguestelecom.fr

Jean-Francois Puget
ILOG
puget@ilog.fr

Abstract This chapter discusses current and future research directions for the integration of AI and OR techniques in Constraint Programming. We describe the challenges driving this research area, outlining some of the areas where much remains to be done to meet real-world needs. We then investigate the task of mapping problems to solutions, listing the steps to be taken and the choices facing the problem solver. Next we review some new techniques which build on the hybrid algorithms approach presented in this book. Finally we list the new applications areas which will both result from these advances, and inspire further research.

Keywords: future, constraints, mathematical programming, hybrid solvers

1. Motivations, Challenges and Applications

1.1 Overview

The fields of constraint programming and combinatorial optimisation are extremely motivating, from a science and research perspective, because they are still full of varied challenges. We can expect a lot of activity in both fields in the next ten years and, probably, new results and new approaches.

We first look at some problems and requirements that are not well handled in the current state of the art. These include scalability, uncertainty, support for the problem solving process and integration in the IT environment. Our aim is to set the context for this chapter on open perspectives.

In the next section we follow the process of mapping a problem to an algorithm that solves it, describing the different steps and the research issues that are thrown up by each step in turn. We consider problem decomposition and algorithm composition; problem transformations; search methods; and control of inference.

Next we investigate new techniques which are starting to be explored by researchers building on previous successes in hybrid algorithms research. We consider stochastic optimisation; overconstrained problems; integration of offline planning and online control; simulation; automated algorithm adaptation and learning; packaging of constraint technologies and components.

Finally we list some of the application areas which will come within the range of our techniques, and drive forward the frontiers of our research.

1.2 Challenges

The first challenge comes from the very nature of combinatorial optimisation, which means that, unless someone finds that $P = NP$, there are many problems that are very hard to solve. These problems exist both in the academic world, where very small and elegant problems are still open, and in the industrial world, where there are real and practical problems that are too large or complex to be solved by current computational techniques and systems. For many combinatorial optimisation problems, solutions have been found using incomplete methods, but there is no guarantee of their quality in the sense of how far they fall short of the optimal solution.

The second challenge comes from problems that are, in different ways, incompletely, imprecisely or ill-defined. There may be multiple optimisation criteria, uncertain data, or conflicting constraints that should be satisfied as far as possible. For these problems our technology is required to provide not just an optimal solution, but information relating different assumptions or properties of the problem to properties of the solutions. One context where this requirement arises is when the specified problem is a part of a larger problem, and any solutions to the specified problem will need to be integrated into solutions of the larger problem.

The third challenge is to support the problem solving process. There is usually little value in developing a system that solves a specific problem very efficiently, if the development takes many man-years. Hybrid algorithms exploiting CP, AI and OR may be very complex and sophisticated, and finding the "right" hybrid algorithm for a problem could be like looking for a needle in a haystack. Not only is this process time consuming, but it also requires a great deal of expertise from the algorithm developer. Arguably, in the current state of the art, the development of each hybrid algorithm corresponds to a full PhD. To make an impact, our technology must be made useable so that

1 highly qualified experts can develop solutions very quickly

2 less expert users can also exploit the technology successfully

The fourth challenge is to make our technology integrate into the current and emerging IT world. This challenge arises naturally from both the second and third challenges. Our technology can be exploited as a "black box" or, better, it could be built to run as multiple, possibly distributed, components with standard interfaces. Our technology can be applied by experts trained to use our systems or, better, it could support standard application programmer interfaces that enable it to be exploited by a wide range of users competent with standard IT tools and techniques.

The remainder of this first Section will take a closer look at these different challenges.

1.3 Focus on Hard to Solve Problems

1.3.1 Global Constraints. Mathematical Programmers have made a huge industrial impact over the last 50 years, and have built up not only an armoury of techniques, but also a body of experience that enables them to gauge whether a given combinatorial problem is accessible to their approach. Constraint Programming and Stochastic Search have appeared more recently, both achieving success on a wide range of benchmarks and industrial applications.

Constraint programming also offers a language for combining these techniques, and building hybrid algorithms which exploit the best aspects of each. Indeed the integration of many Operations Research algorithms into a constraint programming framework has already been achieved using global constraints. Since global constraints support both modelling and solving of complex, application-specific constraints, their exploitation within constraint programming seems very promising. Indeed for practical applications such as scheduling, timetabling and configuration, this technology has proven highly successful by

- producing solutions of higher quality than was previously possible
- taking many aspects of the problem into account (flexibility)
- enabling the participation of the end user in building the solution

However, there are at the same time many occurrences of "horror stories", where regular users have failed to produce satisfying solutions using these tools.

1.3.2 Scalability Limitations of Tree Search. To date constraint programming, like mathematical programming, has used tree search (for optimisation, branch-and-bound which is an extension of tree search) as its default method of exploring the space of potential solutions. Indeed global constraints are designed specifically to enhance tree search, and their potential for integration into other forms of search is so far unknown. The drawback of tree search is its unpredictability. The search can dwell at length in a region of the search space where there are no (good) solutions. For large, complex problems complete exploration of the search space is not viable. An incomplete tree search may terminate without ever escaping from such a barren region.

The example of vehicle routing is symptomatic of the strengths and weaknesses of constraint programming in its current form. The strengths are without any doubt the flexibility and the expressivity which are necessary for industrial vehicle routing problems for two reasons: the diversity of problems (optimising the number of trucks, the amount of work, the mileage, the location of warehouse, the satisfaction of customers, etc.) and the large number of industry-specific side-constraints (labour regulation, traffic regulation, customer wishes, truck-driver preferences, etc.). On the other hand, vehicle routing problems are often large-scale problems for which a partial tree search does not produce high-quality solutions. The consequence is that, after a positive experience of modelling her or his problem, the user is left waiting for the system to produce high quality solutions that never come. This is corroborated by the poor relative ranking of CP technology on classical benchmarks such as Solomon's (Solomon, 1987).

The example of call centre scheduling (Rottembourg, 2000) is quite similar. It is also a large-scale problem, with a strong underlying structure. The abstraction of the problem is the coverage of a load curve with intervals, which is a polynomial problem that may be solved with a flow algorithm. Obviously, real problems are more complex since additional constraints (such as breaks or lunches) make them NP-complete. Here we also find a lot of complexity with the rules and practices for labour scheduling, and the set of specialized global constraints in that field is second-to-none as far as size and complexity is concerned (cf. the sequence global constraint that was proposed in CHIP (Chan et al., 1998) for this type of application). Since, therefore, constraint programming has been reasonably successful for employee scheduling, it would appear to be an appropriate technology for call centre scheduling as well.

However, due to the large-scale nature of these problems, branch-and-bound search has proven a weak approach for producing high quality solutions. In practise, call centre scheduling software is built using complex non-tree search algorithms, whose use of constraints is reduced to passive checking as opposed to active solving.

1.3.3 Limitations of Global "Soft" Constraints. The other weakness of constraint programming and global constraints is in the handling of soft constraints. Active solving of global constraints is only applicable on the assumption that the constraint must be satisfied in any solution. Soft constraints may be violated in an admissible solution, and cannot therefore be handled by the usual techniques for global constraints.

Our third example, Frequency Assignment, reveals the consequences of this weakness. The ROADEF 2001 challenge (ROADEF, 2001) problem was the assignment of radio frequencies and polarities to links in a cellular network. The problem is a combination of a frequency assignment problem with distance constraints (to avoid interferences) that may be thought of as a disjunctive scheduling problem and of a 2-SAT problem for deciding the polarities. The intricate combination of two problems (distances depend on polarities) and the introduction of soft constraints (each constraint has different levels of satisfaction) makes this a very hard problem, "the hardest of any industrial problem that I have ever seen" according to one of the authors. Although many teams subscribed to the challenge, no solution was ever proposed that was built on CP tools.

The best solutions, in practice, were again ones that used non-tree search, handling the soft constraints by incorprating them into the optimisation criteria.

These illustrations show that the standard hybrid form, based on global constraints, is no magic bullet that defeats all combinatorial dragons. Does this mean that those hard problems are not suited to CP technology ? Quite the opposite, since these problems, because of their richness and the wide variety of side constraints, are begging for a constraint-based approach, from a user and a modelling perspective.

Rather, there is an urgent need for the integration of different constraint handling and search techniques to address the scale and complexity of industrial problems.

1.4 Problem Analysis vs. Resolution

The second challenge is to deal with problems that are incompletely, imprecisely or ill-defined. These problems can be broadly divided into three classes, discussed in the following subsections.

1.4.1 Multi-criteria problems. For many, if not most, optimisation problems it is unclear what exactly should be optimised. For scheduling problems is it maximising throughput, minimising tardiness, balancing workloads or some mixture of all these? It is common, especially in the industrial context, for different groups whose share a global problem to have conflicting goals: for example the marketing, sales and production divisions of a company. Moreover many problem "constraints" are typically preferences rather than necessary requirements on all solutions. In other words practical problems typically include soft constraints as well as hard ones. Minimising the violation of soft constraints introduces more optimisation criteria which must be balanced with the others. The number of different optimisation criteria may grow very large if each soft constraint is an orthogonal optimisation criterion.

We may optimise successively, in various orders, against different criteria producing Pareto-optimal solutions. We then characterize the solution space with a few singular solutions that are each of them optimal with respect to a given combination of criteria: a solution with the best throughput, one with the minimal tardiness, the lowest tardiness for an optimal throughput solution, the most balanced solution with this lowest tardiness etc. As long as this concerns an interactive process with one user and one problem, exploration of Pareto-optimal solutions may be an acceptable process.

But if the goal is to perform collaborative work on the problem, a more abstract and practical characterization of the solution space is needed. This need arises, for instance, in exploring "what-if" scenarios, with multiple players such as the different divisions of a company.

The difficulty of addressing these issues has had an impact on the deployment of computer-aided-decision tools in a typical company. Such tools are rarely deployed globally across the company, but rather they belong to a few of the players - those who can produce a well-defined objective. For applications involving different companies the obstacles to deploying computer-aided-decision tools on the global problem are even greater.

The multiple-player scenario is also becoming more visible in "e-business" type of applications, such as supply-chain management, where different companies try to collaborate over the Internet (a topic that will also be mentioned in Section 1.6.1 below).

1.4.2 Uncertain Problems. Again, most practical problems cannot be precisely specified because the data is uncertain. For instance, in the field of construction scheduling, all task durations are estimates, and the real duration will depend on many external factors such as the weather, the terrain, the materials and the workforce. The uncertainty of journey times, in transportation, is of course notorious. In ship loading, the order in which containers arrive at the dock is unpredictable.

If the data is uncertain, the exact optimisation is often meaningless. Instead of a precise solution, we think about classes of solutions, which represent structural choices. In a construction scheduling application, for example, do I start the next floor of the building or do I finish the non-structural elements of the current one ?

The questions that must be answered by the delivered system are

- How robust is the ordering of tasks in the schedule?
- What external events could impact the schedule, and how?
- What can be done to protect the schedule?

Here again, we see that the object is not a solution but rather an abstraction of the solution space, and the issue is not to produce the object but to reason about it.

The systems we build might compute sets of solutions covering all possibly data values (the "uncertainty closure" (Yorke-Smith and Gervet, 2002)), or "robust" solutions that cover as many data values as possible (Ben-Tal and Nemirovski, 1998)). Often the uncertainty is incrementally resolved during the implementation of the solution (for example task durations take specific values while a schedule is being carried out). In this case the interaction between problem solving and the implementation of the solution raises a host of new challenges.

1.4.3 Probabilistic Data. If the uncertainty can be associated with probabilities, then precise optimisation crieria can be specified, such as optimising the expected value, or achieving a given objective with a certain probability (i.e. achieving a throughput of 20 jobs per day with 90% probability). Finding an "optimal" solution, in this sense, would appear to be easier than analysing classes of solutions, as required for handling problems with uncertain data without probabilities.

Actually, the challenge of finding the optimal solution to a probabilistic combinatorial problem, with either of the objectives descibed above, is extremely hard. We do not, yet, have any methods of computing it. Even if the set of data points is finite, we have to perform some kind of exhaustive exploration of all solutions against all data points, weighting them by their probabilities. The state-of-the art (Benoist et al., 2001) is often to apply crude stochastic and simulation techniques to the combinatorial problem, or to simplify the problem to reduce it to a form where stochastic analysis may be applied.

Consequently the possible approaches for solving uncertain and probabilistic problems share a common thread: the need to think about how to represent and to reason about a solution space.

1.5 Supporting the Problem Solving Process

Faced with a combinatorial problem and a toolset of algorithms, the user has a very difficult task in trying to find the combinations of algorithms that best solves his problem. This "meta-problem" is, itself, combinatorial. Moreover, even testing a solution (i.e. a particular hybrid algorithm) uses an amount of computing resource that is exponential in the size of the problem!

The challenge is to make the user's task more manageable. It can be addressed in three ways:

- Identify general problem features that correlate with suitable algorithms. Define simple rules about when to use one problem solving method and when to use another one. Categorise which forms of hybridisation work best for which kinds of problems and algorithms. Some hints along these lines were published in (Chamard et al., 1995) and (Gervet, 2001).

  It is unrealistic to expect a complete categorisation of problems and algorithm hybrids that can be applied mechanically to each problem. However the mathematical programming community has established some folklore along these lines, and there is a very valuable opportunity to try and extend this folklore to hybrid algorithms and to capture it in terms of guidelines or even rules.

- Build formalisms, languages and systems that support mapping of problems to hybrid algorithms which can solve them. This requires both primitives for defining algorithms, operators for combining them, and constructs for mapping problems onto these primitives and operators. This is an area of active research, and major steps forward are achieved each year (e.g. (Fourer, 1998; ILOG, 2003; Cheadle et al., 2002; Michel and Van Hentenryck, 1997; Laburthe and Caseau, 2002; Van Hentenryck et al., 1999; Guéret et al., 2002; Arbab and Monfroy, 1998)).

  There is no "best" solution in sight. Some researchers seek a minimal set of algorithm primitives and operators, that are in some sense powerful enough. Keeping this set small ensures the user's combinatorial meta-problem also remains small and manageable. Other researchers seek rather to explore the field of algorithms in all its diversity, so as to ensure the resulting system offers all the tools necessary for solving the user's problem efficiently.

- Build systems that partially or fully automate the process of solving the combinatorial meta-problem, e.g. (Minton, 1996; Caseau et al., 2001) This challenge becomes much harder as the number of primitives and operators (algorithms and ways of combining them) grows.

These three approaches to supporting the user are highly dependent on each other. It has often been remarked that the holy grail of a system that automatically maps a problem specification to the hybrid algorithm best suited to solving it, is probably not achievable. Such an ideal system would as a side-effect resolve the open question $P = NP$. Nevertheless, the ideal system can still serve as a target, towards which we seek to converge, even though we may not expect to reach the holy grail itself.

Until we approach the ideal system, users will face a difficulty which is peculiar to solving combinatorial problems. This is the unpredictable behaviour of any algorithm on a class of combinatorial problems. When the last line of code is written, the user does not know whether success or failure should be expected. Modelling is actually easy with regular CP tools, the user has a very positive experience from design to code production. Then there is the empty screen syndrome: the search starts and nothing happens. This may sound mundane to a seasoned researcher, but it is a rare behaviour from a user's perspective.

The reason is obviously the exponential nature of the search paradigm underneath the tools, but this is no consolation to our user. The matter is made worse by the sensitivity to anything, from data, to constraints, to heuristic tuning. A program that worked perfectly well may start producing an empty screen with a new set of data, with the addition of a small insignificant side constraints or if the user tries another value ordering heuristic.

Support for the user therefore includes not only support for developing a suitable algorithm, but also for analysing the behaviour of algorithms during development. This is no mundane piece of engineering, but a real and testing challenge. It becomes even harder when hybrid approaches are introduced, since the number of levers that have this dramatic effect increases. How do we augment an algorithm and especially a hybrid algorithm, to give relevant comprehensible feedback to the end user about its (lack of) progress.

1.6 Software Engineering Issues

Last, there is a software engineering issue, about the integration of CP and hybrid technology with the rest of the software world. Today software development is very different from what it was 10 years ago. This world is evolving rapidly, with the normalization of exchanges (XML), the emergence of new concepts such as Web Services and the reinforcement of old ones such as business objects or component frameworks. The consequence is that integration issues are changing and they are becoming more and more important. There are obviously exceptions, but most companies are turning their IT department into integration department, where most of the action is the integration of existing software.

1.6.1 Software Engineering in Support of Combinatorial Problem Solving. As a consequence, the integration dependencies, or constraints, that apply to a decision-analysis tool are more stringent, both in terms of data and user interaction. Where the standard was to read and write data into a SQL database, and propose a graphical GUI for single users, today's requirements are more complex and many applications require support for complex interaction (both for data and control) with multiple other components. This translates into another challenge for CP technology in order to become a standard engineering practice: it must propose integration tools, methodologies and metrics to control the quality of the development, monitoring and debugging tools.

As a programming tool discipline, CP has a good track record, measured by the use of modern programming languages, component-oriented libraries, and introduction of frameworks. CP program development environments are the subject of continuing research. These techniques serve the development and exploitation of hybrid algorithms as "black-box" modules interfacing to other modules, and end users. The technology is presented to the user as a programing system, or set of libraries available to the user's programming system.

Some researchers have a more ambitious vision, of problem data and constraints, and algorithms, hybridisation forms and search methods encapsulated in different modules, objects or agents. This architecture allows different versions of a problem to be constructed at will by changing the components. It also enables different versions of the problem to be seen or owned or solved by different users. Some sub-domains, such as the encoding of constraints that express a consumer's wishes are very hot, both from a scientific (Laburthe and Caseau, 2003) and industrial perspective. When successfully implemented, this architecture will enable our technology to be presented to the user as a set of components to be "integrated" into a solution which is fully integrated in the IT environment.

This architecture motivates, and depends upon, major advances towards the ideal system described in the previous challenge "Supporting the Problem-Solving Process". The requirement is for standard interfaces enabling constraints, and results elicited from (partial) problem solving, to be passed between different solvers, and the environment. This argues for some research to focus on common formalisms expressive enough to support our existing understanding of hybrid algorithms, rather than on the development of new techniques and forms of hybridisation.

1.6.2 Combinatorial Problem Solving in Support of Software Engineering. There is a huge opportunity for CP technology in support of software engineering, since many of the integration issues may be thought of constraint satisfaction/optimisation problems. If one steps back from the everyday life of an integration shop, one sees configuration problems, routing problems, resource allocation problems. These problems are not necessarily complex ones, thus the previous challenges do not need to be resolved, but the integration issues must be ironed out before the technology will be used: which services can a distributed solver component offer ?

2. Transforming Models to Algorithms

2.1 Conceptual and Design Models

In many areas of computer science, there is a long term aim - or perhaps it is merely a dream - to automate the mapping of the problem specification to its solution. Both the specification and the implemented solution are computer programs, but the programs are very different. The specification is an uncomplicated, unambiguous, clear statement of the problem, in a form that can be verified by the person, or organisation, that "owns" the problem. The implemented solution is a program encoding an algorithm that solves the problem. This program must be efficient in its use of computing resources, and it may well embody some compromise between:

- a program that runs in reasonable time with data requirements that can realistically provided by the computing environment
- a program that computes the very best answer to the exact problem specified

The CHIC-2 methodology for solving large scale combinatorial optimisation (LSCO) problems (Gervet, 2001) uses the term *conceptual model* for the specification and *design model* for the implemented solution. The conceptual model for an LSCO problem specifies:

- the problem variables
- their range of possible values
- the problem constraints, stating for each constraint
  - the definition of the constraint
  - which variables it constrains

The conceptual model is the starting point for the problem solving process. The end point - the implemented solution - is the design model. The idea is that every possible hybrid algorithm should be expressible as a design model, and that LSCO problem solving is the process of mapping the conceptual model down to the best solution, expressed as a design model.

The hybrid algorithms research community is currently designing and developing new hybrid algorithms and new forms of hybridisation. We need to capture the new algorithms and forms in a way that enables them to be reused and combined with other techniques. The design model is - or rather would be - the result of successfully meeting this need. Building and implementing all the constructs that go into the design modelling language, is an essential part of our hybrid algorithms research. Indeed it is the essence of our research.

In the next few sections we will discuss the steps involved in mapping a conceptual model down to an implemented solution. As the precise constructs of a design model emerge, it will become possible to make these steps more concrete. A conceptual model is mapped to a design model by

- decomposing it
- transforming it
- adding search
- adding inference
- handling symmetries

As we pointed out above, these steps constitute the process of solving LSCO problems. One property of this process must be correctness: when we map the conceptual model to a design model, the process should ensure that the resulting solution indeed solves the very same problem formalised in the design model.

The second requirement on this process is flexibility. Just as the design modelling language must be powerful enough to express each algorithm and each hybridisation form, in the same way the process must be flexible enough to map a single conceptual model to "all" the hybrid algorithms that could possibly be used to handle it.

2.2 Decompositions

Hybrid algorithms enable us to exploit the structure of the problem we are solving. LSCO problems typically have components that can be efficiently handled by particular algorithms. For example there might be a linear subproblem, or a job-shop scheduling subproblem. The complete problem is best solved by applying the most suitable algorithm to each component subproblem, as long as cooperation can be achieved between these different algorithms.

The solving behaviour associated with a component is itself expressed as a design model, possibly involving its own decomposition. Thus the problem decomposition may include more than one layer, generating not just a set of components, but a tree.

A problem can be decomposed into subproblems by the variables, or by the constraints. Moreover they can be divided statically, or dynamically during problem solving. If the subproblems were independent, then solving the problem via the decomposition would be simple. However, the subproblems are invariably linked through shared variables, or linking constraints. The challenge is to solve the subproblems in a way that takes into account the dependencies between them.

To map a conceptual model into a design model that takes advantage of problem decomposition, the user requires a natural and if possible lightweight syntax for:

- defining the components of the problem
- associating a solving behaviour with different components
- stating how the solvers which handle the different components should interact

Hybrid algorithms research is all about the ways in which solving behaviour for different components interact, so the following list is both speculative and incomplete. Three ways in which components interact are

- Propagating logical consequences to other components.
  Often the medium through which the components interact are dynamically growing constraint sets associated with particular constraint solvers.

  In constraint programming, global constraints are often used to handle specific problem components. Domain tightenings are logical consequences derived by the global constraints, which are communicated to domain solver. In mathematical programming linear constraints ("cuts") are derived, and added to the linear problem component. When search fails, some algorithms learn "nogoods". These are also logical consequences, posted to a pool of nogood constraints.

  These dynamically growing constraint stores are visible to other problem component handlers, which react to the added constraints by performing further logical inferences, or some other behaviour.

- Propagating the results of sensitivity analysis to other components.
  Information under this heading includes reduced costs, and shadow prices, computed by linear solvers.

  Reduced costs measure the influence on the optimal cost of changing the value assigned to certain variables. This can be used for domain pruning and search heuristics. They enable information that the mathematical programming solver extracts from the cost function to be exploited by other solvers and the search engine.

  The shadow price offers a measure of "tighteness" of each constraint. In particular this can be used as a penalty to add to the cost function in place of the constraint in subproblems where this constraint is relaxed.

- Propagating heuristic information to other components.
  The heuristic information is very often a "good" value associated with a variable. If all the values have heuristic values, then further heuristic

information can be derived, such as conflicts between these values. A typical example is the deployment of a linear solver to find specific values for the variables at which the linear relaxation of the problem has an optimal solution.

Heuristics can also be based on meta-information, such as the number of values remaining in the domain of a (finite domain) variable, the number of constraints on a variable, or the number of times during search that a variable has been involved in a violated constraint.

Traditionally this information was communicated between solvers of the same type. Domain constraints propagated information in the form of tightened domains, which was used by other domain constraints. Integer/linear programming methods propagated "global cuts" in the form of linear constraints added to the underlying linear solver. Reduced costs and shadow prices were passed between integer/linear master and subproblems. Heuristics were passed from one type of solver to its associated search engine.

Now we are allowing this information to be communicated between solvers of different kinds, and between the different kinds of solvers and the search engine. This communication has proven very useful in the combination of different techniques into tightly integrated, efficient, tailored hybrid algorithms.

2.3 Transformations

2.3.1 Separating Modeling for Performance Improvements. Experts often claim that it is unrealistic, or indeed useless, to build a conceptual model of a problem without taking the intended design model into consideration. The "wrong" conceptual model may use the wrong variables, domains and constraints in its formulation, that simply don't map down into the best algorithm for solving the problem.

Mathematical programmers would claim that the success of an application owes as much to the way it is represented (the model that is used) as to the efficiency of the algorithms used for solving it. In order to ease the work of practitioners, a set of software, called modeling languages have been developed, such as (Fourer et al., 2002; GAMS; B.V. ; Software).

Progress in LSCO problem solving is, we believe, eating away the substance of this claim. We believe that design models include variables, domains and constraints (as well as search and inference), and that therefore there exists a transformation algorithm that maps the variables, domains and constraints of the conceptual model onto those of the design model. The challenge here is to identify a (small) set of transformations that cover (almost) all the different cases that come up in practical LSCO problem solving.

2.3.2 Motivations and Examples. Constraint Programming offers a concise formalism for conceptual modeling. The types of variables and constraints that can be used as modeling components are of a wider variety than in Mathematical Programming. They include symbolic variables, compound terms and other complex data types; non linear constraints such as polynomials, rational functions, and exponents; global constraints such as all different (Regin, 1994) and cardinalities (Regin, 1996); conjunction, disjunction and negation of constraints. The resulting model is often closer to the original problem, more compact and easier to understand.

As an example, let's consider a resource allocation problem where we need to assign n persons to n machines. Because of the properties of the problem at hand, not all assignments are equivalent. This can be represented by a 2 dimensional cost matrix C where $C[i, j]$ represents the cost of assigning person i to machine j.

The Mathematical programming model for this problem employs just linear and integer constraints, thus:

$$\min \sum_{i,j} C[i,j] * Y_{ij} \tag{10.1}$$

$$s.t \quad Y_{ij} \in 0..1, Y_{ij} \text{ integer} \tag{10.2}$$

$$\sum_{j} Y_{ij} = 1, \forall\, i \in 1..n \tag{10.3}$$

$$\sum_{j} Y_{ij} = 1, \forall\, j \in 1..n \tag{10.4}$$

The constraint programming model uses the *alldifferent* constraint to achieve both clarity and conciseness:

$$\min C[i, Xi] \tag{10.5}$$

$$s.t \quad X_i \in 1..n, X_i \text{ integer} \tag{10.6}$$

$$\textit{alldifferent}(X_i) \tag{10.7}$$

The MIP model has n^2 variables and $2n$ constraints whereas the CP one has only n variables and 1 constraint. However, it turns out that the MIP model can be solved exactly by an LP algorithm as it is a simple flow problem.

This example prompts us to look for means of combining the expressiveness of CP with the power of MP algorithm. This has been the motivation for a variety of work over the past decade on hybrid approaches between CP an MIP. Instead of devising hybrid algorithms, it is also possible to transform a CP model into an equivalent MIP model. Each finite domain variable in the CP model is transformed to set of $0-1$ variables into the MIP model, one for each possible value of the original CP variable. Disjunction and negation of constraints are transformed to an MIP representation that includes an extra $0-1$ variable in each constraint (Rodosek et al., 1999) Many global constraints of C can then be expressed by a set of linear constraints on those 0-1 variables (Refalo, 2000).

A CP model can be transformed to many different alternative MIP models. Choosing the best one may be quite tricky, and is subject of active research. One potential drawback of this approach is that it generates MIP models with a very large number of 0-1 variables and lots of dense constraints (constraints involving many variables).

Although the transformations can yield difficult models for MIP algorithms, this approach seems promising enough to interest the MP community. For instance, some CP constraints have been added to the AMPL modeling language (Fourer, 1998).

CP could be used as a modeling tool for other kind of algorithms. For instance, we could envision the automatic translation of a CP model into a Propositional Satisfiability (SAT) model. A SAT model only involves binary variables and logical clauses (logical disjunction of variable or their negations). It remains to be seen if this yields good computation results.

Over the past years, several methods have been devised to transform models into ones which will ultimately produce more efficient design models. We will review two of them. It would be interesting to study their automatic application to a given model.

A first technique is the so-called dual model. Let's explain the underlying idea on our simple resource allocation problem. In our model we choose to use one variable X_i per person i, whose value will be the machine assigned to that person. We could also chose to have one variable Y_j per machine j, whose value will be the person assigned to it. This yields a second model, equivalent to the first one. In this example, the two models yield equivalent computational results. This is not true in general for more complex problems, and it may difficult to predict which one would be the best. It is possible to combine both models using so-called channeling constraints (Cheng et al., 1996) to relate the two models. The channeling constraint merely states that $X_i = j$ holds if and only $Y_j = i$ holds.

The second one is the use of additional constraints that do not modify the set of solutions of the model, but that improve the efficiency of the search for solutions. Such constraints are often called implied constraints, or redundant constraints.

2.3.3 Transforming Conceptual Models into Design Models. A variety of such transformations have been identified:

- Mapping constraints to variables, and variables to constraints. This can be achieved by CSP and linear dualisation, and by reification (of constraints)
- Mapping more expressive models to simpler ones. Examples are n-ary constraints to binary ones, mapping CLP variables and constraints to integer/linear ones, and mapping an arbitrary combinatorial problem to a 3-sat problem
- Mapping values to variables and variables to values. This is often used for problems which involve permutations (travelling salesman; matching tasks to discrete resources; competition scheduling)
- Adding multiple representations, and chanelling constraints between them
- Adding redundant constraints
- Mapping constraints into the optimisation function, and optimisation requirements into constraints. Examples are conflict minimisation problems; overconstrained problems and multi-criteria optimisation problems

Capturing such transformations in the conceptual-to-design mapping process is relatively unproblematic. The challenge here is to then apply further mappings on the transformed model. The new variables, domains and constraints are implicit, until the transformation has been applied to a problem instance. It is hard to choose, and to express, the best search and inference behaviour for a set of variables, domains and constraints that are merely implicit.

2.4 Search

Search has to be introduced in the mapping from the conceptual model, which involves no search whatsoever, to the design model, where search is a key aspect.

2.4.1 Local and Constructive Search. Historically two forms of search have been kept quite distinct:

- Local search, such as hill climbing, which typically involves a sequence of "moves" from one complete assignment to another
- Constructive search, where solutions are created incrementally by instantiating the variables one by one, or adding other constraints to the problem one after another.

In the first section of this chapter, we reported on the scalability problems that can result from a "pure" tree-search approach, where alternative solutions are constructed incrementally down each branch of the tree.

Integrating constraint handling with local search, is very different from integrating with constructive search. Ideally, the form of search could be specified in the design model, independently of the problem constraints. The constraint behaviour - detecting violations, inference and so on - should adapt itself to the chosen form of search.

2.4.2 Combining Different Search Methods. Our research community has recognised the importance of combining different search methods and exploiting incomplete search to achieve scalability with good problem solving behaviour.

The system should support hybrid forms of search comprising both local and constructive components. The hybridisation may be based upon problem decompositions. For example certain problem variables may be labelled by local search, and others by constructive search. The decomposition may be temporal: all variables first labelled using constructive search and then the solution may be improved using local search. The different kinds of search may be interleaved, for example by applying local search to the partial assignments produced during a constructive search procedure.

Local search may be used once a first solution has been found with a global tree search, a most common pattern since local search is usually based on neighbourhoods that apply to complete solutions. Local search may also be used within the global tree search to speed up the search or increase the quality of the solution, as soon as we have a neighbourhood structure that applies to partial solutions. ILO (Incremental local optimisation) is such an example (Caseau and Laburthe, 1999) where local optimisation moves are used to fix the inconsistencies that were made during the global tree search, when two nearby regions are modified by decisions that were made at very different times during the search. Another approach is to use global tree search to define neighbourhoods, such as the Large Neighbourhood Search (LNS) (Shaw, 1998) method. All these methods can be combined sequentially, in parallel or with a meta-search approach. For instance, most LNS implementations today use very sophisticated neighbourhoods, and very simple meta-strategies, based on hill-climbing. This means that once a local move is computed, if it yields an improvement it is applied and discarded otherwise. One could imagine that a more sophisticated approach, such as a Tabu search, is applied as a meta-search strategy. The "meta-level" of hybridisation is a promising avenue for new research, since many successful specialized algorithms, such as GSAT (Selman et al., 1992), may be described as a smart meta-strategy applied to a simple problem solving approach.

2.4.3 Concurrent Search. Most of the hybrid algorithms described in this book employ different solvers to handle different problem variables and constraints, but exploit a single global search algorithm to synchronise the problem solving behaviour.

There are performance advantages to be gained from running more than one search procedure in a hybrid algorithm. In such a case each search procedure in effect explores a different subproblem (though the subproblems may overlap).

One way to use a local search procedure is for looking ahead. Testing the consistency of a value for a variable can be achieved by instantiating the variable to that value, and searching for compatible values for all the other variables occurring in the constraint. N-ary consistency can be enforced by labelling sets of N variables. Probing in MIP is also achieved by, effectively, labeling variables.

Searching to a restricted depth is not only useful for ruling out (combinations of) values for variables, but it also enables heuristic information about a subproblem to be accumulated. The labelling which is most promising, can be used as a value-choice heuristic for a global search routine.

These examples employ local search in the service of a (more) global search routine. However multi-agent approaches to solving combinatorial problems may associate a search routing with each agent. In this case there is typically no global search routine.

For managing an algorithm where the search is distributed concurrently over separate subproblems the main issues are:

- How to achieve and maintain consistency between search procedures
- How to ensure completeness of the global search routine in case two search trees are explored asynchronously

2.5 Inference

2.5.1 Global and Local Control of Inference. Traditionally a single form of inference was applied to the whole problem. For example in the AI literature arc-consistency is a property of the whole constraint network, and it is achieved by applying propagation on all the problem constraints. In the mathematical programming literature, the linear constraints are all added to the same matrix, and the algorithm is applied to the complete set of linear constraints.

This book is about combining different algorithms to solve a problem. This can be achieved, to a useful extent, by problem decomposition. However a more radical approach is to choose a (possibly different) solver for each separate problem constraint. Moreover a single constraint can be sent to several solvers.

It is sometimes useful to employ not only different types of solvers, but even different instances of the same solver. In the case of a master/subproblem decomposition, it is natural to use a different matrix for each distinct linear subproblem. As another approach, several different relaxations can be handled by different (possibly linear) solvers, whose different solutions support alternative heuristics for guiding search.

2.5.2 Controlling the Local Inference Procedure. Orthogonal to the choice of which solver each constraint is sent to, there is a choice of what inference the solver should perform on the constraint. For example when a constraint is handled by a domain solver, the solver may infer only tighter bounds or it may additionally infer "holes" in domains.

The most powerful form of inference is problem solving. The information inferred is the complete set of solutions to the problem. In "perfect relaxation" (Montanari and Rossi, 1991), the subproblem is simply replaced by its set of solutions. However, to enable this information to be efficiently exploited by other solvers, this set of solutions is often represented intensionally using appropriate constraints (such as the convex hull of the set of solutions, as in Dantzig-Wolfe decomposition or as the tightest "box" containing all solutions, which corresponds to generalised arc-consistency).

The subproblem itself can be solved by any algorithm, even a hybrid algorithm exploiting the decomposition, transformation, inference and search techniques outlined in this section.

For most purposes less powerful forms of inference are used. It is not necessary to compute all the solutions to a subproblem to extract useful information. There are a wide range of "lookahead" techniques that explore partial search trees, and extract information that can be inferred from all the branches of the partial search tree. Indeed the information extracted from looking ahead may be heuristic, based on an assessment of the quality of the deepest nodes reached on the different branches of the partial search tree.

2.5.3 Controlling Communication of Inferences. The inference performed by a solver becomes more complicated to control when it passes information to other solvers. Suppose, for example, a constraint propagates new linear constraints (for example a job-shop scheduling constraint might infer the order of two tasks on the same machine). If there are multiple linear solvers, the user must be able to specify to which solvers the propagated linear constraints should be sent.

As yet there is no method for algorithm developers to control propagation of information between solver instances. Advances in modelling languages which enable users to employ more solver instances and more varied propagation will bring this issue to the fore.

2.6 Symmetries

For many problems, there is a set of solutions from which, by different transformations, it is possible to construct all the solutions. For example from one solution to a map colouring problem, many different solutions can be quickly generated by simply permuting the colours. These other solutions are often termed "symmetric" solutions. Indeed if the problem statement says that an answer is merely a partitioning of the map into regions which share the same colour, then the assignment of an actual colour to a region is an artifact of the solving process, and the symmetric solutions, all represent the same answer as far as the user is concerned.

Exploiting symmetry provides a much more compact way to represent the complete set of solutions (by explicitly representing the core set, and the transformations which can be used to generate all the other solutions). More importantly it can be used to dramatically improve search efficiency. This efficiency gain results from the observation that, if part of the solution space can only produce solutions symmetric to solutions which will necessarily be produced elsewhere by the search procedure, then there is no need to further explore this part of the search space. This optimisation can even be exploited when symmetrical solutions have different costs: in effect each core solution is taken to represent all its symmetric solutions, and during search its cost is taken to be the best cost of any symmetrical solution.

Pruning the symmetric parts of the search space can be done by adding extra constraints to the design model, or by dynamically generating and testing symmetry conditions conditions during search. The first approach is handled as part of the problem transformation process. For example in a map colouring model an ordering could be associated with the regions and colours, and the first regions of each colour could be constrained to be in the same order as the colours themselves. The second approach is handled as part of the inference process, posting a propagating anti-symmetry constraints at each node of the search tree.

3. New Techniques

3.1 Stochastic Optimisation

Stochastic optimisation occurs when the input data that describes a problem has a stochastic nature, that is, is defined with a probabilistic point of view. The stochastic information may be complete, with probability distribution for each input parameter (length of a task duration, expected sales price, consumer demand for a given product, etc.), or it may be incomplete, when we only know the mean and the standard deviation of the input values. We may want to optimise a decision according to an objective function that is defined in a stochastic manner (e.g., maximize the expected revenue) or characterize the probability distribution of the output parameters of the problem. The field of stochastic optimisation is (Birge and Louveaux, 1997) tightly coupled to probability theory and has produced a large set of results and algorithm.

The challenge occurs when the problem is, independently, from its stochastic nature, a hard combinatorial problem. Indeed, the probabilistic methods apply well to close-form solution that may be expressed with linear, quadratic or sometimes polynomial formulas. On the other hand, the stochastic analysis of an exponential algorithm is most often out of reach of mathematical analysis. Therefore, the state-of-the-art for these stochastic combinatorial problem is often to get rid of one dimension of the problem. If the stochastic nature is too important, a simpler relaxation (non-combinatorial) is used, for which stochastic analysis may be applied.

On the other hand, if no such relaxation approximates the problem, the combinatorial algorithm is then itself used on sample problems, like mean values, extremal values, or a set of values generated from the probability distributions using sampling techniques. In (Benoist et al., 2001) an experiment was proposed that studied different trade-offs between a purely stochastic and a purely combinatorial approach. Since then further experiments have been made using quasi random methods for improving the sampling approaches with very positive results. This follows the footsteps of stochastic optimisers who introduced "quasi-Monte-Carlo" approaches as a follow-up to Monte-Carlo methods, since quasi-random distribution are more evenly distributed than truly random sequence and thus may produce faster convergence. One should not underestimate the importance of this stochastic-combinatorial combination. Experience suggests that 90% of industrial problems have indeed a stochastic or uncertain nature. The absence of tools and methods makes it unrealistic to take this aspect into account for many applications, but one may think that it would be different given the proper tools.

Then, there exists a lot of stochastic problems such as reservation systems, which are truly stochastic and where the combinatorial aspects are ignored. Thus, one may say that there exist a large number of "real" problems waiting for new advances in this domain. These advances may come from three directions. First, from a CP angle, we need to develop better sampling techniques that support the use of combinatorial optimising algorithms to tackle stochastic problems. The use of quasi-random sequences is an interesting new development, but there are many other solutions that need to be investigated, depending on the global objective.

Quasi-random sampling may be ideal for optimising expected values, but sensitivity analysis may be a better approach to answer robustness and "what-if" questions. Second, we need to better understand CP global and local search algorithms, to be able to apply stochastic analysis to them. Obviously, this approach which is already difficult with complete branch-and-bound search becomes even harder with hybrid local search approaches. Last, we need to find tools and techniques to reason about solution spaces, as mentioned earlier.

Reasoning about solution space is much more tolerant to the stochastic nature of the problem.

3.2 Overconstrained problems and robustness

ICL's Dave Brunswick, who worked on constraint problems in the airlines industry, said that "there is no such thing as a truly hard constraint". In practice solutions frequently violate constraints, and there is almost always a way to implement such solutions in the real world. In the same vein, an ILOG user asked "What solution does SOLVER return, when there is no solution?" In practise, then, our systems must be able to handle problems whose solutions all violate one or more constraints.

Historically four techniques have been used to handle overconstrained problems.

- Hierarchical Constraints: The system searches for a solution which satisfied the more important constraints, relaxing those lower in the hierarchy
- Penalties: Constraints are extended with a penalty term which are included in the optimisation function
- Dummy Values: The constraints are extended so they can be satisfied by extra "dummy" values associated with certain variables. Solutions without dummy values are preferred.
- Count the violations: Associate a penalty of 1 with each violated constraint, but when the cost function has reached its upper bound, enforce all the remaining constraints. For scheduling applications, a recent algorithm for minimising constraint violations is (Petit et al., 2002).

These methods are augmented by sensitivity analysis techniques that are available from linear solvers. In particular we can elicit:

- The reduced cost of a variable is a conservative estimate of how much worse the optimum becomes under changes to the value of that variable.
- The shadow price measures the rate of improvement in the optimal value as the constraint is loosened.

It is also possible to extract *irreducible infeasible sets*, minimal sets of constraints that are inconsistent. This can be done by analysis of linear constraints, or more generally by a binary-chop search of the constraint set, until a minimal inconsistent set has been found (Junker, 2001).

Explanation of inconsistencies, is a crucial part of the program development environment (Deransart et al., 2000, Chapter 5). Minimal correction of inconsistent constraint sets is a new area of research (Amarel and Barahona, 2002).

3.3 User Support

Hybrid techniques are powerful and successful, but their practice is an art, not a science and definitely not an engineering discipline. This means that, although they can be used successively by seasoned researchers who will obtain very competitive results, they have not yey caught on in the industrial world. Many of the techniques that are described in this book would be very relevant to existing industrial problems, but for those who are brave enough, the complexity of the techniques and the instability of the algorithm performance are a real obstacles. In order to make the hybridisation of local search and constraint programming a successful practice, one needs to focus on ease of use.

In this section, therefore, the aim is not necessarily to provide the user with the tools to build the very best algorithm for the problem at hand, but rather to enable "ordinary" users - who may not be completing a PhD in the area of hybrid algorithms - to solve large scale industrial combinatorial optimisation (LSICO) problems.

To spread the use of hybrid techniques in industry we propose two directions of research

- The development of a relatively understandable and simple environment for building hybrid algorithms to solve LSICO problems
- The automation of some aspects of the process of building and testing hybrid algorithms, using machine learning techniques

3.3.1 A Simple Environment for Solving LSICO Problems. To understand the challenge of making CP and especially hybrid CP techniques user-friendly, we need to look at the difficulties that lie in front of a regular user. The question of when to apply local search, which type of neighbourhood, the opportunity of "shaving" (Martin and Shmoys, 1996), the introduction of restrictive branching scheme, all are difficult choices for which the two necessary virtues are patience and intuition. This does not live well in an industrial world which is often short on both.

There have been numerous attempts in the past 10 years to produce methodologies to help our regular user to solve large-scale industrial combinatorial optimisation (LSICO) problems (Chamard et al., 1995; Gervet, 2001). This wealth of effort should not be under-estimated and is indeed very valuable for the training of new optimisation engineers. On the other hand, we are still far from a user-friendly technology.

To maintain user confidence it is necessary to establish explicitly what techniques may work on a problem, and what techniques probably won't. Combinatorial optimisation is a very competitive field and some of the algorithms have been exploited in sophisticated ways to solve large complex problems. There are many other meta-heuristic techniques but (hybrid) Tabu Search is the leader of the pack, on almost any published benchmarks. Similarly, there are many problems for which MP is the technique of choice, either because the problems are indeed large-scale linear problems, or because our knowledge of the associated polytope has produced hybrid strategies (branch-and-cut, branch-and-price, column generation) that are today's record holders (such as branch-and-cut on TSP). Last, there are problems such as SAT for which specialized algorithms, especially randomised algorithms, have been developed that are far more efficient than what may be accomplished using constraint propagation embedded in a tree search.

There is a regular confusion among non-specialists between CP, AI and rule-based programming, which is especially true in North America and in the industrial world. A commonly accepted list of problems or criterions for selecting problems that are indeed relevant to CP technology would be a big help. Similarly to the methodology issue, there have been many attempts to compile such a list. However, making this knowledge part of standard courses on optimisation is still an open challenge.

Here are four steps to "productising" hybrid algorithm development environments for LSICO problems:

- Develop a solution roadmap, which tells which method to apply for which problem, and an easy test for success/failure. Publish books like this one that present the state of the art.

- Select a small subset of methods by the community, and its implementation in various tools. We need fewer techniques: to push a technique from the world of academic research towards the fields of engineering practices, a little bit of simplification and standardization is necessary.

- Regular users have a legitimate need for explanations, training and metrics (how do I know that this is working?). The community needs to spend time to characterize how to tune these methods, when to and when not to apply them, and produce indicators that tells when the method is working properly. There is still a lack of explanation and debugging tools, though there is an ongoing research effort in this area (Deransart et al., 2000).

- One issue, which has been intensively addressed in the mathematical programming community, is what default algorithm to associate with a conceptual model. A good "out of the box" performance can avoid the shock many users experience when they build and run their first design model for a problem. The current default would be Simplex or barrier solving techniques for linear constraints, and domain tightening propagation for non-linear and symbolic constraints, embedded in a branch and bound search. The drawback is the blank screen syndrome. The question, and the challenge, is what could replace it, delivering acceptable solutions in most cases, and a consistent experience to the user?

3.3.2 Automating Algorithm Development and Testing. An interesting avenue for the future is the principle of auto-adaptative hybrid algorithms, where the tuning part of the hybrid algorithm is done automatically with a learning process. The idea is to represent the space of all possible hybrid algorithms that may be produced with one or many methods with a term algebra, where each term represents a selection of heuristic methods, of heuristic values, and a combination order for global and local search principles. This approach is explained with careful details in (Caseau et al., 2001), and relies on a common idea of representing algorithms with terms from an algebra. This algebraic representation is the foundation for a "meta" adaptative algorithm which, given a sample of problems and an objective function, will either discover a proper algorithm or tune an existing one by creating or modifying the associate term.

The experiments in (Caseau et al., 2001) are encouraging since they show that it is possible to obtain similar and even better algorithm with such an automated process than those created by hand. It is far to early to conclude that this is a perfect solution to the problems that were stated earlier.

First, a learning algorithm takes time and it requires many orders of magnitude (10^4 to 10^5) of experiments, which makes it practical for tuning an reactive system that runs in a second, but unrealistic for a back-office system that runs in a few hours.

Second, we lack the scientific knowledge that will determine what is the ideal combination of meta-heuristic that must belong to the algebra. The more complex the algebra, the more difficult is the learning process, and the more difficult it is to provide explanations about the algorithm that is produced. Therefore, a long sequence of experiments with different problems and resolution span is necessary before this adaptative technique may be imported in the industrial world. Still, there are many reasons why this may be a very promising approach. On the one hand, industrial problems tend to vary during their lifetime. New side-constraints are introduced, the objective function may change, the computer onto which the piece of optimisation software is running may be updated. All these changes do require, at least for most hybrid approaches, a careful rethinking of the algorithm. Our experience is that these updates are not performed in the industrial world (once a system works, there are few incentives to introduce changes), thus an automated tuning is bound to be better than none. On the other hand, the amount of time that is available in the "real" world to perform the careful tuning of an hybrid approach is usually quite small.

Therefore, although producing adaptative algorithms that may compete against the state-of-the-art of the research community may be an impossible challenge, producing algorithms that are better than what may be accomplished with a limited amount of time is much easier. For instance, the ratio of development vs. tuning time for the two ROADEF 2001 and 2003 hybrid methods from the e-Lab (cf. (ROADEF, 2001)) is 20%/80%, where 80% is an unrealistic figure in a true industrial setting.

3.4 Packaging

It can be argued that CP in its current state is too complex for the average engineer in the industry. In order to support this view, let's contrast CP and MP technologies from the standpoint of the user. There are several major differences, some of which have been already discussed.

- Toolkit versus algorithm
  In MP, people produce algorithms, that take as input a model and some data, and produce a result, usually an optimal solution. In CP, people produce a toolkit, in the form of a full fledged programming language, or a library, and users have to build an algorithm out of it. Therefore, more work remains to be done when one uses a CP system than when a MP system is used.

- Technology abstraction
  An MP system can be used by someone who does not understand its basic algorithms. On the contrary, CP system are best used by people that have a deep understanding of how constraints are propagated and how the tree search behaves. This limits the number of people that can effectively use CP.

- Standard for exchanging models
  There is a standard file format for storing and exchanging models in MP It is called the MPS format, after the name of a famous MP system developed in the 60s. This standardization has several good effects, including the following ones.

  - A large set of models is available for researchers. This enables the experimentation and validation of new ideas on a reasonably sized test set.
  - Commercial users can also try several competing MP solvers as they all use the same format for input. This reduces the dependency of customers from their suppliers, increases competition, and finally increases the market size.

  Some initial efforts toward the standardization of model format have begun. For instance, the COIN initiative, whose aim is to replace the

MPS format by a new, more flexible format, is considering the addition of CP modeling constructs in the format. In an unrelated initiative, several researchers use the OPL language (Van Hentenryck et al., 1999) as a solver independent modeling language (Hnich and Flener, 2001).

- Support for analysis of the quality of a solution.
  As we have seen above, MP systems provide various tools for explaining results and for what-if analysis. MP systems also provide an estimate (the MIP gap) of the quality of a solution.

If those 4 features (algorithms, abstracted technology, standard interface, and explanations) are not present in future CP, its industrial penetration may be threatened. In such a case at least two scenarios could be devised. First, CP might be restricted to some niche markets where its value is already proven, such as for detailed production scheduling. Second, CP might be used in some killer application not yet discovered. In such a case CP would be widely used, but its use would be hidden inside the application.

So far we focused on the acceptance of CP as a general technology for engineering. CP's future can be analyzed along other dimensions. First of all, CP is a health research area, that can potentially bring interesting advances in several computer science areas ranging from programming languages design and semantics to combinatorial optimization. The steady increase of he visibility and attendance of the CP conference series clearly show that the future of CP is quite bright in the academic world. The set of issues reviewed in this paper defines a number of challenging research problems that can fuel the research in the CP community for the years to come.

The development of global constraints addresses the objective of ease-of-use. The addition of a small set of truly generic global constraints to CP solvers is still a very promising avenue to expand the range of application of constraint programming.

1 The increase in quality and speed of computers and algorithms means that the range of applicability of a given technique is growing mechanically. True, there will always be problems that are out of reach, but the bulk of industrial problems that we know today are almost stable. This means that the size of the call centres that we need to schedule, the number of trucks that we need to plan, the number of visits that one employee can do in a day, the number of machines in a plant, etc. are not evolving as rapidly as Moore's law. There are counter-examples, due for instance to consolidation, but the principle that a large set of industrial problems form a "fixed target" holds, which is why the notion of CP becoming an industrial practice one day makes sense.

2 On the other hand, the capacity to handle complex techniques, algorithms and methods does not grow, so scalability must be supported by better abstraction tools and simplification of the algorithm development task. thus to extend the range of applicability of a give technique

The status of a given technology to solve a stable industrial problem evolves through 3 phases: cannot do; can do, but it is hard; done, and there exists a software component. Our belief is that the new frontier for research on global constraints is the third stage, not the second, for which other approaches are better suited.

4. New Application Areas

In this section we will briefly discuss new application areas for constraint programming and hybrid approaches, since we believe that there is much more to constraint programming than producing solutions to combinatorial problems. This section is by no means complete: there are many other new application areas, including ones which the authors are not aware of.

4.1 Computer-Aided decision analysis based on simulation

When we have a complex multi-criteria optimization problem, it is quite frequent that optimization is no longer an issue. Although it is always possible to linearize a set of criteria or to search for Pareto optima, truly complex industrial problems such as construction scheduling problems which involve uncertainty and multiple criteria, require for a much more pragmatic approach, which may be described as how to evaluate or propose possible reactions to unforeseen events, for a given plan of action. For a construction engineer, the issue is not to do but mostly what to watch for. This prompts for new tools and new techniques that are centred around two ideas: characterizing a solution space (again) and simulation, which is a way to handle the complexity of uncertainty and multiple criteria.

Characterizing a solution space is precisely what we need to perform what-if analysis. For instance, if the foreman decides to dig the foundation of the second building before raising the walls of the first one, will it be more robust with regard to poor weather conditions ? With the current state of science, this type of characterization will be done mostly through simulation and sampling, so the two ideas are linked, but this is not necessarily the case. Techniques from theorem proving might be imported and used in this context. In fact, a proper solution to a realistic problem should probably involve a lot of different techniques. The inadequacy of today's tools to address the uncertainties of, for example, construction scheduling, means that currently almost no tools are yet deployed in the field.

If we stick to our construction example, the first stage of planning is a pure combinatorial optimization problem, since we use average values to build the initial schedule. Risk analysis on this schedule involve stochastic optimization. In the future, these two steps may be merged, with today's large-scale problems and our simple technology for stochastic problems, it is better to stick to two phases. Last, decision analysis tools for the construction site should use simulation to evaluate options, compute risk and foresee possible complications. This indeed a multiple criteria problem: time of delivery, money that is spent, quality of the construction are obvious criterions but there is also the human resource aspect which is difficult to put in an equation and is often the most critical factor in getting the job done. Thus rather than putting everything in the computer and pressing a key to find a solution, the need is for methods that allow the foreman to understand the structure of the solution space.

4.2 Cooperative Problem Solving

The advance of distributed and cooperative problem solving is one of the major challenges for the next 10 years to come. This challenge is intimately related with what was said in Section 1.4.1 above, since collaborative work does not simply mean solving a problem. The prototype work that was done in the *Wishbone* project at the Bouygues e-Lab (Laburthe and Caseau, 2003) is a good example of this direction: two constraint engines collaborate, one represent the customer wishes and the other represent the product database. They exchange, over a SOAP connection and using XML encoding, constraints and abstractions of the solution space. The depth and the richness of the information that is exchanged supports an interaction protocol between the customer and the product database that goes beyond current electronic commerce applications. In this example, the same team developed the two solvers; the next step is to make this type of deep interaction possible between components that come from different sources.

4.3 Interleaved Planning and Execution

Most industrial combinatorial optimisation problems involve uncertainty. The solution of such a problem is very often a plan, or schedule, of operations, tasks or activities. During the implementation of such a solution, the uncertainties typically become resolved. After completing the first task in a schedule, for example, the uncertainty about the duration of that task and the performance of the resources, has been resolved. At this stage, the original problem can be solved again, with data and constraints that are more precise, or more correct, or more certain, than before.

The knowledge that the (uncertain) problem can be repeatedly re-solved during the implementation of the solution, with better and better input, can be used to improve the overall system performance. Instead of generating the optimal solution to the original problem, for example, it may be better to generate the most "robust" solution - the one whose first step will not prevent optimal replanning in case its implementation does not work out as expected. In case the problem solving process is much faster than the implementation, a new optimal plan can be computed at each step of the implementation. An example of this kind of approach is (Fowler and Brown, 2003).

Often, however, there is not time to compute a new optimal solution between implementation steps. This is the situation for applications such as network routing, control of autonomous vehicles and mission control in aerospace.

In this case an online decision method has to be employed. The recomputation steps must be completed within a fixed amount of computational resource. Problems of this kind require the initial LSICO problem to produce not a single solution, but an automaton to perform the online replanning during execution. The automaton might be, for example, a finite satae machine.

We have a new kind of LSICO solution therefore, that produces not an assignment of values to decision variables (as described in the last section), but generates a finite state machine. If there is scope for somewhat more recomputation, then the combination of off-line problem solving, at the beginning, and on-line resolving during implementation (e.g. during the mission) becomes more complex and more difficult to capture in a single framework, or specification.

References

Amarel, P. and Barahona, P. (2002). On optimal correction of inconsistent linear constraints. In *Principles and Practice of Constraint Programming, CP 2002*, pages 33–46. Springer.

Arbab, F. and Monfroy, E. (1998). Coordination of heterogeneous distributed cooperative constraint solving. In *Applied Computing Review*, volume 6, pages 4–17. ACM SIGAPP.

Ben-Tal, A. and Nemirovski, A. (1998). Robust covex optimisation. *Mathematics of Operations Research*, 23.

Benoist, T., Bourreau, E., Caseau, Y., and Rottembourg, B. (2001). Towards stochastic constraint programming: A study of online multi-choice knapsack with deadlines. In *Principles and Practice of Constraint Programming, CP 2001*, pages 61–76. Springer.

Birge, J. R. and Louveaux, F. (1997). *Introduction to Stochastic Programming.* Operations Research. Springer.

B.V., P. D. T. AIMMS: Advanced integrated multidimensional modeling software.
Online documentations at: `www.aimms.com`.

Caseau, Y. and Laburthe, F. (1999). Heuristics for large constrained vehicle routing problems. *Journal of Heuristics*, 5(3).

Caseau, Y., Silverstein, G., and Laburthe, F. (2001). Learning hybrid algorithms for vehicle routing problems. *Theory and Practice of Logic Programming (TPLP)*, 1(6):779–806.

Chamard, A., Fischer, B., Guinaudeau, and Guillaud, A. (1995). CHIC lessons on CLP methodology.
`www.icparc.ic.ac.uk/eclipse/reports/CHIC_Methodology.html`.

Chan, P., Heus, K., and Weil, G. (1998). Nurse scheduling with global constraints in CHIP: GYMNASTE. In *Proc of Practical Application of Constraint Technology (PACT98)*, London.

Cheadle, A., Harvey, W., Sadler, A., Schimpf, J., Shen, K., and Wallace, M. (2002). ECLiPSe: an introduction. Technical Report IC-Parc-03-01, IC-Parc, Imperial College London.

Cheng, B., Lee, J., and Wu, J. (1996). Speeding up constraint propagation by redundant modeling. In *Principles and Practice of Constraint Programming, CP 1996*, pages 91–103. Springer.

Deransart, P., Hermenegildo, M. V., and Maluszynski, J. (2000). *Analysis and Visualisation Tools for Constraint Programming*, volume 1870 of *LNCS*. Springer.

Fourer, R. (1998). *Extending a General-Purpose Algebraic Modeling Language to Combinatorial Optimization: A Logic Programming Approach*, pages 31–74. Kluwer.

Fourer, R., Gay, D. M., and Kernighan, B. W. (2002). *AMPL: A Modeling Language for Mathematical Programming*. Duxbury Press.

Fowler, D. W. and Brown, K. N. (2003). Branching constraint satisfaction problems and Markov decision problems compared. *Annals of Operations Research*, 118(1-4):85–100.

GAMS. GAMS: The general algebraic modeling system. Online documentation at: `www.gams.com`.

Gervet, C. (2001). Large scale combinatorial optimization: A methodological viewpoint. In *Series in Discrete Mathematics and Theoretical Computer Science*, volume 57, pages 151–175. American Mathematical Society.

Guéret, C., Prins, C., and Sevaux, M. (2002). *Applications of optimization with Xpress-MP*. Dash Optimization. Translated and revised by S. Heipcke.

Hnich, B. and Flener, P. (2001). High-level reformulation of constraint programs. In Codognet, P., editor, *Proc. of JFPLC'01*, pages 75–89.

ILOG (2003). ILOG Concert technology. `www.ilog.com/products/optimization/tech/concert.cfm`.

Junker, U. (2001). QUICKXPLAIN: Conflict detection for arbitrary constraint propagation algorithms. In *IJCAI'01 Workshop on Modelling and Solving problems with constraints*.

Laburthe, F. and Caseau, Y. (2002). SALSA: A language for search algorithms. *Constraints*, 7(3-4):255–288.

Laburthe, F. and Caseau, Y. (2003). On the use of constraints for exploring catalogs. In *Proc. JFPLC*.

Martin, D. and Shmoys, P. (1996). A time-based approach to the jobshop problem. In Queyranne, M., editor, *Proc. IPCO'5*, volume 1084 of *LNCS*. Springer.

Michel, L. and Van Hentenryck, P. (1997). Localizer: A modeling language for local search. In *Proc. Principles and Practice of Constraint Programming - CP97*, pages 237–251. Springer, LNCS 1330.

Minton, S. (1996). Automatically configuring constraint satisfaction problems: A case study. *Constraints*, 1:7–44.

Montanari, U. and Rossi, F. (1991). Perfect relaxation in constraint logic programming. In *Proc. ICLP*, Paris.

Petit, T., Regin, J.-C., and Bessiere, C. (2002). Range-based algorithm for max-CSP. In *Principles and Practice of Constraint Programming, CP 2002*, pages 280–294. Springer.

Refalo, P. (2000). Linear formulation of constraint programming models and hybrid solvers. In *Principles and Practice of Constraint Programming, CP 2000*, pages 369–383. Springer.

Regin, J. (1994). A filtering algorithm for constraints of difference in CSPs. In *Proceedings of the 12th National Conference on AI*, pages 362–367.

Regin, J.-C. (1996). Generalized arc consistency for global cardinality constraint. In *Proc. of the Fourteenth National Conference on Artificial Intelligence (AAAI-96)*.

ROADEF (2001). Roadef challenges.
`www.prism.uvsq.fr/~vdc/ROADEF/CHALLENGES`.

Rodosek, R., Wallace, M., and Hajian, M. (1999). A new approach to integrating mixed integer programming and constraint logic programming. *Annals of Operations Research*, 86:63–87.

Rottembourg, B. (2000). Call center scheduling. *AIRO Bulletin*, 2000(1). Spring issue of the bulletin of the Italian Operations Research Association (AIRO).

Selman, B., Levesque, H., and Mitchell, D. (1992). A new method for solving hard satisfiability problems. In *Proc. of the National Conference on Artificial Intelligence AAAI'92*, pages 440–446.

Shaw, P. (1998). Using Constraint Programming and Local Search Methods to Solve Vehicle Routing Problems In *Proceedings of the 4th International Conference on Principles and Practice of Constraint Programming - CP98*, pp 417-431.

Software, M. MPL: Mathematical programming language.
Online documentations at: `www.maximal-usa.com/mpl/`.

Solomon, M. (1987). Algorithms for the vehicle routing and scheduling problems with time window constraints. *Operations Research*, 35(2).

Van Hentenryck, P., Lustig, I., Michel, L., and Puget, J.-F. (1999). *The OPL Optimization Programming Language*. MIT Press.

Yorke-Smith, N. and Gervet, C. (2002). On constraint problems with incomplete or erroneous data. In *Principles and Practice of Constraint Programming, CP 2002*. Springer.

Index

Algorithm composition, 332
Alldifferent constraint - consistency, 114
Alldifferent constraint - filtering, 115
Alternating path or cycle, 113
AND function, 17
Application areas, 359
Arc-consistency, 8, 94
Backtracking heuristics, 309
Benders Decomposition, 43
Benders decomposition, 177
Best-bound search, 260
Big-M values, 23
Black box, 333
Boolean Satisfiability (SAT), 234
Boolean Satisfiability, 236, 260
Bounded heavy tailed distribution, 245
Bounded heavy tails, 241
Bounding, 23
Bounding constraint, 7
Branch-and-bound, 22, 208
Branch-and-cut, 225, 355
Branch-and-price, 209, 355
Branch and price, 207
Branching, 22, 72, 209, 234
Branching constraints, 10
Branching rule, 72
Branching strategy, 224
Breakpoints of a piecewise linear function, 194
Brelaz heuristic, 256
Capacity constraint, 80
Car-sequencing problem, 13
CHIP, 6, 41, 334
Chronological backtracking, 10
CLP(R), 41
Coefficient matrix, 16
Column generation, 39, 42, 47, 177, 209, 355
Comb inequalities, 76
Combinatorial optimization problem, 1, 294
Compositionality, 77
Conceptual model, 174, 341
Concurrent search, 348
Configuration, 333
Conflict set, 121
Consistency, 94
Constraint definition, 182
Constraint imposition, 182
Constraint Logic Programming, 2
Constraint network, 94
Constraint Programming, 2, 35, 37, 40, 59
Constraint Programming on Finite Domains, 2
Constraint Propagation, 8, 35
Constraint Satisfaction Problem, 4, 60
Constraint
  alldifferent, 9, 98, 114, 345
  atleast, 6
  atmost, 6
  channeling, 346
  conditional, 64
  conjunction, 345
  cumulative, 62, 65, 96, 191
  cycle, 98
  diff-n, 65, 97
  disjunction, 345
  element, 6
  exactly, 6
  global cardinality, 100
  global minimum distance, 103
  hierarchical, 353
  implied, 346
  k-diff, 103
  linear, 349
  negation, 345
  number of distinct values, 104
  permutation, 98
  sequence, 102
  soft aldifferent, 124
  sort, 98
  stretch, 102
  sum and binary inequalities, 102
  symmetric alldifferent, 99
Control of inference, 332
Cooperation, 342, 360
CP program, 6
Crew Assignment Problem, 218
Crew Pairing Problem, 218
Crew Rostering, 42
Crew Scheduling, 42
Cuts, 194

Cutting plane generation, 141
Cutting planes, 24, 177
Debugging tool, 355
Decision variable, 182
Decomposition, 36–37, 39, 41–42, 44, 47–48, 342, 350
Demon, 180
Depth-first search, 10
Depth Bounded Discrepancy Search, 12
Design model, 174, 341
Deterministic algorithm, 237
Directed graph - digraph, 113
Discrepancy-Bounded Depth First Search, 12
Discrete tomography, 63
Diversification, 37, 47
Dominance Rules, 35, 50
Dual model, 346
Dual value, 21
Dynamic Programming, 39
E-business, 336
Eclipse, 6, 170
Edge-finder algorithm, 97
Edge-Finding, 37, 49
Efficiency, 34, 41
Efficiency, 60
Encoding of over-constrained problems, 123
Eplex, 184
Explanation, 358
Expressiveness, 60
Fat tails, 264
Feasible flow problem, 113
Filtering algorithm, 94, 5
First fail, 10
Fitness landscape, 27
Fixed-point property, 127
Flexibility, 333
Flow, 44–45
Frequency Assignment, 48, 335
GAC schema, 106
Global cardinality constraint - consistency, 117
Global cardinality constraint - filtering, 117
Generalized Arc-Consistency, 8, 106, 350
Generalized Assignment Problem, 210
Genetic Algorithms, 36
Global Constraint, 41, 62, 333, 345, 37, 95, 177
Global cuts, 344
Global soft constraint, 122
Global vs local view of a relaxation, 145
Gomory cuts, 25
Good relaxations, 140
Graph-coloring problem, 116, 262
Graph theory, 95
Greedy algorithm, 308
GSAT, 348
Heavy-tailed distribution, 248
Heavy-tails, 235
Heuristic, 344
Heuristic solution, 26
Hoist Scheduling, 42
Ic, 183
ILOG Solver, 6, 110
Incomplete tree search, 334
Infeasible set, 353
Inference, 342, 350
Inference rule, 72
Information provided by a relaxation, 144
Integer linear programming, 1, 59
Integer Programming, 15, 36, 40
Intelligent backtracking, 238
Intensification, 37
Interface, 340
Job-Shop Scheduling, 44, 49
Knapsack constraints, 24
Labelling, 349, 10
Large-scale problems, 354
Large Neighborhood Search, 44–45, 47, 49, 316, 348
Large scale problems, 334
Latin-square, 245
Learning, 356
Limited Discrepancy Search, 50
Linear constraint, 16, 61
Linear Programming, 2
Linear relaxation, 208, 257
Linearization, 42
Linearize global constraints, 147
Local branching, 47
Local Search, 27, 36, 234, 294, 348
Logical constraints, 16
Longest Path, 45
Look-ahead techniques, 238
Look-ahead, 350
Look-back techniques, 238
LP-diving, 26
Map colouring, 351
Master problem, 208, 344
Maximal Constraint Satisfaction Problem - MAX-CSP, 121
Maximum flow problem, 113
Maximum matching problem, 113
Metaheuristic algorithms, 294, 28
Mixed Integer Linear Programming, 2
Modelling in constraint programming, 5
Modelling in integer programming, 16
Modelling, 339
Monte Carlo methods, 352
Multi-criteria problems, 359, 335
Multidimensional knapsack, 47
Multiple Search, 37, 40, 47
Neighborhood, 27, 295, 348
Neighborhood function, 27
Network Design, 38, 40
Network flow, 113
Nogoods, 50, 343

Non linear constraint, 345
Non primitive constraints, 67
OPL, 214, 358
Optimization constraints, 70
OR function, 17
Over-constrained problem, 120
Partial Solutions, 50
Path, 113
Path Relinking, 36
Penalties, 353
Piecewise linear function, 187
Pivot and Complement, 26, 46
Portfolio design, 279
Portfolio strategy, 269, 279
Preemptive Job-Shop Scheduling, 49
Preferences, 336
Pricing problem, 208, 220
Pricing subproblem, 22
Primal graph, 124
Primal graph based violation cost, 124
Primitive constraints, 67
Probabilistic data, 337
Probability distribution, 237
Probe, 191
Prober, 192
Probing, 190, 349
Probing problem, 189
Problem decomposition, 332
Problem solving, 332
Problem transformation, 332
Project Scheduling, 38
Prolog III, 41
Propagation, 95
Propagation algorithms, 175
Propagation mechanism, 90
Pruning based on optimality reasoning, 156
Quasigroup completion problem, 245
Random jump, 278
Random tie-breaking, 239
Random walk, 241
Randomization, 236
Randomized algorithm, 237
Randomized restart, 236
Randomized Rounding, 46
Randomness, 234
Rapid restarts, 275
Re-compute the relaxation, 152
Reduced cost, 41, 177, 208, 343, 353
Redundant constraints, 65, 303
Relaxation, 20, 51
Relaxation
  bound, 21
  continuous, 41, 45
  Lagrangian, 40, 42, 177
  linear, 21
  perfect, 350
Repair, 49, 185
Residual graph, 117
Resource allocation, 345
Resource constraint, 79, 188, 17
Resource reduction, 198
Restart strategy, 269
Restricted Candidate Lists, 309
Robustness, 35, 41, 337, 361
Rounding Heuristics, 46
Routing constraints, 18
Satisfiability Problem, 346
Scalability, 332
Scheduling, 333
Search, 350, 8, 72, 342
Search methods, 332
Sensitivity analysis, 177
Separation problem, 24, 141, 225
Set Covering, 39, 220
Set partitioning, 220, 223
Shadow price, 343, 353
Shifting Bottleneck, 44
Shortest Path, 38
Shuffle, 44, 49
Side Constraints, 35–37, 48, 334
Simplex, 355, 62
Simulated Annealing, 36, 234
Social Golfers Problem, 216
Soft constraint, 335
Software engineering, 339
Solution quality, 358
Solving in constraint programming, 8
Solving in integer programming, 20
SOS-Type 2 constraints, 25
SOS-Type 2, 194
Sport-scheduling problem, 118
Sport scheduling problem, 256
Stochastic search, 333
Stochastic optimisation, 351
Strong branching, 223
Subgradient optimization, 222
Subproblem, 344
Subtour elimination, 75
Supply chain planning, 77
Support, 94
Symbolic constraint, 5, 62
Symbolic constraint abstraction, 75
Symbolic constraints in integer programming, 74
Symmetries, 342, 351
Tabu Search, 36, 48
Tail truncation, 264
Temporal constraints, 187
Temporal variable, 189
Tentative assignment, 179
Tentative value, 186
Timetabling, 333
Toolkit, 357
Transformation, 350
Transition rule, 72

Transportation Problem, 42
Travelling salesman problem (TSP), 48, 51, 75, 157
Travelling Salesman Problem with Time Windows, 161
Travelling Tournament Problem, 212
Tree search, 334
Triggering events, 9
Tuple, 94
Two dimensional packing, 65
Uncertainty, 332, 336, 352
Valid tuple, 94
Value-selection heuristics, 234
Value graph, 114
Value network, 117
Variable-selection heuristics, 234
Variable attribute, 179
Variable domain, 182
Variable fixing, 22, 152, 177
Vehicle Routing, 39, 42, 45, 47–48, 334
Violation cost, 120
What-if, 336
Workforce Scheduling, 44